MATHEMATICS
1001

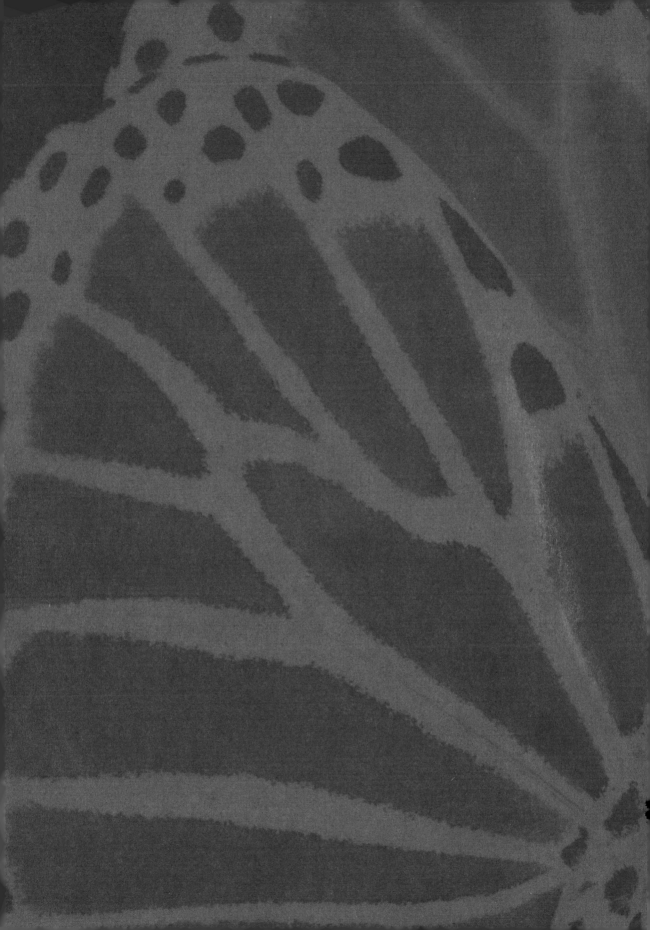

MATHEMATICS
1001

ABSOLUTELY EVERYTHING THAT MATTERS IN MATHEMATICS IN 1001 BITE-SIZED EXPLANATIONS

DR RICHARD ELWES

FIREFLY BOOKS

CONTENTS

INTRODUCTION

THE FIRST MATHEMATICIAN KNOWN BY NAME is Ahmes, an Egyptian scribe who around 1650 BC copied out and studied a list of complex mathematical problems he called the 'ancient writings'. Today, Ahmes' text is known as the *Rhind papyrus*. From this, and older stone tablets, we know that the scholars of ancient Egypt and Babylon had sophisticated numerical notation and a taste for challenging problems in algebra, geometry and number theory.

The study of mathematics, then, is as ancient as civilization; but it also represents the modernity of today's world. In the millennia since Ahmes' work, we have seen scientific and technological progress of which he could not have dreamt. Central to this advance has been the march of mathematics, which has contributed the basic language used in all scientific contexts. Probably mathematics' most fundamental contribution has been in the sphere of physics. Galileo's revolutionary insight in the early 17th century that the universe might yield to a purely mathematical description set the direction towards the world-changing theories of quantum mechanics and relativity.

This reliance on mathematics is not confined to the physical sciences. The social sciences depend on techniques of probability and statistics to validate their theories, as indeed do the worlds of business and government. More recently, with the emergence of information technology, mathematics became entangled in another love-affair, with computer science. This too has had a profound impact on our world.

As its influence has broadened, the subject of mathematics itself has grown at a startling rate. One of history's greatest mathematicians, Henri Poincaré, was described by Eric Temple Bell as 'the last universalist', the final person to have complete mastery of every mathematical discipline that existed during his lifetime. He died in 1912. Today, no-one can claim to have mastered the whole of topology, let alone geometry or logic, and these are just a fraction of the whole of mathematics.

Poincaré lived through a turbulent period in the history of mathematics. Old ideas had been blown away, and new seeds planted which flourished during the 20th century. The result is that the mathematical world we know today is rich and complex in ways that even the greatest visionaries of the past could not have imagined. My aim in this book is to give an overview of this world and how it came to be. I might have tried to sketch a low-resolution map of the entire mathematical landscape, but this would be neither useful nor entertaining. Instead, I have presented 1001 short 'postcards' from interesting landmarks around the mathematical world that nonetheless give a feel for the bigger picture of mathematics.

In the scheme of things, 1001 is a very small number (see the **frivolous theorem of arithmetic**). My challenge has been to select the real highlights: the truly great theorems, the outstanding open problems and the central ideas. I have also sought to represent the surprises and quirks that make the subject truly fascinating.

This book is organized thematically, on three levels. It is divided into ten chapters, each covering a broad subject, beginning with 'Numbers'. Each chapter is subdivided into sections, which are more narrowly focused on a single topic, such as 'Prime numbers'. Each section comprises a series of individual entries, such as the one on **the Riemann hypothesis**.

How you should read Mathematics 1001 depends on what you want from it. If you are interested in prime numbers you can read through that section. If you want a quick explanation of the Riemann hypothesis, you can jump straight there; but, because 'a quick explanation of the Riemann hypothesis' is an impossibility, you will then need to rewind a little, to take in the preceding few entries where the necessary prerequisites are laid out. Alternatively you can dip in and out, perhaps finding a new story by following the bold cross-references to different entries in the book.

Who is this book aimed at? The answer is: anyone with a curiosity about mathematics, from the novice to the informed student or enthusiast. Whatever the reader's current knowledge, I'm sure that there will be material here to enlighten and engage. Some parts of the book undoubtedly cover highly complex subjects. That is the nature of the subject; shying away from it would defeat the purpose. However, the book is structured so that the relevant basic concepts precede the complex ones, giving a foundation for understanding. My job in writing has been to discuss all ideas, from the basic to the most abstract, in as direct and focused a way as possible. I have done my best, and have certainly relished the challenge. Now I can only hope that you will enjoy it too.

Richard Okura Elwes

THMS • THE LAWS OF LOGARITHM
ADDITION BY HAND • SUBTRACTI
METHOD • MULTIPLICATION BY HA
VISION • DIVISIBILITY TESTS • DIV
BY 2 AND 4 • DIVISIBILITY BY
OTHER PRIMES • DIFFEREN
TING OUT NINES • TRACH
BY IT • NUMBER SYSTEMS
TORY OF ZERO • B
NU

NUMBERS

WHAT IS MATHEMATICS? The 'science of numbers' would be many people's first guess, and it is by no means wrong. However, our understanding of what numbers are has evolved over time. Today there are several different number systems which merit attention, each with its own characteristics and mysteries. The subject we call *number theory* concerns the most ancient of all: the natural numbers, comprising 0, 1, 2, 3, 4, 5, . . .

      The first thing to do with these is combine them arithmetically, through addition, subtraction, multiplication and division. There are several time-honoured ways of calculating the results, which rely on our decimal place value notation.

ION • SUMS AND PRODUCTS • POWERS
ERS • ROOTS • FRACTIONAL POWERS •
• SLIDE RULES • LOGARITHMIC SLIDE
N BY HAND • MULTIPLICATION BY HAND,
, COLUMN METHOD • SHORT DIVISION •
BILITY BY 3 AND 9 • DIVISIBILITY BY 6 •
DIVISIBILITY BY 7 • DIVISIBILITY BY 11 •
E OF TWO SQUARES • ARITHMETIC USING
ENBERG ARITHMETIC • TRACHTENBERG
MATHEMATICAL DISCIPLINES • NATURAL
HMAGUPTA'S ZERO • THE NATURALNESS
ERS • INTEGERS • RATIONAL NUMBERS
COMPLEX NUMBERS • QUATERNIONS
RECIPROCALS • EQUIVALENT
RECURRING DECIMALS
THE

At a deeper level, questions about the natural numbers fall into two main categories. The first concerns the *prime numbers*, the atoms from which all others are built. Even today, the primes guard their secrets. Major open questions include Landau's problems and the Riemann hypothesis.

The second principal branch of number theory considers *Diophantine equations*. These encode the possible relationships between whole numbers. Catalan's conjecture (Mihăilescu's theorem), for example, says that 8 and 9 are the only two positive powers you will ever find sitting next to each other.

THE BASICS

Addition

Today, as for thousands of years, numbers are principally used for *counting*. Counting already involves addition: when you include a new item in your collection, you have to *add* one to your total.

More general addition extends this: what happens if you add three items to a collection of five? Efficient methods for adding larger numbers first required the development of numerical notation (see **addition by hand**). Mathematicians have many terms for addition. The *sum* of a collection is the total when everything is added together. As more and more numbers are added together, objects called **series** emerge. Today, addition extends beyond plain numbers to more colourful objects such as **polynomials** and **vectors**.

Subtraction

The mathematical perspective on subtraction may seem strange on first sight. Since the dawn of **negative numbers**, every number (such as 10) has an opposing *additive inverse* (-10). This is defined so that when you add the two together they completely cancel each other out, to leave 0. Subtraction is then a two-step process: to calculate '20 – 9' first you replace 9 with its negative, and then you *add* the numbers 20 and -9. So '20 – 9' is really short-hand for '20 + (-9)'.

This settles a matter which often troubles children: why does the order of addition not matter ($3 + 7 = 7 + 3$), but the order of subtraction does ($3 - 7 \neq 7 - 3$)? When understood as addition, the order does not matter after all: $3 + (-7) = (-7) + 3$.

So how do you subtract negative numbers such as $-7 - (-4)$? The same rules apply, first replace -4 with its inverse, 4, and then add: $-7 + 4 = -3$.

Multiplication

Multiplication's first appearance is as repeated addition: if each member of a family has two beads, and there are four people in the family, how many beads are there altogether? The answer is $2 + 2 + 2 + 2$, or 4×2 for short. From this definition, it is not immediately obvious that $n \times m$ should always equal $m \times n$. This is true of course, as can be seen in a rectangular array of m columns and n rows. If you count this as n rows of m beads each, we get a total of $n \times m$. But it is $m \times n$ when seen as m columns of n beads. Since the total cannot depend on our counting method, these must be equal.

There are several terms to describe multiplication, and several symbols. 'Times' is nicely descriptive. A more everyday term is the English word 'of': three sets *of* seven make 21, for example. This remains valid for fractions: half *of* six is three, which translates as $\frac{1}{2} \times 6 = 3$. (It is common to mistake 'of' as meaning division here.)

The most common symbol for multiplication is '×', though mathematicians often prefer '·', or even nothing at all: $3 \cdot x$ and $3x$ mean the same as $3 \times x$. For multiplying lots of numbers together, we use the product notation (see **sums and products**).

Since $3 \times 5 = 15$, we say that 3 is a *factor* of 15, and 15 is a *multiple* of 3. The prime factors of a number are its basic building blocks, as affirmed by the **fundamental theorem of arithmetic**.

Sums and products

Suppose you want to add together the numbers 1 to 100. Writing out the whole list would take up too much time and paper, so mathematicians have devised a short-hand. A capital Greek sigma (standing for *sum*) is used:

$$\sum_{j=1}^{100} j$$

The numbers on top and bottom of the sigma show the range: j starts at 1 and takes each value successively up to 100. After the sigma comes the formula we are adding. Since in this case we are just adding the plain numbers, the formula is just j.

Instead if we wanted to add the first 100 square numbers, we would write:

$$\sum_{j=1}^{100} j^2 = 1^2 + 2^2 + 3^2 + \cdots + 100^2$$

(These expressions have easily computable answers; see **adding up 1 to 100** and **adding the first hundred squares**.)

If we want to multiply instead of add, we use a capital Greek pi (standing for *product*):

$$\prod_{j=1}^{100} j = 1 \times 2 \times 3 \times \ldots \times 100$$

(This is also written as '100!', that is 100 **factorial**.)

Powers

Just as multiplication is repeated addition ($4 \times 3 = 3 + 3 + 3 + 3$), so *exponentiation*, or *raising to a power*, is repeated multiplication: $3^4 = 3 \times 3 \times 3 \times 3$. (Here 3 is the *base* and 4 is the *exponent* or *power*.) Some powers have their own names: raising to the power 2 is called *squaring* because a square with sides x cm long has area x^2 cm^2 (16 is 4 squared for example). Similarly, raising to the third power is called *cubing*.

We can understand x^n as meaning $1 \times \underbrace{x \times x \times x \times \ldots \times x}_{n \text{ times}}$.

This suggests that $x^1 = x$ and $x^0 = 1$. (This is a useful convention rather than a profound theorem, and students are often resistant to it, insisting that x^0 should mean 0.)

The laws of powers

What happens when we multiply powers together? If we unpack the calculation $2^3 \times 2^4$ it becomes $\underbrace{2 \times 2 \times 2} \times \underbrace{2 \times 2 \times 2 \times 2}$, which is 2^7.

It is no coincidence that $7 = 3 + 4$. This is an example of the *first law of powers*:

$$x^a \times x^b = x^{a+b}$$

For the *second law of powers*, consider $(5^2)^3 = \underbrace{5 \times 5} \times \underbrace{5 \times 5} \times \underbrace{5 \times 5} = 5^6$.

The important observation is that $2 \times 3 = 6$. In general:

$$(x^a)^b = x^{a \times b}$$

Exponentiation can easily be extended to negative powers. It can also be extended to broader classes of number, such as **complex numbers**. This is subtler, and relies on the **exponential function**.

Negative powers

To give meaning to negative powers, such as 10^{-2}, it makes sense to look at the pattern for positive powers first, and then try to extend it. The rule for getting larger positive powers of 10 is 'keep multiplying by 10'. Starting with 10^1 (which is 10), multiplying by 10 tells us $10^2 = 10 \times 10 = 100$. Multiplying by 10 again gives $10^3 = 10^2 \times 10 = 100 \times 10 = 1,000$, and so on.

We can turn this on its head. If we start with 10^6 (which is 1,000,000) and count down, to get to the next power down, we divide by 10. That is, $10^5 = 10^6 \div 10 = 1,000,000 \div 10 = 100,000$. Then to get to 10^4 we divide by 10 again, and so on, until we get back to 10^1 (which equals 10).

But there is no reason to stop here. To reach the next power down, 10^0, we divide by 10 again, so $10^0 = 10^1 \div 10 = 1$. If we continue, we arrive among the negative powers: $10^{-1} = 10^0 \div 10 = 1 \div 10 = \frac{1}{10}$. Then $10^{-2} = 10^{-1} \div 10 = \frac{1}{10} \div 10 = \frac{1}{100}$, and so on. The pattern is:

$$10^{-n} = \underbrace{\frac{1}{10} \times \frac{1}{10} \times ... \times \frac{1}{10}}_{n \text{ times}}$$

What works for 10 works for every other number (other than 0). Negative powers of x are defined as the corresponding positive power of the **reciprocal** of x. That is $x^{-n} = (x^{-1})^n$.

Roots

What number when squared gives 16? The answer of course is 4. To put it another way, 4 is the *square root* of 16. In the same way, 3 cubed is 27, so 3 is the *cube root* of 27. Similarly $2^5 = 32$, so 2 is the *fifth root* of 32. The symbol we use to denote this is $\sqrt{}$. So $\sqrt[5]{32} = 2$ and $\sqrt[3]{27} = 3$. For square roots the little 2 is usually omitted, and we just write $\sqrt{16} = 4$. Roots can also be written as fractional powers.

Fractional powers

What might a fractional power mean? If 3^4 is '3 multiplied by itself 4 times' then $3^{\frac{1}{2}}$ should be '3 multiplied by itself $\frac{1}{2}$ a time', which does not seem very meaningful. Just as negative powers are made comprehensible by incorporating the reciprocal, fractional powers can make sense when interpreted as roots.

If notation such as $32^{\frac{1}{5}}$ is to mean anything at all, then it must satisfy the second law of powers: $(x^a)^b = x^{a \times b}$. In particular $\left(32^{\frac{1}{5}}\right)^5$ should equal $32^{\frac{1}{5} \times 5} = 32^1 = 32$. That is, $32^{\frac{1}{5}}$ should be a number which when raised to the fifth power gives 32. This must mean $32^{\frac{1}{5}} = \sqrt[5]{32} = 2$. Similarly we can write $27^{\frac{1}{3}} = \sqrt[3]{27} = 3$ and $16^{\frac{1}{2}} = \sqrt{16} = 4$. In general, $x^{\frac{1}{n}} = \sqrt[n]{x}$.

What meaning can we give to notation like $32^{\frac{4}{5}}$? Again the second law of powers helps: it should be equal to $(32^{\frac{1}{5}})^4 = 2^4 = 16$. Similarly, $27^{\frac{4}{3}} = 3^4 = 81$, and in general, $x^{\frac{m}{n}} = (\sqrt[n]{x})^m$.

Logarithms

As subtraction is to addition, and division is to multiplication, so *logarithms* are to powers. $2^3 = 8$ can be rephrased as $\log_2(8) = 3$, where $\log_2(8)$ is pronounced 'the logarithm of eight, to base 2'. To evaluate $\log_3(81)$ we need to answer $3^? = 81$ (the answer, of course, is 4).

You can take logarithms to any base number, as long as it is positive. But two bases are particularly common. Because powers of 10 are such a convenient way to represent numbers, logarithms to base 10 are very useful for measuring the order of magnitude of a number: $\log_{10} N$ is approximately the number of digits in the decimal representation of N.

Logarithms to base e are called **natural logarithms**, and are the most commonly occurring within pure mathematics.

The laws of logarithms

The first law of powers says $x^a \times x^b = x^{a+b}$. Corresponding to this is the *first law of logarithms*, which states that, for any c and d:

$$\log(cd) = \log c + \log d$$

(All the logarithms must be taken to the same base here, but every positive base works.) To see why this holds, define $c = x^a$ and $d = x^b$ in the first law of powers. Translating these, we get $a = \log c$ and $b = \log d$. (Since all the logarithms are to base x, we take the x as read.) Now, $cd = x^{a+b}$, and so $a + b = \log(cd)$, giving the result above.

The *second law of logarithms* corresponds to the second law of powers. It says that, for any c and d:

$$\log(c^d) = d \times \log c$$

Slide rules

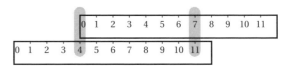

A rudimentary slide rule for addition might work as follows: take two rulers marked in centimetres, and place them side by side. If you want to calculate 4 + 7, slide the top ruler so that its start aligns with the '4' on the bottom ruler. Find the '7' on the top ruler and read off the value aligned with it on the bottom. With a slight modification, this simple idea produces a logarithmic slide rule, which can manage multiplication instead of addition.

Logarithmic slide rules

In the days before calculators, multiplying large numbers was time consuming and error-prone. A *logarithmic slide rule* is a device which uses logarithms to produce a quick and easy method. The crucial ingredient is the first law of logarithms:

$$\log(cd) = \log c + \log d$$

This says that a logarithm converts multiplication into addition: if two numbers are multiplied (cd), then their logarithms are added: $\log c + \log d$.

A logarithmic slide rule works as an ordinary one, with one important difference. Instead of standard rulers where the '4' is marked 4 cm from the end, it uses logarithmic rulers where the '4' is marked $\log 4$ cm from the end. (One consequence is that the logarithmic ruler starts at '1' instead of '0', because $\log 1 = 0$.) Following exactly the same procedure as for an ordinary slide rule, you will have arrived at a point $\log 4 + \log 7$ cm along the bottom ruler, which will be marked 28.

Logarithmic slide rules will work to any base, but were first designed using natural logarithms, by William Oughtred in the 1620s.

ARITHMETIC

Addition by hand

The advantage of a good system of numerical notation is that it allows shortcuts for arithmetic. To add 765 and 123, we can do a lot better than starting at 765 and counting up by one, 123 times. The basic idea is simple enough: write the two numbers one over the other, keeping the place values aligned, and then add up each column.

$$\begin{array}{r} 765 \\ +123 \\ \hline 888 \end{array}$$

The difficulty arises when the numbers in a column add up to more than 9. Suppose we want to add 56 and 37. We always begin on the right, with the units column. This time, 6 and 7 give us 13.

The final number of units will certainly be 3, so we can write this as the answer in the units column. This leaves us with the extra 10 to cope with. Well, the next stage is to add up the tens column anyway, so we just need to add one more ten to the pile. So we carry the 1 to the top of the tens column before adding that up:

$$\begin{array}{r} \overset{\scriptstyle 1}{5}6 \\ +37 \\ \hline 93 \end{array}$$

This method easily generalizes to adding more than two numbers:

$$\begin{array}{r} \overset{\scriptstyle 1}{3}\overset{\scriptstyle 2}{3}\overset{\scriptstyle 2}{9} \\ 389 \\ +273 \\ \hline 1001 \end{array}$$

Subtraction by hand
As with addition by hand, the basic idea for subtraction is to align the numbers in columns and proceed column by column, starting with the units:

$$\begin{array}{r} 96 \\ -34 \\ \hline 62 \end{array}$$

This time, we may encounter the problem that we need to take a larger digit from a smaller:

$$\begin{array}{r} 73 \\ -58 \\ \hline ? \end{array}$$

Here we seem to need to take 8 from 3, which we cannot do without heading into the negative numbers (which is best avoided until inevitable). The way around this is to split up 73 differently. Currently it is split into 7 tens (T), and 3 ones, or units (U), which is proving inconvenient. So instead we will write it as 6 tens, and 13 units. Essentially we are rewriting the calculation as:

T	U
6	13
−5	8

Now we can proceed as before. What this looks like when written normally is:

$$\begin{array}{r} \overset{\scriptstyle 6}{7}{}^{1}3 \\ -58 \\ \hline 15 \end{array}$$

This process of 'borrowing 1' from the next column might have to be repeated several times in one calculation.

Multiplication by hand, table method

If you know your times tables, then multiplying two single-digit numbers should be straightforward. Once we can do this, the decimal system makes it fairly simple to multiply larger numbers.

To calculate 53×7, as usual we split 53 into 5 tens and 3 units (ones). The key fact is that we can multiply each part separately. That is: $(50 + 3) \times 7 = (50 \times 7) + (3 \times 7)$. Some people use a grid method:

	7
50	350
3	21

To finish the calculation, we add up everything in the inner part of the table: $350 + 21 = 371$. This easily extends to calculations with more digits, such as 123×45:

	40	5
100	4000	500
20	800	100
3	120	15

To finish this calculation, we add up: $4000 + 800 + 120 + 500 + 100 + 15 = 5535$.

Multiplication by hand, column method

An alternative method to the table method of multiplication (which uses less ink) is column by column, adding as we go along, instead of at the end:

$$\begin{array}{r} 13 \\ \times 5 \end{array}$$

Starting with the units column, we get 15. So the answer's digit in that column will be 5, and we have also gained an extra ten. So, after we perform the multiplication in the tens column, we need to add that on:

$$\begin{array}{r} \overset{\scriptstyle 1}{13} \\ \times 5 \\ \hline 65 \end{array}$$

Multiplying by 40 just involves multiplying by 4, and then shunting everything one column to the left, suffixing with a zero:

$$\begin{array}{r} \overset{\scriptstyle 1}{13} \\ \times 40 \\ \hline 520 \end{array}$$

To multiply 13 by 45, we multiply by 40 and 5 separately and then add up. Writing these two calculations one under the other makes this addition quicker to do:

$$\begin{array}{r} 13 \\ \times 45 \\ \hline 65 \\ 520 \\ \hline 585 \end{array}$$

Short division

Short division is a method for dividing a large number (the *dividend*) by a single-digit number (the *divisor*). The basic idea is to take the dividend digit by digit starting on the left. Above each digit we write the number of times the divisor fits in:

$$\begin{array}{r} 132 \\ 3\overline{)396} \end{array}$$

The complication comes when the divisor does not fit exactly into one of the digits, but leaves a remainder. In the next example, 3 into 7 goes 2 times, with a remainder of 1. The 2 goes above the 7 as before, and the 1 is carried to the next step, placed before the 8, which is then considered as 18:

$$\begin{array}{r} 2\ 6 \\ 3\overline{)7\ ^18} \end{array}$$

Often we will have to carry more than once:

$$\begin{array}{r} 0\ 3\ 2 \\ 9\overline{)2\ ^28\ ^18} \end{array}$$

Short division can work with small double-digit divisors such as 12 too (as long as you know their times table). In this example, 12 cannot go into 9, so the whole 9 is carried at the start:

$$\begin{array}{r} 0\ 7\ 6 \\ 12\overline{)9\ ^91\ ^72} \end{array}$$

Long division

Long division is essentially the same procedure as short division. As the divisors become larger, however, more digits have to be carried, and calculating the remainders becomes lengthier. Rather than cluttering up the division symbol with carried remainders, they are written out underneath. So instead of writing

$$\begin{array}{r} 0\ 4\ 7 \\ 18\overline{)8\ ^84\ ^{12}6} \end{array}$$

we write them out underneath

$$\begin{array}{r} 0\ 4\ 7 \\ 18\overline{)8\ 4\ 6} \\ -\ 7\ 2\downarrow \\ \hline 1\ 2\ 6 \end{array}$$

Since 18 cannot divide the 8 in the hundreds column, 84 is the first number to be divided by 18, corresponding to the tens column. It goes in four times since $4 \times 18 = 72$, but $5 \times 18 = 90$. 4 is written on top, and 72 is written below 84 and then subtracted from it to find the remainder, 12. If we were doing short division, this would be the number to carry to the next column. The equivalent here is to bring down the next digit from 846 (namely 6) and append it to the 12 to get 126, the next number to be divided by 18. This goes exactly 7 times, so we write 7 on the top and we have finished.

Divisibility tests
How can you tell when one whole number is divisible by another? In general there is no easy method. But for small numbers there are various tricks which exploit quirks in the decimal system, our ordinary method of writing numbers. Some are so easy that we do them without thinking:
- A number is divisible by 2 if it ends in 0, 2, 4, 6 or 8.
- A number is divisible by 5 if it ends in 0 or 5.
- A number is divisible by 10 if it ends in 0.

This last one easily extends to divisibility tests by 100, 1000, and so on.

Divisibility by 3 and 9
A whole number is divisible by 3 if its digits add up to a multiple of 3. So 123 is divisible by 3 because $1 + 2 + 3 = 6$, and 6 is a multiple of 3. But 235 is not divisible by 3, as $2 + 3 + 5 = 10$, which is not a multiple of 3.

This trick works because the number written as xyz is really $100x + 10y + z$. This is equal to $99x + 9y + x + y + z$. Now, $99x + 9y$ is certainly divisible by 3. So the whole thing is divisible by 3, if and only if $x + y + z$ is divisible by 3. This proof also shows that the same trick works for 9. So 972 is divisible by 9, as $9 + 7 + 2 = 18$, a multiple of 9. But 1001 is not divisible by 9, as $1 + 0 + 0 + 1 = 2$.

Divisibility by 6
The test of divisibility by 6 simply amounts to applying the tests for both 2 and 3: a whole number is divisible by 6 if and only if it is even, and its digits add up to a multiple of 3. So 431 is not divisible by 6, as it is not even. Also 430 is not divisible by 6, as its digits add up to 7, not a multiple of 3. But 432 is divisible by 6 as it is even, and its digits add up to 9, a multiple of 3. (Notice that its digits do not have to add up to a multiple of 6.)

Divisibility by 2 and 4
We can tell whether a number is divisible by 2 just by looking at the last digit. The reason this works is that the number written as 'xyz' is really $100x + 10y + z$. Now, $100x + 10y$ is always divisible by 2. So the answer depends only on z.

Similarly, we can tell whether a number is divisible by 4 just by looking at the last two digits. If they form a number divisible by 4, then the whole thing is divisible by 4. So 1924 is divisible by 4, because 24 is divisible by 4. On the other hand 846 is not divisible by 4, because 46 is not divisible by 4. Again, '*wxyz*' is short-hand for $1000w + 100x + 10y + z$. This time, $1000w + 100x$ is always divisible by 4, so whether the whole thing is divisible by 4 depends only on whether $10y + z$ is.

Divisibility by 8

The divisibility test for 4 easily extends to 8, 16, 32, and so on. Looking at the last three digits of a number is enough to determine divisibility by 8. So 7448 is divisible by 8, as 448 is.

Admittedly, this divisibility test relies on knowing your 8 times table up to 1000, but is still useful when analysing very large numbers. For smaller numbers, it may be more practical to divide by 2, and then apply the divisibility test for 4. Similarly the last four digits determine divisibility by 16, and so on.

Divisibility by 7

The trickiest of the numbers up to 10 for divisibility is 7. One test works as follows: chop off the final digit, and double it. Then subtract the result from the shortened number. If the result is divisible by 7 then so was the original number. For example, starting with 224, we remove the final 4 and double it to get 8. Then we subtract this from 22, to get 14. Since this is divisible by 7, so is 224.

For larger numbers, we might need to apply this trick more than once. Starting with 3028, remove the 8 and double it to get 16. Now subtract that from 302 to give 286, and we repeat. Remove the final 6, double it to get 12, and subtract that from 28 to leave 16. That is not divisible by 7, so neither is 286, and therefore neither is 3028. This works because every number can be written as $10x + y$, where y is the last digit (and so between 0 and 9), and x is the result of chopping off y. In the example of 224, $x = 22$, and $y = 4$. Next, $10x + y$ is divisible by 7 if and only if $20x + 2y$ is divisible by 7 (multiplying by 2 does not affect divisibility by 7). Now, $20x + 2y = 21x - x + 2y$. Of course $21x$ is always divisible by 7, so whether or not the original number is divisible by 7 depends on $-x + 2y$, or equivalently its negative $x - 2y$.

Divisibility by 11

11 has an elegant divisibility test. It works by taking the *alternating sum* of the digits: add the first, subtract the second, add the third, and so on. If the result is divisible by 11, then so is the original number. More precisely, taking a five-digit number '*vwxyz*' as an example, this is divisible by 11 if and only if $v - w + x - y + z$ is divisible by 11. (If $v - w + x - y + z = 0$, that is classed as divisible by 11.) So, to test 5893, we calculate $5 - 8 + 9 - 3 = 3$, which is not divisible by 11.

This works because the following numbers are all divisible by 11: 99, 9999, 999999, and so on. On the other hand 9, 999, 99999, etc. are not, but 11, 1001, 100001 etc. are. Writing '*vwxyz*' as $10000v + 1000w + 100x + 10y + z$, this is equal to $9999v + 1001w + 99x + 11y + v - w + x - y + z$

From the above observation $9999v + 1001w + 99x + 11y$ will always be divisible by 11. So whether or not the whole number is divisible by 11 depends on $v - w + x - y + z$.

Divisibility by other primes

For composite numbers the best approach to divisibility is to test for every constituent prime individually. Other prime numbers can all be tested in a way similar to 7. They involve chopping off the final digit, multiplying it by some suitable constant, and then adding or subtracting that from the curtailed number.

- To test for divisibility by 13, chop off the last digit, multiply it by 4, and add that to the shortened number. So, to test 197, chop off the 7, multiply by 4 to give 28, and add that to 19 to give 47. As this is not divisible by 13, neither is 197.
- For 17, chop off the final digit, multiply by 5, and subtract that from the curtailed number. For example, starting with 272, chop off the 2, multiply by 5 to get 10, and subtract that from 27 to leave 17, which is divisible by 17, so 272 is too.
- For 19, chop off the last digit, double it, and add that to the curtailed number.

Similar tests work for larger primes too.

Difference of two squares

One of the simplest and most useful algebraic identities is the *difference of two squares*. This says that for any numbers a and b, $a^2 - b^2 = (a + b)(a - b)$. The proof is simply a matter of expanding brackets:

$$(a + b)(a - b) = a^2 + ab - ab - b^2$$

This works equally well with any combination of numbers and algebraic variables. For instance, $15^2 - 3^2 = (15 + 3)(15 - 3)$ and $x^2 - 16 = (x + 4)(x - 4)$, because 16 is 4^2. One of many uses for this identity is as a technique for speeding up mental arithmetic.

Arithmetic using squares

One of the first tasks for people training for speed arithmetic is to memorise the first 32 square numbers. As well as being useful on their own, they can be used to multiply other pairs of numbers. The trick is to exploit the difference of two squares.

If the two numbers are both odd or both even, then there will be another number directly in the middle of them. For example, if we want to multiply 14×18, we note that 16 is in the middle. So we can rewrite the problem as $(16 - 2) \times (16 + 2)$. This is now the difference of two squares: $16^2 - 2^2$. Since we have memorized that $16^2 = 256$, the answer is 252.

If the two numbers are not both odd or both even, we can do it in two steps. For example, to calculate 15×18, split it up as $(14 \times 18) + 18$. We calculated $14 \times 18 = 252$ above, so $15 \times 18 = 252 + 18 = 270$.

Casting out nines

Casting out nines is a useful technique for checking for errors in arithmetic. The basic idea is to add up the digits of the number, and subtract 9 as many times as possible, to get an answer between 0 and 8. So starting with 16,987 we add 1 and

6 to get 7. We can ignore the next 9. Then add 8 to get 15, and subtract 9 to get 6, add 7 to get 13, and subtract 9 to get an answer of 4. We can write $N(16{,}987) = 4$.

The point of this is that if we have calculated $16{,}987 + 41{,}245$ as $58{,}242$, we can check it as follows: $N(16{,}987) = 4$ and $N(41{,}245) = 7$. Adding these together, and casting out nines again gives an N-value for the question of $4 + 7 - 9 = 2$. However, our answer produces $N(58{,}242) = 3$. As these do not match, we know we have made a mistake. In fact, $16{,}987 + 41{,}245 = 58{,}232$.

The same trick works for subtraction, multiplication and integer division. For example, if we have calculated 845×637 as $538{,}265$, we work out $N(845) = 8$ and $N(637) = 7$. Multiplying these together gives 56. Repeating the process, we get a result for the left-hand side of $N(56) = 5 + 6 - 9 = 2$. Since $N(538{,}265) = 2$ too, the two sides match, and the test is passed.

This technique amounts to checking the answer in arithmetic modulo 9 (see **modular arithmetic**). It is useful for detecting errors, but also gives false negatives (for instance, it cannot detect swapping two digits in the answer).

Trachtenberg arithmetic
Jakow Trachtenberg was a Russian mathematician and engineer, who fled to Germany after the 1917 revolution. A Jew and an outspoken pacifist, after the rise of Nazism he was captured and imprisoned in a concentration camp. During his seven-year incarceration, he developed a new system for doing mental arithmetic, with an emphasis on speed. These techniques form the basis of modern speed arithmetic. An example is Trachtenberg multiplication by 11.

In 1944, aided by his wife, Trachtenberg evaded a death sentence by escaping to Switzerland. There he founded the Mathematical Institute in Zürich, and taught his methods to generations of students.

Trachtenberg multiplication by 11
An example of Trachtenberg arithmetic is in multiplying a large number, such as 726,154, by 11. We take the digits of 726,154 from right to left. To start with, copy down the first digit: 4. Next, we add the first two digits: 5 and 4, to get 9. So far we have 94. Next we add the second and third 1 and 5, to get 6. This takes us to 694. We carry on adding digits in pairs, until we reach 7 and 2 which make 9. This takes us to 987,694. The final step is to copy down the last number, 7. So the answer is 7,987,694.

The only complicating factor is when a pair of digits makes 10 or more. To multiply 87 by 11, for example, we first write down 7. Now, 8 and 7 sum to 15. So we write down 5, and carry 1 to the next step. So far we have 57. The final step is usually to write down the final digit: 8. But this time we must also add the carried 1, so the final answer is 957.

This method can be summarized as 'add each digit to its neighbour', where 'neighbour' means the digit to its right. With a little practice, this makes multiplying by 11 almost instantaneous. Jakow Trachtenberg devised similar methods for multiplying by all the numbers from 1 to 12. The corresponding rule for multiplying by 12, for example, is 'double each digit and add its neighbour'.

NUMBER SYSTEMS

Number systems

The most ancient number system is the one that humans have used to count objects for millennia: the system **N** of **natural numbers** consisting of 0, 1, 2, 3, 4, 5, ... As civilizations advanced, more sophisticated number systems became necessary. To measure **profit and debt**, we need to incorporate **negative numbers**, giving **Z**, the system of **integers**.

Not everything can be measured using whole numbers. Half a day, or two thirds of a metre, show the need for a system extending beyond the integers. Today, the system which unites the fractions and the integers is known as **Q**, **the rational numbers**. As the Pythagoreans discovered, the rational numbers are not adequate for measuring every length. By plugging the gaps between rational numbers, we arrive at the system **R** of **real numbers**. In the 16th century, Italian algebraists working on solving equations realized that this was still not enough. The system that results from introducing a square root of −1 is **C**, the **complex numbers**.

Mathematical disciplines

Each number system can be studied and investigated on its own terms, and mathematicians have come to know the characters and idiosyncrasies of each. **N** and **Q** seem straightforward and welcoming on first meeting, but are highly secretive, and downright awkward to work with. This is the realm of **number theory**.

At the other end of the spectrum, **C** perplexes those who see it from a distance, but rewards anyone brave enough to get to know it, with its incredible simplicity and power. **Complex analysis** was one of the triumphs of 19th-century mathematics.

In between, **R** is the right arena for understanding lengths, as the ancient Greeks first realized. Much of geometry and topology is built from **R**.

Natural numbers

In a cave among the Lebombo mountains of Swaziland, archaeologists in the 1970s found a baboon's leg bone, with 29 notches carved into it. It had been used as a tally, and the number 29 suggests a lunar calendar. Dating from around 35,000 BC, the *Lebombo bone* is the oldest mathematical artifact that we have. It illustrates the first number system, the one that humans have used to count, for millennia: *the natural numbers*.

$$0, 1, 2, 3, 4, 5, \ldots$$

The system of natural numbers is known as **N**. From a theoretical perspective, the defining feature of **N** is **mathematical induction**. This says, roughly, that by starting at 0 and repeatedly adding 1, every number will eventually be reached. In fact, mathematicians are divided on whether **N** comprises 0, 1, 2, 3, 4, 5, ... or 1, 2, 3, 4, 5, ..., The point at issue is the naturalness (or otherwise) of 0.

The prehistory of zero

The entity we know as *zero* took many years to be accepted as a number in its own right. It required a leap of imagination to start thinking of 0, which represents nothing, as being something. The trigger for the ascent of zero was the development of place value notation. There are, of course, infinitely many numbers. But we don't want to have to invent ever more symbols to describe them. Today, we use only the symbols 1–9, as well as 0, to describe any number, with the place of the number imparting as much information as its value. So in '512', the '5' means 'five hundreds', while in '54' it means 'five tens'.

This is an ingenious system, but what happens when you have no tens, as in two hundred and three? The ancient Babylonians simply left gaps. So they might have written two hundred and three as '2 3' (of course, they did not use Arabic numerals or a decimal base, nevertheless this illustrates the idea). The problem is obvious: this can be easily mistaken for 23. By the third century BC, the Babylonians, in common with other cultures, had got around this by using a place holding symbol to indicate an empty column. Ancient Chinese mathematicians had a mathematical concept of zero, and indeed negative numbers, but their primitive notation lagged behind this deeper understanding.

Brahmagupta's zero

It was in India that the Babylonian notation and Chinese conception of zero finally came together, in the development of 0 as a full-blown number. In AD 628, Brahmagupta formally defined it as:

the result of subtracting any number from itself

His arithmetical insights may seem obvious today, but they represent a real breakthrough in the history of human thought: 'When zero is added to a number or subtracted from a number, the number remains unchanged; and a number multiplied by zero becomes zero.'

Brahmagupta's work also laid out the basic theory of negative numbers, though these took longer to gain widespread acceptance.

The naturalness of 0

The history of mathematics contains many notable disagreements. Gottfried Leibniz and Isaac Newton feuded bitterly over the development of **calculus**. One of the most enduring disputes, and also one of the most uninteresting, with least prospect of resolution, is whether or not **0** should be classed as a natural number. The carver of the Lebombo bone would not have thought so; modern mathematicians are split. This book generally adopts the Zen-like philosophy that 0 is the epitome of naturalness. In some contexts, however, this leads to the annoyance that 0 is the *first* natural number, 1 is the *second*, 2 the *third*, and so on. So, when working with **sequences**, for example, it is more convenient to exclude zero.

Profit and debt

If numbers are principally used for counting, then what do negative numbers mean? How can you have −3 apples? The likely origin of negative numbers is in trade, where positive numbers represent profit, and negative numbers debt. Ancient Chinese mathematicians represented numbers with counting rods, and used red and black rods respectively to distinguish between positive and negative numbers. (These colours have been reversed in the western world, with the phrases 'in the black' and 'in the red' meaning 'in credit' and 'in debit', or debt, respectively.)

Negative numbers

Despite their ancient pedigree, negative numbers were long viewed with suspicion, even until the early 19th century. Many saw them as short-hand for something else (a positive quantity of debt instead of a negative quantity of profit), rather than legitimate numbers in their own right. Many mathematicians were content to employ them as tools for calculating, but if the final answer came out negative, it would often be abandoned as invalid. (Complex numbers held a similarly indeterminate status for several years.) However, the direction had been set in 628, when the Indian mathematician Brahmagupta wrote his treatise on the combined arithmetic of positive and negative whole numbers, and 0. We now call this the system of integers.

Integers

The *integers* are the whole numbers: positive, negative and zero. The system of integers is known as **Z**, standing for *Zahlen*, meaning 'numbers' in German. The advantage of **Z** is that quantities, such as temperature, which naturally come either side of 0 can be measured. Similarly, profits and debts can be measured on a single scale.

From a mathematical perspective, **Z** is better behaved than the narrower system of natural numbers **N**. While natural numbers can be added without leaving **N**, they cannot always be subtracted. To solve $2 - 3$ you have to leave the natural numbers. In the integers, however, we can subtract any two numbers, $b - a$.

This also allows more equations to be solved. For example, $x + 3 = 2$ is an equation built solely from natural numbers, but it has no solution in that system. In the integers, however, it does. Indeed any equation $x + a = b$, where a and b are integers, can now be solved without leaving **Z**.

Rational numbers

Any number which can be expressed as a fraction of integers (whole numbers) is *rational* (meaning that they are *ratios*, rather than that they are logical or cerebral). Examples are 2 (since it equals $\frac{2}{1}$), $\frac{17}{8}$ and $\frac{-3}{4}$. The reasons for the development of fractions are self-evident. For measuring time, distance or resources, quantities such as half a month, a third of a mile, or three-quarters of a gallon are obviously useful.

Mathematicians denote the system of all rational numbers by **Q** (standing for quotient). This system augments the integers, and also brings mathematical benefits. Among the whole numbers, division is not well-behaved. You can divide 8 by 2, but not by 3. The rational numbers form a system called a **field**, where any numbers can be divided, as well as added, subtracted or multiplied. The solitary exception is 0, by which you can never divide (see **division by 0**).

Real numbers

Integers are a fixed distance apart. From one integer to the next is always a distance of 1. For the rational numbers this is no longer true; they can measure shorter distances. Starting at 1, there is no 'next' rational number. There are $1 + \frac{1}{2}$, $1 + \frac{1}{3}$, $1 + \frac{1}{10}$, $1 + \frac{1}{20,000}$, and rational numbers as close to 1 as you like. It seems strange, then, to insist that there are nevertheless 'gaps' among the rational numbers. But this is true, as the **irrationality of** $\sqrt{2}$ demonstrates. Drawing the graph of $y = x^2 - 2$, at the place where it should cross the x-axis, it sneaks through a gap in the rational numbers.

Filling in these gaps is a technical procedure, which results in **R**, the system of *real numbers*. Also known as the *real line*, **R** can be thought of as all the points on an infinite line, which is now *complete*, meaning that there are no gaps in it.

Another way to think of **R** is as the collection of all possible infinite decimal expansions. Start with an integer, such as -14, then put a decimal point, followed by any infinite string of digits: $-14.6936027480\ldots$ Beware, however: some numbers, such as $0.99999999\ldots$, have more than one decimal expansion.

Imaginary numbers

In 16th-century Italy, the mathematical community was focused on solving **cubic** and **quartic equations**. An obstacle to this work was that some simpler equations such as $x^2 = -1$, do not have any solutions among the real numbers. It is a consequence of the rules for multiplying negative numbers that no real number, when squared, produces a negative answer. However, the Italian algebraists discovered that if they simply *imagined* a solution to this equation, and called it 'i', they could then use this in their calculations, and it would produce accurate, real results for solving other equations.

Multiples of i by real numbers, such as 3i, πi and $-\frac{1}{2}$i, are called *imaginary numbers*. This terminology of *imaginary numbers* contrasting with *real numbers* is misleading and regrettable, if historically understandable. The home of imaginary numbers is the system of complex numbers, which was set on a firm foundation by Rafael Bombelli in 1572. In fact, we know that the system of complex numbers can be built from the real numbers in a straightforward and explicit way. In truth, i is no more imaginary than π or -3. Unfortunately, however, the name has stuck.

Complex numbers

The system (**C**) of *complex numbers* is constructed by formally adding the imaginary numbers to the real numbers. Indeed, an individual complex number is nothing more than a real number (such as $\frac{1}{2}$) added to an imaginary number (such as 3i). So $\frac{1}{2} + 3i$ is a complex number. These can then be added, subtracted, multiplied and divided according to the rules of complex arithmetic.

The *Argand diagram* is the standard way to represent complex numbers, as a 2-dimensional plane. Then complex numbers look like familiar Cartesian coordinates, with the real axis horizontal, and the imaginary axis vertical. This complex plane is a wonderful setting for geometry, as geometric and algebraic ideas mesh perfectly.

In many ways, the complex numbers form the endpoint for the evolution of the concept of *number*. For the purposes of solving polynomial equations, the **fundamental theorem of algebra** says that they do everything that could possibly be required.

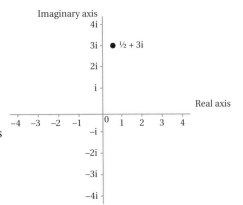

Quaternions

At key moments in history, mathematicians have reconsidered what it means to be a *number*. Important moments were the introduction of 0, negative numbers and irrational numbers. The complex numbers form a natural stopping point for this expansion. The fundamental theorem of algebra says that every equation that we would want to solve can indeed now be solved.

However, we do not *have* to stop there. It is possible to extend the complex numbers to a still larger system. The complex numbers are built from pairs of real numbers (a, b), with a new ingredient i, the square root of -1. Every complex number can be written in the form $a + ib$. In 1843, Sir William Hamilton discovered a system based on quadruples of real numbers (a, b, c, d) with three new ingredients: i, j, k. Every *quaternion* can be written as $a + ib + jc + kd$. The fundamental rule of the quaternions is $i^2 = j^2 = k^2 = ijk = -1$. Hamilton was so excited with his discovery that he carved this equation onto Broom Bridge in Dublin. (A plaque still commemorates the spot.)

There is a cost to this expansion. Multiplication is now *non-commutative*: it is not always true that $A \times B = B \times A$. Technically, then, the quaternions do not form a **field**. Following their discovery, quaternions became a major topic in mathematics. Though their popularity later waned, they remain in use today as they neatly capture the behaviour of *rotations* of 3- and 4-dimensional spaces.

Octonions

When Sir William Hamilton had discovered his system of quaternions, he wrote to his friend John Graves explaining his breakthrough. Graves replied 'If with your alchemy you can make three pounds of gold, why should you stop there?' As good as his word, Graves came up with an even larger system, now known as the *octonions*. It is built from octuples of real numbers: $(a_0, a_1, a_2, a_3, a_4, a_5, a_6, a_7)$ along with seven new ingredients: $i_1, i_2, i_3, i_4, i_5, i_6, i_7$, each of which squares to -1. So a general octonion is of the form $a_0 + i_1 a_1 + i_2 a_2 + i_3 a_3 + i_4 a_4 + i_5 a_5 + i_6 a_6 + i_7 a_7$.

Hurwitz's theorem

Hamilton produced the quaternions, and Graves the octonions; how much further can this generalization be pushed? In 1898, Adolf Hurwitz proved that this really is the limit now. The real and complex numbers, the quaternions and the octonions are the only four *normed division algebras*: structures containing the real numbers which allow multiplication and division in a geometrically sensible way. The mathematical physicist John Baez described this family in 2002:

The real numbers are the dependable breadwinner of the family, the complete ordered field we all rely on. The complex numbers are a slightly flashier but still respectable younger brother: not ordered, but algebraically complete. The quaternions, being non-commutative, are the eccentric cousin who is shunned at important family gatherings. But the octonions are the crazy old uncle nobody lets out of the attic: they are non-associative.

Non-associative means that, if you multiply octonions A, B and C, you might find $A \times (B \times C) \neq (A \times B) \times C$. This is a violation of one of the most basic algebraic laws. Despite (or because of) their craziness, the quaternions and octonions are useful for explaining other mathematical anomalies, such as the exceptional **Lie groups**.

RATIONAL NUMBERS

Reciprocals

Also known as the *multiplicative inverse*, the *reciprocal* of a number is what you need to multiply it by to get 1. So the reciprocal of 2 is $\frac{1}{2}$, and vice versa. The reciprocal of a fraction is easily found: just turn it upside down. So the reciprocal of $\frac{5}{7}$ is $\frac{7}{5}$.

The reciprocal of x is denoted $\frac{1}{x}$ or x^{-1}, although for mathematicians (if not historians) reciprocation is a more fundamental notion than either division or taking negative powers. Reciprocation also provides a mirror between large numbers and small numbers. The reciprocal of one million is one millionth, and vice versa. There is one number which, uniquely, has no reciprocal. There is no value of x which satisfies $x \times 0 = 1$, so 0 does not have a reciprocal.

Division

Division measures how many times one number fits into another. So $15 \div 3 = 5$ because 3 fits into 15 exactly 5 times: $5 \times 3 = 15$. This equally applies to fractions, so $\frac{1}{2} \div \frac{1}{4} = 2$ because $\frac{1}{4}$ fits into $\frac{1}{2}$ exactly 2 times: $2 \times \frac{1}{4} = \frac{1}{2}$.

An alternative view is to see division as being built from the more basic operation of reciprocation. (Of course the notation of the reciprocal is already highly suggestive of 1 being *divided by* 4.)

We can then understand $m \div n$ as meaning m multiplied by the reciprocal of n, that is $m \times \frac{1}{n}$. So $15 \div 3 = 15 \times \frac{1}{3} = 5$, and $\frac{1}{2} \div \frac{1}{4} = \frac{1}{2} \times 4 = 2$. This easily extends to more general fractions. So $\frac{2}{5} \div \frac{3}{4} = \frac{2}{5} \times \frac{4}{3} = \frac{8}{15}$.

Equivalent fractions
In the world of whole numbers, there is just one way to denote any number: 7 is 7, and you cannot write it in any other way. (James Bond might fairly object that it can be prefixed with zeroes, but this presents no real ambiguity.) When we enter the realm of the rational numbers, that is to say fractions, something inconvenient happens. There are now several genuinely different ways to write the same number. For instance, $\frac{2}{3}$ represents the same number as $\frac{4}{6}, \frac{6}{9}, \frac{14}{21}$, and a host of other fractions which do not immediately seem the same. These are called *equivalent fractions*. If you start with a number, multiply it by 2, and then divide the result by 2, you would expect to arrive back where you started. This leads to the rule for equivalent fractions: if you multiply the top by some number (not including zero) and you do the same to the bottom, you get an equivalent fraction. So $\frac{4}{6} = \frac{2}{3}$, because $\frac{4}{6} = \frac{2 \times 2}{3 \times 2}$.

Taking this backwards, if the top and bottom of the fraction are each divisible by some number, you can always *cancel down* to produce an equivalent fraction: $\frac{12}{15}$ can be cancelled down since 12 and 15 are both divisible by 3. This is called *simplifying* and gives the equivalent fraction $\frac{4}{5}$. For most purposes, the best (simplest) representation of a fraction is when all possible cancelling has been done. So the top and bottom of the fraction have no common factors (or are *coprime*). It is a consequence of the **fundamental theorem of arithmetic** that such a representation always exists, and is unique.

Multiplying fractions
The basic rule for multiplying fractions is simple: multiply the top, and multiply the bottom. So $\frac{2}{3} \times \frac{4}{5} = \frac{2 \times 4}{3 \times 5} = \frac{8}{15}$. A shortcut is to do any possible cancelling at the beginning, instead of the end. So instead of calculating $\frac{2}{15} \times \frac{21}{8}$ as $\frac{42}{120}$ and then simplifying, we spot that the top and bottom are both divisible by 2 and 3. Cancelling these gives $\frac{1}{5} \times \frac{7}{4} = \frac{7}{20}$.

Adding fractions
One of the commonest mistakes young students make is to think $\frac{1}{4} + \frac{2}{3} = \frac{3}{7}$. The reasoning here is obvious, but a little thought-experiment quickly dismisses it: $\frac{1}{2}$ a cake with another $\frac{1}{4}$ of a cake clearly amount to $\frac{3}{4}$ of a cake, not $\frac{2}{6}$.

Some fractions are easy to add: those which have the same number on the bottom, or a *common denominator*. It is intuitive enough that $\frac{1}{5} + \frac{2}{5} = \frac{3}{5}$. To add other pairs of fractions we first have to find an equivalent pair with common denominator. When one denominator is a multiple of the other, this is straightforward. To evaluate $\frac{3}{7} + \frac{5}{14}$, first convert $\frac{3}{7}$ into 14ths, namely $\frac{6}{14}$. Then we can add: $\frac{6}{14} + \frac{5}{14} = \frac{11}{14}$.

When faced with $\frac{3}{4} + \frac{1}{6}$, we need to find a common multiple of the two denominators, that is, a number into which both 4 and 6 divide. One possibility is to multiply them: $4 \times 6 = 24$. This will work perfectly well. But it will reduce the cancelling later on if we use their **lowest common multiple**, 12, instead. Now we convert each into 12ths, and add: $\frac{3}{4} + \frac{1}{6} = \frac{9}{12} + \frac{2}{12} = \frac{11}{12}$.

Recurring decimals

Recurring decimals are decimal expansions which repeat for ever without end. For example $\frac{1}{12} = 0.083333333...$ This is conventionally written as $0.08\dot{3}$. Others have longer repeating patterns: $\frac{2}{7} = 0.285714285714285714...$, which is written as $0.\overline{285714}$ (or $0.\dot{2}8571\dot{4}$). Recurring decimals can always be rewritten as exact fractions, and so represent rational numbers.

Irrational numbers such as $\sqrt{2}$ have decimal expansions which continue for ever without repeating, and so are not recurring decimals.

Some numbers have both terminating and recurring representations: the number 1 is an example.

$0.\dot{9} = 1$

The fact that the recurring decimal $0.\dot{9}$ (that is, $0.99999999...$) is equal to 1 is one of the most resisted facts in elementary mathematics. Students often insist that the two numbers are 'next to each other', but not the same. Or they say that something should happen 'after all the 9s'. Here are three different proofs that the two are equal. (Of course any one of them is enough.)

1 Switching between fractions and decimals, $\frac{1}{3} = 0.\dot{3}$ (that is, $0.333333...$) Multiplying both sides by 3, we get $1 = 0.\dot{9}$.

2 $1 - 0.\dot{9} = 0.\dot{0}$ Obviously, $0.\dot{0} = 0$. In other words the distance between 1 and $0.\dot{9}$ is 0, so it must be that they are the same.

3 Let $x = 0.\dot{9}$. Multiplying by 10, we get $10x = 9.\dot{9}$. Subtracting the first equation from the second shows that $9x = 9$, and so $x = 1$.

The last of these is the template for converting any recurring decimal into a fraction. Let $x = 0.\dot{4}$. Multiplying by 10, we get $10x = 4.\dot{4}$. Subtracting the first equation from the second shows that $9x = 4$, and so $x = \frac{4}{9}$.

Irrational numbers

'Rational number' is the mathematicians' term for a fraction. Numbers, such as π, $\sqrt{2}$ and exponential e, which can never be written exactly as fractions, are therefore *irrational*. (Again this has nothing to do with being illogical or stupid.) The Pythagorean cult in ancient Greece attached mystical significance to the integers, and believed that all numbers were rational. According to legend, when Hippasus of Metapontum first proved the irrationality of $\sqrt{2}$ around 500 BC, he was drowned for heresy.

The irrationality of $\sqrt{2}$

The fact that $\sqrt{2}$ is irrational is a famous example of **proof by contradiction**. Being irrational means that $\sqrt{2}$ can never be written exactly as a fraction. So the proof begins by assuming that it can, say $\sqrt{2} = \frac{a}{b}$. The aim is now to produce a contradiction from this assumption.

If a and b have any common factors then the fraction can be cancelled down: we will assume this has been done. So a and b have no common factors. Now, the definition of $\sqrt{2}$ means that when you multiply it by itself you get 2. So $\frac{a}{b} \times \frac{a}{b} = 2$. That is, $\frac{a^2}{b^2} = 2$, and so $a^2 = 2b^2$. So a^2 is even. Now it must be that a is itself even, because the square of any odd number is odd. So a is a multiple of 2. Say $a = 2c$. Then $(2c)^2 = 2b^2$, and $4c^2 = 2b^2$. But then $b^2 = 2c^2$, and so b^2 is even, and therefore b is even and so divisible by 2. But we assumed that a and b had no common factors, and now we have shown that they are both divisible by 2, which is the required contradiction.

This proof, often attributed to Hippasus of Metapontum around 500 BC, can be adapted to show that the square root of any prime number (in fact any non-square whole number) is irrational.

The problem of the ray

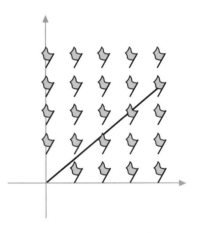

We work on the plane equipped with Cartestian coordinates, and we plant a flag at all the points with integer coordinates $(1, 1)$, $(2, 5)$, $(-4, 7)$, and so on. Now imagine that a laser beam is fired from the origin out across the plane. Will it eventually hit a flagpole or not?

The answer depends on the gradient m of the straight line followed by the ray. The equation of this line is $y = mx$. If the ray hits the post at (p, q), then it must be that $q = mp$, and so $m = \frac{q}{p}$. Since q and p are whole numbers, this means that m is a rational number. So the answer is that if m is rational the ray will hit a post, and if m is irrational it will not.

The problem of the reflected ray

König and Szücs considered an interesting adaptation of the problem of the ray. Instead of posts, they imagine a square whose internal walls are mirrors. A laser is fired from one corner into this mirrored box. What sort of path will it follow? (If the ray ever hits a corner of the box, we assume it bounces back in the direction it came.)

Again the answer depends on the initial gradient m of the ray. If m is rational, then after some time the ray will start retracing its former path, and will then repeat the same loop over and over again. On the other hand, if m is irrational, then the ray never repeats itself. The resulting line will be *dense* inside the box. This does not mean that it will literally pass through every point of the interior of the box (so it is not quite a **space-filling curve**).

But if you choose any point inside the box, and specify some distance, no matter how tiny, then eventually the ray will pass within that distance of your chosen point.

FACTORS AND MULTIPLES

Multiplying negative numbers

Suppose Anna pays Bob $3 every day, and has been doing so for some time. So, each day, Bob's balance changes by $+3$, and Anna's by -3. In 2 days from now, Bob's balance will be $+6$, compared with today, illustrating that $2 \times 3 = 6$. On the other hand -2 days from now (that is two days ago) Bob had $6 less, that is -6, relative to today.

Writing these in a table we get:

$2 \times 3 =$		6
$1 \times 3 =$		3
$0 \times 3 =$		0
$-1 \times 3 =$	-3	
$-2 \times 3 =$	-6	

(The middle row reflects that we are comparing Bob's balance to today's level. After 0 days, of course it has changed by 0 dollars!)

Now consider Anna's money, which changes by -3 dollars each day. So, in $+2$ days time, it will be -6, compared with today's level, illustrating that $2 \times -3 = -6$. What about -2 days from now? Anna had $6 more, that is, a relative level of $+6$. Putting these in a table we get the multiplication table for -3:

$2 \times -3 =$	-6	
$1 \times -3 =$	-3	
$0 \times -3 =$		0
$-1 \times -3 =$		3
$-2 \times -3 =$		6

Another way to think of this is that multiplying something by a negative number always changes its sign between $+/-$, but multiplying by a positive number always leaves the sign unchanged. So, starting with 3 and multiplying by -2 changes its sign: $+3$ becomes -6. But this applies to negative numbers too: starting with -3 and multiplying by -2 changes its sign: -3 becomes $+6$.

Division by 0

Dividing by 0 is probably the commonest mathematical mistake of all. Even experienced researchers can tell horror stories of finding division by 0 lurking within their proofs. There is a good reason why division by 0 is forbidden, coming straight from the meaning of 'divide'. We write that $8 \div 2 = 4$ because 4 is the number which, when multiplied by 2, gives 8. So to calculate $8 \div 0$ we would need to find a number which, when multiplied by 0 gives 8. Obviously, there is no such number. On the other hand, to calculate $0 \div 0$, we would need a number which, when multiplied by 0, gives 0. But every number satisfies this!

Differential calculus studies fractions of the form $\frac{x}{y}$ as x and y both get closer and closer to 0. Depending on the precise relationship between x and y, this fraction can approach any fixed number, or explode out to infinity, or cycle around seemingly at random.

A proof that 1 = 2

Let $a = b$. Multiplying both sides by b, we get $ab = b^2$. Subtracting a^2 from both sides gives $ab - a^2 = b^2 - a^2$. Factorizing each side gives $a(b - a) = (b + a)(b - a)$. Cancelling $(b - a)$ from both sides shows that $a = b + a$. But $a = b$, so this says that $a = 2a$, and so $1 = 2$. The algebra here is pure camouflage, and endless variations are possible (the more apparently sophisticated, the better). The basic argument is $1 \times 0 = 2 \times 0$ (which is undeniably true), therefore $1 = 2$. The false step in the 'proof' is the cancellation of $(b - a)$ from both sides, which amounts to hidden division by 0.

The fundamental theorem of arithmetic

In Proposition 7.30 of **Euclid's** *Elements*, an important property of prime numbers is recognized: if a prime number p divides $a \times b$, then it must also divide either a or b. This is not true of composite numbers such as 10, which divides 5×4, but neither 5 nor 4.

A consequence of this, also known to the ancient Greeks, is the *fundamental theorem of arithmetic*, which says two things:

1 Every positive whole number can be broken up into prime factors.

2 This can happen in only one way.

So, 308 can be broken down into $2 \times 2 \times 7 \times 11$, and the only other ways of writing 308 as a multiple of primes are re-orderings of this (such as $11 \times 2 \times 7 \times 2$). So we know immediately, without checking, that $308 \neq 2 \times 2 \times 2 \times 3 \times 13$.

As its name suggests, this fact is a foundation for the whole of mathematics. It is peculiar to the system of natural numbers, however. In the rational numbers, nothing similar holds, as there are many different ways to divide up a number. For example $2 = \frac{4}{3} \times \frac{3}{2}$ and $2 = \frac{7}{8} \times \frac{16}{7}$.

Highest common factor

The highest common factor (or *greatest common denominator*) of two natural numbers is the largest number which divides each of them. For example, the highest common factor of 18 and 24 is 6: there is nothing bigger which divides both.

The *hcf* of two numbers can be found by dividing them up into primes, and multiplying together all their common prime factors (including any repetitions). For example, $60 = 2 \times 2 \times 3 \times 5$ and $84 = 2 \times 2 \times 3 \times 7$. So the hcf of 60 and 84 is $2 \times 2 \times 3 = 12$. The same method extends to finding the hcf of more than two numbers. Numbers like 8 and 9, whose highest common factor is 1, are called *co-prime*.

Lowest common multiple

The *lowest common multiple* (or *lowest common denominator*) of two natural numbers is the smallest number into which they both divide. So the lowest common multiple of 4 and 6 is 12, since this is the first number appearing in both the 4 and 6 times tables. The *lcm* of two numbers can be found by multiplying them together, and then dividing by their highest common factor. So the lcm of 60 and 84 is $\frac{60 \times 84}{12} = 420$.

Perfect numbers

A whole number is *perfect* if all its factors (including 1 but not the number itself) add together to make the original number. The first perfect number is 6: its factors are 1, 2 and 3. The next is 28 whose factors are 1, 2, 4, 7 and 14. Perfect numbers have aroused curiosity since the Pythagoreans, who attached a mystical significance to this balancing out of additive and multiplicative components. Perfect numbers continue to attract the attentions of mathematicians today, and their study is divided between the even perfect numbers and the odd perfect numbers.

Even perfect numbers

It was around 300 BC, in Euclid's *Elements*, that the first important result on perfect numbers was proved. Proposition 9.36 proved that if $2^k - 1$ is prime, then $\frac{(2^k - 1)\,(2^k)}{2}$ is perfect. Centuries later this result would be revisited, once prime numbers of the form $2^k - 1$ had come to be known as **Mersenne primes**. So, to rephrase Euclid's result, if M is a Mersenne prime, then $\frac{M(M+1)}{2}$ is an even perfect number.

The converse to Euclid's theorem is also true, as the 10th-century scientist Ibn al-Haytham noticed. If an even number is perfect, then it must be of the form $\frac{M(M+1)}{2}$ where M is a Mersenne prime. For example, $6 = \frac{3 \times 4}{2}$ and $28 = \frac{7 \times 8}{2}$. However Ibn al-Haytham was not able to prove this result fully. That had to wait for Leonhard Euler, around 800 years later.

This result, sometimes known as the *Euclid–Euler theorem*, establishes a one-to-one correspondence between even perfect numbers and Mersenne primes. So discoveries of new Mersenne primes immediately give new even perfect numbers. Similarly, questions on one side translate to questions on the other. In particular, this holds for the biggest question in the area: whether the list of even perfect numbers is finite or infinite.

Odd perfect numbers

No-one has yet found an odd perfect number and most experts today doubt their existence. However, no-one has managed to prove that such a creature cannot exist either. In any case, their likely non-existence is no obstacle for mathematicians to study them in depth. By investigating what an odd perfect number will look like if it does exist, mathematicians hope either to pin down where one might be found, or gather ammunition for an eventual **proof by contradiction**.

In the 19th century, James Sylvester identified numerous conditions which will have to be met by any odd perfect number. Sylvester believed that for such a number to exist 'would be little short of a miracle'. Certainly, if there are any, they will have to be extremely large. In 1991, Brent, Cohen and te Riele used a computer to rule out the existence of any odd perfect number shorter than 300 digits long.

Amicable pairs

Most numbers are not perfect. When you add up their factors you are likely either to come up short (in the case of a *deficient* number) or overshoot (for an *abundant* number). But sometimes abundant and deficient numbers can balance each other out. For example, 220 is abundant: its factors are 1, 2, 4, 5, 10, 11, 20, 22, 44, 55 and 110, which sum to 284. The factors of 284 are 1, 2, 4, 71 and 142, which total not to 284, but back to 220.

Mathematicians of the classical and Islamic worlds were fascinated by *amicable pairs* of numbers such as this. In the 10th century, Thâbit ibn Kurrah discovered a rule for producing amicable pairs, which was later improved by Leonhard Euler.

At the time of writing, 11,994,387 different amicable pairs are known, the largest being 24,073 digits long. As with perfect numbers, it remains an open question whether there are really infinitely many of them. Every known pair consists either of two odd or two even numbers, but it has never been proved that this must always be the case.

Sociable numbers

An amicable pair consists of two numbers where the process of adding up all the factors for each number takes you from one to the other. But longer cycles can exist too: adding up the factors of 12,496 gives 14,288. Next we get 15,472, then 14,536 and 14,264, before getting back to 12,496. So these form a cycle of length 5. Numbers like this are known as *sociable*. At the time of writing, the longest known cycle of sociable numbers has length 28. It is: 14316, 19116, 31704, 47616, 83328, 177792, 295488, 629072, 589786, 294896, 358336, 418904, 366556, 274924, 275444, 243760, 376736, 381028, 285778, 152990, 122410, 97946, 48976, 45946, 22976, 22744, 19916, 17716.

Aliquot sequences

Start with any number S_1, and add up all its proper factors to get a new number S_2. If $S_1 = S_2$, then we have a perfect number. Otherwise, repeat the process: add up all the proper factors of S_2, to get a new number S_3. If $S_1 = S_3$, then S_1 and S_2 form an amicable pair. Otherwise we can continue, to get S_4, S_5, S_6 and so on. This is called an *aliquot sequence* (*aliquot* is the Latin word for 'several'). The aliquot sequence of 95 is 95, 25, 6, 6, 6, … This has landed on a perfect number, where it stays. Similarly, if an aliquot sequence ever lands on a member of an amicable pair, or any sociable number, then it will simply cycle around for ever.

An alternative is for the sequence to hit a prime number, which marks its death. The sequence starting with 49, for example, proceeds next to 8, and then to 7. But the next number must be 1 (since 7 has no other factors), and after that 0.

In 1888, Eugène Catalan conjectured that every aliquot sequence will end in one of these ways. But some numbers are throwing this conjecture into serious doubt: 276 is the first of many numbers whose eventual fate is unknown. Around 2000 terms of its aliquot sequence have been calculated to date and so far it is simply growing and growing. After 1500 terms, the numbers are over 150 digits long.

INDUCTION

Proof by induction
How can you prove infinitely many things in one go? One prized technique is *mathematical induction*, which is used to prove results involving natural numbers. Suppose I want to prove that every number satisfies some property, call it X. Induction attacks the natural numbers in order. The *base case* of the argument is to show that 0 is an X-number. The *inductive step* is to show that *if* 0, …, k are all X-numbers, then $k + 1$ must be one too. If this can be done, then no number can ever be the first non-X-number, so there cannot be any non-X-numbers at all. So all numbers are X-numbers.

Induction resembles a mathematical domino-effect: the base case pushes over the first domino. Then the inductive step shows that the first domino must knock over the second, and the second will knock over the third, and so on, until no domino is left standing. Induction is a defining feature of the natural numbers: it does not apply directly to the real numbers for example. Adding the numbers 1 to 100 is an example of induction in use, and light-hearted takes on it include the **bald man paradox**, and the **proof that every number is interesting**.

1729 is interesting

$$1729 = 1^3 + 12^3 = 9^3 + 10^3$$

When G.H.Hardy visited the extraordinary mathematical virtuoso Srinivasa Ramanujan, he had travelled in a taxi numbered 1729. He remarked to his friend that the number seemed rather a dull one. 'No,' replied Ramanujan immediately, 'it is a very interesting number; it is the smallest number expressible as the sum of two cubes in two different ways'. A lesser mind than Ramanujan's could instead have argued that every natural number is interesting, a fact for which there is a famous light-hearted 'proof' by induction.

A proof that every number is interesting

The *base case* is the number 0: unquestionably among the most interesting of all numbers. The *inductive step* assumes that the numbers 0, 1, 2, ..., k are all interesting. The next one is $k + 1$ which must be either interesting or not. If not, then it is the first non-interesting number, but this would make it of unique interest: a contradiction. So, it must be interesting. This completes the inductive step, and so by induction, every number is interesting.

This is, of course, a parody of a proof, rather than the real thing. *Interestingness* is not rigorously defined, and more accurately operates on a subjective sliding-scale. In reality some numbers are more interesting than others (and it depends on who you ask). This 'proof' requires an artificial inflexibility whereby every number is deemed either absolutely interesting or not.

Adding up 1 to 100

In a classroom in 18th-century Germany, primary school teacher Herr Büttner set his class the task of adding up all the numbers from 1 to 100. Hoping for a nice, quiet lesson, he was surprised when a boy put up his hand within seconds, to give the answer: 5050. The boy grew up to become a towering figure in the history of mathematics: Carl Friedrich Gauss. The young Gauss must have worked out the formula for adding up successive numbers: $1 + 2 + \cdots + n = \frac{n \times (n+1)}{2}$. For $n = 3$, we add up the first three numbers: $1 + 2 + 3 = 6$: the same answer given by $\frac{3 \times (3+1)}{2}$. Gauss realized that all he needed to do was substitute $n = 100$ into this formula to get his answer: $\frac{100 \times (100+1)}{2}$

There is a geometric way to see this, by considering the nth **triangular number**. It can be proved more rigorously by induction.

Adding 1 up to n, proof by induction

We shall prove by induction that $1 + 2 + \cdots + n = \frac{n \times (n+1)}{2}$. The *base case* involves adding up the first zero numbers. Obviously the answer is 0. The right-hand side is $\frac{0 \times (0+1)}{2}$, so the formula holds for $n = 0$. Now, the inductive step: we suppose that the formula holds for some particular value of n, say k. So $1 + 2 + \cdots + k = \frac{k \times (k+1)}{2}$. We want to deduce that the corresponding thing holds for the next term, $n = k + 1$. It follows that $1 + 2 + \cdots + k + (k + 1) = \frac{k \times (k+1)}{2} + (k + 1)$. After a little algebraic sleight of hand, the right-hand side comes out as $\frac{(k+1) \times (k+2)}{2}$, which is the correct formula for $n = k + 1$.

Adding the first hundred squares

The formula for adding up 1 to n can be written neatly in sum notation as:

$$\sum_{j=1}^{n} j = \frac{n(n+1)}{2}$$

There is a similar formula for the sum of the first n square numbers, which can also be proved by induction:

$$\sum_{j=1}^{n} j^2 = \frac{n(n+1)(2n+1)}{6}$$

So if we want to know $1^2 + 2^2 + 3^2 + \cdots + 100^2$, all we have to do is put in $n = 100$ into this formula: $\frac{100 \times 101 \times 201}{6}$, which is 338,350.

To sum cubes, we have the formula:

$$\sum_{j=1}^{n} j^3 = \frac{n^2(n+1)^2}{4}$$

(Notice that this is just the first formula squared.) For higher powers, we get:

$$\sum_{j=1}^{n} j^4 = \frac{1}{5}n^5 + \frac{1}{2}n^4 + \frac{1}{3}n^3 - \frac{1}{30}n$$

$$\sum_{j=1}^{n} j^5 = \frac{1}{6}n^6 + \frac{1}{2}n^5 + \frac{5}{12}n^4 - \frac{1}{12}n^2$$

$$\sum_{j=1}^{n} j^6 = \frac{1}{7}n^7 + \frac{1}{2}n^6 + \frac{1}{2}n^5 - \frac{1}{6}n^3 + \frac{1}{42}n$$

To give a general formula for higher powers requires more work, and involves the *Bernoulli numbers*, an important sequence in number theory.

REPRESENTATIONS OF NUMBERS

Place value and decimal notation

Counting from 1 to 9 is easy; all we need to do is remember the correct symbols. But something strange happens when we arrive at 10. Instead of a new symbol, we start recycling the old ones. This is a deceptively sophisticated system, which took many centuries to evolve. A crucial moment was the arrival of a symbol for 0.

In *place value* notation, the symbol '3' does not just represent the number 3. It can also stand for 30, 300 or 0.3. The position of the symbol carries as much meaning as the symbol itself. Whole numbers are represented as digits arranged in columns. On the right are the units (ones), and with each step left we go up a power of 10. In a place value table, the number 1001 is shown as:

Thousands	Hundreds	Tens	Units
1	0	0	1

This is called the *decimal* system, because 10 is its base. Other choices of base are perfectly possible.

Bases

The fact that 10 is the base of our counting system is surely due in part to evolution gifting us 10 fingers rather than 8 or 12. From a mathematical perspective, it is an arbitrary choice. You can form an equally good counting system based on any number. Indeed a binary system, that is base 2, has some advantages (at least in the computer age). It should be stressed that this discussion is only concerned with how to *represent* numbers using symbols. Whether written as 11 in decimal notation, 1011 in binary, B in hexadecimal, or 10 in base 11, the fundamental object is the same throughout. It is no more affected by these cosmetic alterations than a human is by a change of hat.

Today we mostly work with decimals, but not entirely. Telling the time is not decimal: there are 60 seconds in a minute, 60 minutes in an hour. This is a hangover from ancient Babylon, whose mathematicians and bureaucrats worked in base 60. The ancient Chinese divided the day into 100 *ke* (around $\frac{1}{4}$ of an hour), until the adoption of the western system in the 17th century. Plans to replace hours and minutes with a decimal system have come and gone several times since then, only ever meeting with fleeting success, such as during the French revolution.

Binary

Binary means 'base 2'. So, counting from the right, the place values represent units (or ones), twos, fours, eights, sixteens, etc. (powers of 2). To translate a decimal number into binary, we break it down into these pieces. For example, $45 = (1 \times 32) + (1 \times 8) + (1 \times 4) + (1 \times 1)$, giving a binary representation of 101101. Going the other way, we translate binary to decimal as follows: 11001 has 1 in the units column, 0 in the twos column, 0 in the fours, 1 in the eights, and 1 in the sixteens. So it represents $1 + 8 + 16 = 25$.

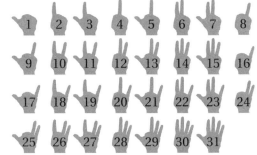

Binary is the most convenient base for computers, since the 1 and 0 can be stored as the 'on' and 'off' settings of a basic component. These binary digits are known as *bits*. Eight bits make a *byte*, which is used to measure computer memory. The way in which strings of bits can carry data is the subject of **information theory**. Like so much of modern mathematics, binary was first conceived by Gottfried Leibniz in the 17th century.

Binary is not only useful for computers. Using your fingers to represent bits, it is possible to count to 31 on one hand, or 1023 on two.

Hexadecimals
Binary may be the easiest numerical representation for a computer but, to the human eye, a long string of 1s and 0s is not easy to decipher. For the most part, we deal in decimals instead. These have the disadvantage of not being easy to convert into binary. For this reason, some computer scientists prefer to work in *hexadecimals*, base 16. The digits 0–9 have their ordinary meanings, and A, B, C, D, E, F stand for 10, 11, 12, 13, 14 and 15 respectively. The decimal '441' would be written in hexadecimals as '1B9'.

Translating between binary and hexadecimals is much easier than between binary and decimals. We just group the binary expression into clumps of four digits, and translate each in turn. So 1111001011 gets grouped as '(00)11 1100 1011', which translates to hexadecimals as 3CB.

Standard form
Thanks to our decimal system, the number 10 has several uniquely useful properties. Multiplying or dividing by 10 has the effect of sliding the digits one place left or right, relative to the decimal point. For example, $47 \div 10 = 4.7$, and $0.89 \times 10 = 8.9$. Exploiting this, every number can be expressed as a number between 1 and 10, multiplied or divided by some number of 10s, which can be written as a positive or negative power of 10. This is called *standard form*. For example, 3.14×10^6 and 2.71×10^{-5} are both written in standard form.

To convert an ordinary number, such as 14,100, into standard form, first slide the decimal point so that it gives a number between 1 and 10. In this case it gives 1.41. To get back to 14,100 we need to multiply by 10 four times. So in standard form 14100 becomes 1.41×10^4. For small numbers such as 0.00173, the procedure is the same. First slide the decimal point (right this time) to give 1.73. This time, to get back to the original number, we need to divide by 10 three times. So we get a negative power: 1.73×10^{-3}.

Standard form is useful as it allows the order of magnitude of the number to be gauged quickly (by the power of 10), and meshes well with the metric system.

Surds
Because $\sqrt{2}$ is an **irrational number**, it cannot be written exactly as a fraction. Worse, because its decimal expansion goes on for ever and never repeats it cannot be written down exactly as a terminating or recurring decimal. In fact, the best way to write $\sqrt{2}$ *exactly* is as '$\sqrt{2}$'. So when the result of a calculation involves $\sqrt{2}$, there is a good argument for maintaining precision by keeping it in this form, as for example in $3 + \sqrt{2}$. Expressions like this are called *surds*. (This word comes from the Latin *surdus*, meaning 'voiceless' reflecting al Al-Khwarizmi's take on irrational numbers, against the more 'audible' rationals.)

What goes for $\sqrt{2}$ applies equally to the square root of any non-square natural number. Although it makes sense to keep these expressions as surds rather than approximating them with decimals or fractions, we usually want to simplify these expressions as far as possible. The key mathematical ingredient in working with surds is the observation that $\sqrt{a \times b} = \sqrt{a} \times \sqrt{b}$, and in particular $\sqrt{a^2} = a$. So, rather than writing $\sqrt{12}$ we simplify it: $\sqrt{4 \times 3} = \sqrt{4} \times \sqrt{3} = 2\sqrt{3}$.

Rationalizing the denominator

If we have an expression like $\frac{2}{1+\sqrt{3}}$, we often want to simplify to it a form where the square roots occur only on the top of the fraction. To achieve this, we can always multiply the top and bottom of any fraction by the same thing without changing its value. This trick is in choosing the right multiplier. In this case choose $1 - \sqrt{3}$. To see that this works:

$$\frac{2}{1+\sqrt{3}} = \frac{2 \times (1-\sqrt{3})}{(1+\sqrt{3}) \times (1-\sqrt{3})} = \frac{2 - 2\sqrt{3}}{1 + \sqrt{3} - \sqrt{3} - \sqrt{3} \times \sqrt{3}}$$

$$= \frac{2 - 2\sqrt{3}}{1 - 3} = \frac{2 - 2\sqrt{3}}{-2} = \sqrt{3} - 1$$

In general, if $a + \sqrt{b}$ is the denominator of a fraction, it can be rationalized by multiplying top and bottom by $a - \sqrt{b}$. This will transform the denominator into the whole number $a^2 - b$ (see **difference of two squares**).

Large numbers

Large numbers have always held a great fascination for humans (or at least humans of a certain disposition). Jaina mathematicians of ancient India attached deep mystical significance to enormous numbers. They defined a *rajju* as the distance a God travels in six months. (Gods travel a million kilometres in every blink of the eye.) Building on this, they imagined a cubic box whose sides are one rajju long, filled with wool. A *palya* is then the time it will take to empty the box, removing one strand per century. The Jains also developed a theory of different denominations of infinity, anticipating Georg Cantor's **set theory** by over 2000 years.

Archimedes' Sand Reckoner

In his work the *Sand Reckoner* of around 250 BC, Archimedes estimated the number of grains of sand needed to fill the universe. His solution was that no more than 10^{63} grains should be required. This number is of no great interest, marred as it is by the heliocentric cosmology of Archimedes' time, where the stars were assumed a fixed distance from the sun. Nevertheless the text marks an important conceptual distinction: that between very large natural numbers and infinity.

The *Sand Reckoner* was no idle game. Archimedes was correcting a common misconception of his time: that there is no number which can measure anything so huge, that sand is essentially an infinite quantity. To do this, he had to invent a whole system of notation for large numbers. He got as far as 'a myriad myriad units of a myriad myriad numbers of the myriad myriadth period' or, as we would write it, $10^{8 \times 10^{16}}$. Modern notation, such as taller towers of exponentials, can take us far beyond this.

Towers of exponentials

How can large numbers best be described, using modern mathematics? We could start by trying to write them down in ordinary decimal place value notation. Following this strategy, a *googol* is:

$$10,000,000,000,000,000,000,000,000,000,000,000,000,000,000,000,000,$$
$$000,000,000,000,000,000,000,000,000,000,000,000,000,000,000$$

It is much easier to say '1 followed by a hundred zeroes', or 10^{100}. This illustrates that exponentiation (raising to a power) is excellent at capturing large numbers. (The term 'googol' was coined by Milton Sirotta, aged 9, in 1938.)

For most purposes, exponentiation is all we need. The number of atoms in the universe is around 10^{80}, and the number of possible games of chess is estimated at 10^{123}. But we can always concoct larger numbers. A *googolplex* is 1 followed by a googol zeroes. In exponential notation that is:

$$10^{10,000,000,000,000,000,000,000,000,000,000,000,000,000,000,000,000,000,000,000,000,000,000,000,000,000,000,000,000,000,000,000,000,000,000,000}$$

It is more satisfactory to write $10^{10^{100}}$ or $10^{10^{10^2}}$. (The term 'googolplex' was chosen by Milton Sirotta's uncle, the mathematician Edward Kasner.) This suggests how to extend the system, by building ever taller *towers* of exponentials, such as $10^{10^{10^{10^{10^{10}}}}}$. To build larger numbers than towers can give us, we need Knuth's arrow notation.

Knuth's arrows
Donald Knuth occupies a place in every mathematician's heart. As the creator of the typesetting programming *TeX*, he is largely responsible for what modern mathematics looks like, in the pages of countless books and journals. In 1976, Knuth also devised an efficient notation for writing down very large numbers. It is based on iteration. To start with, multiplication is iterated (repeated) addition: $4 \times 3 = 4 + 4 + 4$. Then exponentiation is iterated multiplication: $4^3 = 4 \times 4 \times 4$. Exponentiation is the first arrow, so $4\uparrow3$ means $4^3 = 64$. The second arrow is the first arrow iterated: $4\uparrow\uparrow3 = 4\uparrow4\uparrow4$, which means 4^{4^4}. This is considerably more than a googol, while $4\uparrow\uparrow4$ is $4^{4^{4^4}}$, which dwarfs a googolplex.

Similarly the third arrow is the second iterated: $4\uparrow\uparrow\uparrow3 = 4\uparrow\uparrow4\uparrow\uparrow4$, which is $4^{4^{.^{.^{.^4}}}}$ where the tower is 4^{4^4} storeys high. This is a stupendously large number, which is already almost impossible to express in any other way. We continue by defining the fourth arrow as the third iterated, and so on. The next problem is that the number of arrows might grow unmanageable. To counter this, we may write $4\{n\}3$ as short-hand for $4\uparrow\uparrow\uparrow\dots\uparrow3$ where there are n arrows. For still larger numbers, we need more powerful notation, such as Bowers' operators.

Bowers' operators
The Texan amateur mathematician Jonathan Bowers has devoted a great deal of time into finding and naming ever larger numbers. At time of writing, his largest is a colossus he has called *meameamealokkapoowa oompa*. Bowers' basic idea is a process to far extend Knuth's arrows. His first operator is $\{\{1\}\}$ defined by:

$$m\{\{1\}\}2 = m\{m\}m$$
$$m\{\{1\}\}3 = m\{m\{m\}m\}m$$
$$m\{\{1\}\}4 = m\{m\{m\{m\}m\}m\}m$$

and so on. This is enough to locate one of the largest of all mathematical constants, **Graham's number**. But we can continue with a second operator {{2}} defined by:

$$m\{\{2\}\}2 = m\{\{1\}\}m$$
$$m\{\{2\}\}3 = m\{\{1\}\}(m\{\{2\}\}2)$$
$$m\{\{2\}\}4 = m\{\{1\}\}(m\{\{2\}\}3)$$

and so on. Then the operators {{3}}, {{4}}, etc. can all be defined analogously.

We begin the next level with {{{1}}}, which behaves in relation to {{·}} as {{·}} does to {·}, and so on. We can press on, with a new function which counts the brackets: we write $[m, n, p, q]$ to mean $m\{\{\ldots\{p\}\ldots\}\}n$ where there are q sets of brackets. Of course, Bowers does not stop here, pushing this line of thought to ever more outrageous heights. But some numbers will always remain out of reach, such as Friedman's *TREE*(3).

Graham's number

Graham's number is often cited as the largest number ever put to practical use. The previous record holder was *Skewe's number*, a puny $10^{10^{10^{34}}}$. (Whether Graham's number still holds the crown depends on whether you class the likes of *TREE*(3) as useful.) While Skewes' number can easily be written as a short tower of exponentials, it is impossible to describe Graham's number without the aid of some heavy machinery, such as Bowers' operators. In these terms, Graham's number lies between 3{{1}}63 and 3{{1}}64.

To give some idea of its magnitude, we start with 3^{3^3} (which is 7,625,597,484,987). Next, we build a new tower of threes, 3^{3^3} storeys high. Call this number A_1 (this is already unimaginably large). Then we build A_2 as a tower of 3s with A_1 many storeys, and A_3 a tower of 3s with A_2 many storeys. We keep repeating this process, all the way to A_{A_1}. This number is B_1 (in Knuth arrow notation it is $3\uparrow\uparrow\uparrow\uparrow3$). Then we form B_2 as $3\uparrow\uparrow\ldots\uparrow3$ where there are B_1 arrows. In short-hand that is $B_2 = 3\{B_1\}3$. Next, $B_3 = 3\{B_2\}3$, $B_4 = 3\{B_3\}3$, and so on. Graham's number lies between B_{63} and B_{64}. Employed by Ronald Graham in 1977, it is an upper bound for a problem in **Ramsey theory**, for which a conjectured true solution is 12.

Friedman's *TREE* sequence

In the 1980s, the logician Harvey Friedman discovered a rapidly growing sequence he dubbed *TREE*, deriving from a problem in Ramsey theory. At first glance, the problem does not seem too troublesome, and the sequence starts innocuously enough with *TREE*(1) = 1, *TREE*(2) = 3. But then comes *TREE*(3), where we hit a wall. *TREE*(3) renders even Bowers' operators powerless. Friedman realized that any attempt to describe *TREE*(3) in ordinary mathematical language would necessarily involve 'incomprehensibly many symbols' (e.g more than Graham's number). Essentially, it is indescribable. You could write out Bowers' operators at higher and higher levels until the end of the universe, without making the slightest impact on it.

Friedman's *TREE* sequence grows so fast that ordinary mathematics (as formalized in **Peano arithmetic**) simply cannot cope, making this one of the most concrete established examples of **Gödelian incompleteness**.

Continued fractions

A *continued fraction* is an object such as:

$$a + \cfrac{b}{c + \cfrac{d}{e + \cfrac{f}{g + \cdots}}}$$

This can be written more briefly as: $a + \dfrac{b}{c+} \dfrac{d}{e+} \dfrac{f}{g+}$

Usually, $a, b, c, d, e, f, g \ldots$ are integers. It is a difficult problem to decide whether a sequence $a, b, c, d \ldots$ produces a continued fraction which converges to some fixed number or one which diverges to infinity.

However, plenty of examples of convergent continued fractions are known. The simplest infinite continued fraction is that for the **golden section**:

$$\phi = 1 + \frac{1}{1+} \ \frac{1}{1+} \ \frac{1}{1+} \ \frac{1}{1+} \ \cdots$$

Another nice example is that for the square root of 2:

$$\sqrt{2} = 1 + \frac{1}{2+} \ \frac{1}{2+} \ \frac{1}{2+} \ \frac{1}{2+} \ \cdots$$

Leonhard Euler showed that if a continued fraction continues for ever and converges, it must represent an **irrational number**. He then deduced for the first time that the number e is irrational by showing that it is equal to:

$$e = 2 + \frac{1}{1+} \ \frac{1}{2+} \ \frac{1}{1+} \ \frac{1}{1+} \ \frac{1}{4+} \ \frac{1}{1+} \ \frac{1}{1+} \ \frac{1}{6+} \ \cdots$$

These are all *simple continued fractions*, because their numerators (top rows) are all 1.

Non-simple continued fractions

A non-simple continued fraction for e is:

$$e = 2 + \frac{1}{1+} \ \frac{1}{2+} \ \frac{2}{3+} \ \frac{3}{4+} \ \frac{4}{5+} \ \frac{5}{6+} \ \cdots$$

One of the earliest continued fractions was discovered by Lord William Brouncker, in the early 17th century:

$$\frac{4}{\pi} = 1 + \frac{1^2}{2+} \ \frac{3^2}{2+} \ \frac{5^2}{2+} \ \frac{7^2}{2+} \ \cdots$$

This can be manipulated to produce a continued fraction for π. But π's **simple continued fraction** remains mysterious. Ramanujan's continued fractions are not only non-simple, but are not even built from whole numbers.

Ramanujan's continued fractions

According to his friend and mentor G.H.Hardy, the Indian virtuoso Srinivasa Ramanujan, had 'mastery of continued fractions ... beyond that of any mathematician in the world'. Ramanujan discovered numerous spectacular formulas involving continued fractions, many of which were discovered in unorganized notebooks after his death. Not only did Ramanujan not provide proofs of his formulas, often he did not even leave hints as to how he performed these astonishing feats of mental acrobatics. It was left to later mathematicians to verify his formulas, and some of his highly individual notation remains undeciphered to this day.

An example of his work on continued fractions is this, involving the golden section ϕ:

$$\cfrac{1}{1+}\ \cfrac{e^{-2\pi\sqrt{5}}}{1+}\ \cfrac{e^{-4\pi\sqrt{5}}}{1+}\ \cfrac{e^{-6\pi\sqrt{5}}}{1+}\ \cdots = e^{\frac{2\pi}{\sqrt{5}}}\left(\frac{\sqrt{5}}{1+\left(5^{\frac{3}{4}}(\phi-1)^{\frac{5}{2}}-1\right)^{\frac{1}{5}}}-\phi\right)$$

(The numerical value of this is around 0.999999, which may be why Ramanujan found it intriguing.) This is a special case of the celebrated *Rogers–Ramanujan fraction*. Discovered independently by Leonard Rogers, this is an ingenious system for calculating the value of

$$\cfrac{1}{1+}\ \cfrac{q}{1+}\ \cfrac{q^2}{1+}\ \cfrac{q^3}{1+}\ \cdots$$

for different values of q.

Forming continued fractions

Suppose we want to convert the ordinary fraction $\frac{43}{30}$ into a simple continued fraction. There are two basic steps: first we separate the integer part from the fractional part, which means splitting $\frac{43}{30}$ into $1+\frac{13}{30}$. Next we turn the fractional part upside down, below a 1. So our fraction becomes:

$$1+\cfrac{1}{\left(\frac{30}{13}\right)}$$

Now we repeat the process with $\frac{30}{13}$. First separate out the integer part: $2+\frac{4}{13}$. Then turn the fractional part upside down, $2+\frac{1}{\left(\frac{13}{4}\right)}$. Putting this together with the previous step takes us to:

$$1+\cfrac{1}{2+\cfrac{1}{\left(\frac{13}{4}\right)}}$$

Separating out the integer part of $\frac{13}{4}$ produces the final result:

$$1+\cfrac{1}{2+\cfrac{1}{3+\frac{1}{4}}}$$

Because we started with a rational number, the process ended after finitely many steps. For an irrational number, this process will produce an infinitely long continued fraction. This is a *simple* continued fraction, because all the components are whole numbers, and all the numerators are 1. Every real number can be expressed as a simple continued fraction, and this can be done in exactly one way.

π's simple continued fraction

A *simple continued fraction* looks like this:

$$a + \cfrac{1}{b + \cfrac{1}{c + \cfrac{1}{d + \cdots}}}$$

We can write the resulting sequence as $(a, b, c, d, \ldots)$.

A perplexing question involves the simple continued fraction for π. The sequence begins: 3, 7, 15, 1, 292, 1, 1, 1, 2, 1, 3, 1, 14, 2, 1, 1, 2, 2, 2, 2, 1, 84, 2, … and has been calculated to 100 million terms by Eric Weisstein, in 2003. The underlying pattern, however, remains mysterious. By truncating the sequence after a few steps, good fractional approximations to π can be found: $3, \frac{22}{7}, \frac{333}{106}, \frac{355}{113}, \frac{103993}{33102}$, etc.

Khinchin's constant

Here is a recipe for producing a new number K from any real number x.

1 x can be expressed as a simple continued fraction, in a unique way. So, just as for π, we get a sequence $(a, b, c, d, e \ldots)$ which encapsulates x.

2 Next we can write a new sequence: $\sqrt{ab}, \sqrt[3]{abc}, \sqrt[4]{abcd}, \sqrt[5]{abcde}, \ldots$

3 This sequence will get closer and closer to some fixed number. This is K, the *geometric mean* of the sequence $(a, b, c, d, e, f, \ldots)$.

This may seem a convoluted process, but the punch-line is truly astounding: almost every value of x produces the same value of K. This amazing number is approximately 2.685452… and is known as *Khinchin's constant*, after its discoverer Aleksandr Khinchin, in 1936.

It is not true that literally every value of x reveals the same value. Rational numbers do not produce K, for example, and neither does e. However, these exceptions are infinitely outnumbered by those which do have Khinchin's constant hiding behind them. If we were able to pick a real number at random, the probability that it would yield K would be 100% (exactly). It is all the more surprising then, that no-one has yet managed to prove that any individual value of x does produce K. π appears to (as does K itself), but no full proofs have yet been found. K itself is very secretive and is not even known to be irrational.

TRANSCENDENTAL NUMBERS

Transcendental numbers

An *irrational number* is one which cannot be written as a fraction of whole numbers. $\sqrt{2}$ is an example. Although $\sqrt{2}$ is irrational, there is a sense in which it is not too far away from the safety of the whole numbers. There is a quick route back from $\sqrt{2}$, using only multiplication: $\sqrt{2} \times \sqrt{2} = 2$, a whole number.

Transcendental numbers such as π are not like this. $\pi \times \pi$ is not a whole number, and neither is $3\pi \times \pi \times \pi$, or $1001\pi^5 + 64\pi$. In fact, there is no route from a transcendental number back to the whole numbers, using addition, subtraction, multiplication and division. The technical definition is that a number a is *transcendental* if there is no **polynomial** built from integers which produces an integer when a is substituted in. Every transcendental number is irrational. However, many irrational numbers, such as $\sqrt{2}$ and indeed all roots and powers of rational numbers, are non-transcendental. Non-transcendental numbers are called *algebraic*.

The transcendence of π and e

The first transcendental numbers were discovered by Joseph Liouville in 1844. The most famous of his numbers is 0.110001000000000000000001000… The 1s appear in the 1st, 2nd, 6th, 24th digit, and so on, in the sequence of **factorials**. The appearance of transcendental numbers was certainly striking, although Liouville's numbers seemed more like artificial curiosities than objects of any profound importance.

However, in 1873 Charles Hermite cemented the importance of transcendence within mathematics by showing that e is transcendental. In 1882 Ferdinand von Lindemann added π to the list. This was enough to settle the ancient problem of **squaring the circle**. The importance of transcendental numbers was established, but no-one could have predicted Georg Cantor's revelation of just how many they are.

Cantor's uncountability of transcendental numbers

Georg Cantor's **set theory** split open the old notion of infinity as a single entity. His contrasting proofs of the uncountability of the real numbers and the countability of the rational numbers showed that the irrational numbers infinitely outnumber their rational cousins.

Cantor went beyond this, however. The *algebraic numbers* include the rational numbers but also many of the more common irrational numbers, such as $\sqrt{2}$. Cantor proved that, as well as all the rational numbers, the algebraic numbers are also countable. This had a stunning consequence: transcendental numbers, rather than being exotic anomalies, are the norm. In fact, *almost every* real number is transcendental. The familiar numbers that we use the most, the integers, rationals and algebraic numbers, are just a tiny sliver in a universe of transcendence.

Transcendental number theory

Georg Cantor had shown that almost all real numbers are transcendental. It is all the more surprising then that specific examples are hard to find. Liouville's numbers, e, and π were for some time the only known examples. Hilbert's 7th problem of 1900 first addressed the core of the difficulty: the way that transcendence and exponentiation interact.

In answer, the *Gelfond–Schneider theorem* of 1934 provided the first solid rule of transcendence: it said that if a is an algebraic number (not 0 or 1), and b is an irrational algebraic number, then a^b is always transcendental. So $\sqrt{2}^{\sqrt{2}}$ and $3^{\sqrt{7}}$ are transcendental, for example. Over the 20th century this beginning was built upon, notably in results such as the six exponentials theorem, and Alan Baker's pioneering work in the 1960s, for which he won the Fields medal.

Baker's work investigated sums of numbers of the form $b \ln a$. His results extended the Gelfond–Schneider theorem to *products* of numbers of the form a^b (where a and and b are both algebraic, a is not 0 or 1, and b is irrational). This hugely increased the stock of known examples of transcendental numbers, by including the likes of $2^{\sqrt{3}} \times 5^{\sqrt{7}} \times \sqrt{10}^{\sqrt{11}}$. However, the phenomenon of transcendence remains extremely difficult to pin down. Even today, the status of many numbers, including e^e and $e + \pi$, is unknown. Most of these unresolved questions would follow from Schanuel's conjecture.

Six exponentials theorem

The six exponentials theorem was proved by Siegel, Schneider, Lang and Ramachandra. It attacks the central problem of transcendental number theory: how often exponentiation produces a transcendental result. It states that if a and b are complex numbers, and x, y and z are complex numbers, then at least one of e^{ax}, e^{bx}, e^{ay}, e^{by}, e^{az} and e^{bz} is transcendental. The caveats are that a and b must be *linearly independent*, meaning that neither is a multiple of the other by a rational number (so $a \neq \frac{3}{4} b$, for example). Similarly none of x, y and z can be reached by multiplying the other two by rational numbers and adding the results together (so $x \neq \frac{1}{3} y + \frac{2}{5} z$, for example).

It is an open question, known as the *four exponentials conjecture*, whether the same holds when z is omitted. This would follow from Schanuel's conjecture.

Schanuel's conjecture

Since the beginning of transcendental number theory, a critical problem has always been how transcendence behaves under exponentiation. In 1960, Stephen Schanuel made a sweeping conjecture which, if proved, would transform our understanding of the whole phenomenon. In fact Schanuel's conjecture would subsume almost every known theorem about transcendental numbers, and at the same time settle hundreds of open questions, including the four exponentials conjecture, and the transcendence of e^e and $e + \pi$.

Schanuel's conjecture is phrased in the technical language of **Galois theory**, and says, in essence, that there are no nasty surprises in store: transcendence and exponentiation interact in as simple a fashion as could be hoped. According to the number theorist David Masser,

Schanuel's conjecture 'is generally regarded as impossibly difficult to prove'. However, in 2004 Boris Zilber applied techniques of **model theory** to provide strong, indirect evidence that Schanuel's conjecture should be true. It remains to be seen whether this insight can be captured as a proof of this momentous conjecture.

RULER AND COMPASS CONSTRUCTIONS

Ruler and compass constructions

The world of geometry is filled with exotic figures and shapes. Mostly we consider them from a lofty theoretical perspective, but how can we actually construct them? Taking this question to its limit, which of these shapes can be constructed using just the simplest of tools: a ruler to draw a straight line, and a compass (pair of compasses) to draw a circle? This was a question which fascinated ancient Greek mathematicians.

Sometimes this process is called *straight-edge and compass construction* to make clear that the ruler is not to be used for measuring: it is just a tool for drawing straight lines. Similarly, the compass can only be set to the length of any line which has already been constructed (or to a randomly chosen length).

As specific examples were conquered (such as the construction of **Gauss' heptadecagon**), the underlying algebraic principles began to become clear. It was the work of Pierre Wantzel in the 19th century, and the development of **constructible numbers**, which removed this question entirely from geometry, and placed it in the realm of algebraic number theory.

Bisecting a line

We are presented with two points on a page: the challenge is to find the point exactly half way between them, using only a ruler and compass. The solution, provided as Proposition 1.10 of **Euclid's *Elements*** is as follows:

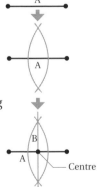

1 Join the two points with a straight line (A).
2 Next, set the compass to more than half the distance between the points. Putting the pin in one of the points, draw an arc crossing the line. Keeping the compass at the same setting, do the same thing at the other point.
3 The two arcs should meet at two places (as long as you drew them large enough). Join these with a straight line (B).
4 The place where the line A and line B meet is the midpoint we want.

In fact this construction does a little more than just find the midpoint: line B is the *perpendicular bisector* of line A.

Constructing parallel lines
A basic axiom of Euclidean geometry is the **parallel postulate**, which says that, for any line L, and any point A not on L, there is another line through A, parallel to L. Can this be constructed by a ruler and compass?

Proposition 1.31 of *Elements* shows that it can. First set your compass to any length greater than that from A to L, and keep it at this setting throughout the whole construction.

1 Draw a circle (X) centred at A. This will cross L at two places; pick one and call it B.

2 Draw another circle (Y) centred at B. This will cross L at two places too. Pick one and call it C.

3 Draw a third circle (Z) centred at C. This will cross the circle X at B, and at a second point. Call this D.

4 The line AD is parallel to L, as we wanted.

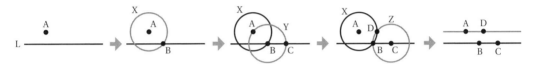

Trisecting a line
Since a line can be bisected using only a ruler and compass, it can also be divided into four, eight or sixteen equal parts, just by repeating that procedure. But can a line be divided into three equal parts? In Proposition 6.9 of his *Elements*, Euclid shows that it can:

1 To divide the line AB into three, first draw any other line (L) through A.

2 Pick any point (C) on L, and then use the compass to find another point D on L so that the distance from A to C is the same as that from C to D.

3 Repeat this, to find a third point E on L so that the distance from A to C is the same as that from D to E, so that C is one third of the way from A to E.

4 Now join E to B with a new line M.

5 Draw a line N through C, parallel to M.

6 The point where N crosses AB is one third of the distance from A to B.

Step 5 requires use of the construction of parallel lines.

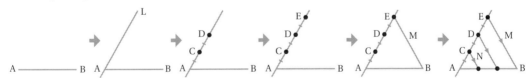

Lines of rational length
The method for trisecting a line easily generalizes to allow a segment to be divided into as many pieces as you like. If we begin with a line of length 1, we can now create a line of length $\frac{x}{y}$, where $\frac{x}{y}$ is any rational number. First divide the line into y equal pieces. Next place x such pieces end to end (by measuring off chunks of a long line with the compass). This shows that all rational numbers are constructible. Some, but not all, irrational numbers are constructible too, since the square root is a constructible operation.

Bisecting an angle

We are given a sheet of paper containing two lines meeting at an angle: the challenge is to divide that angle in half, using only a ruler and compass.

The solution, supplied as Proposition 1.9 of Euclid's *Elements*, is as follows:

1 Position the compass at the vertex (V) of the angle, set to any length. Draw an arc which crosses the two lines, say at points A and B.
2 Ensuring that the compass is set large enough, put the compass at A and draw circle.
3 Keeping the compass at the same setting, draw another circle centred at B.
4 The two circles should meet; call this point C.
5 Join V to C with a straight line: this bisects the angle.

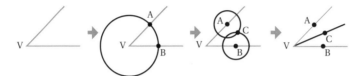

Trisecting an angle

The procedure for bisecting an angle is not especially complicated. Altogether harder is the question of whether a general angle can be trisected: divided into three equal angles. This was a problem tackled by Archimedes, among others. He discovered how to do it using an extra tool which allowed him to draw an **Archimidean spiral**. Neither he nor anyone else could manage the task exactly using just a ruler and compass, however. Approximate methods were found: one such is to draw a chord across the angle, and trisect that line.

The conundrum remained unsettled until 1836 when Pierre Wantzel finally proved algebraically that, in general, angles cannot be trisected using only a ruler and compass. Some specific angles, such as right angles, are trisectable however.

Trisecting an angle using a ruler, compass, and Archimedean spiral

Suppose the angle is centred at the origin O, and is formed by a horizontal line and an inclined line. Draw the spiral centred at O. At some point, call it A, the curve will cross the inclined line.

By definition, at every point on the spiral, the distance to the origin is equal to the angle formed, since $r = \theta$. At this stage, trisecting the angle becomes equivalent to trisecting the distance from A to O. Of course this is a problem which can be solved. Say the resulting distance is X. We set our compass to X, and use this to find a point B on the spiral which is a distance X from the origin. This solves the original problem.

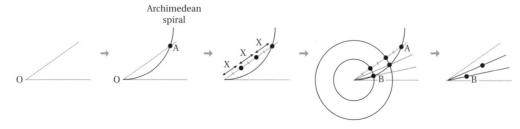

Constructing an equilateral triangle

The very first proposition, 1.1, of Euclid's *Elements* shows that an equilateral triangle can be constructed with ruler and compass. We assume that we are given a segment of straight line (say with ends at A and B). The challenge is to find another point C so that ABC is an equilateral triangle. Euclid's solution is as follows:

1 Set the compass to the distance from A to B. Draw an arc centred at A, passing through B.

2 Without changing the compass setting, draw and arc centred at B passing through A.

3 The two arcs should meet: that place is C.

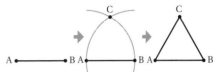

Constructing squares and pentagons

The first regular polygon is an equilateral triangle, which is constructible. The second is a square. Proposition 1.46 of Euclid's *Elements* shows that this is constructible too. The procedure hinges on being able to create a right angle at a point A on a line L. This can be done by taking points B and C an equal distance on either side of A, and then bisecting the segment BC.

The *Elements* also contains instructions for constructing a regular pentagon (see illustration), as well as a hexagon and pentadecagon (15-gon).

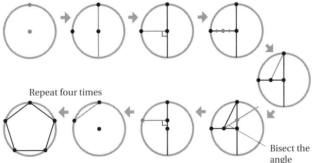

Repeat four times

Bisect the angle

Gauss' heptadecagon
Euclid's methods for constructing regular polygons could be extended to other polygons. Altogether mathematicians were able to construct a regular *n*-gon, where *n* was one of 3, 4, 5, 6, 8, 10, 12, 15, 16, 20, 24, 30, 32, 40, 48, 60, 64, … This situation remained essentially unchanged for 2000 years. It was highly unsatisfactory: could 7, 9 and 11-sided polygons not be constructed? If not, what was the meaning of this sequence? Or had the right methods simply not yet been found?

In 1796, Carl Friedrich Gauss stunned the mathematical community by announcing that he could construct a regular heptadecagon (or 17-gon). The explanation lay in his analysis of constructible polygons via **Fermat primes**. Gauss was so thrilled with his discovery that he requested the shape be carved on his tombstone. The stonemason refused, protesting that it would look like a circle. In his home town of Braunschweig in Germany, Gauss' beloved heptadecagon is represented in a statue to his honour.

Constructible polygons

Gauss' construction of the heptadecagon may make the sequence of constructible polygons seem even more impenetrable: 3, 4, 5, 6, 8, 10, 12, 15, 16, 17, 20, 24, 30, 32, 40, 48, 60, 64, … However, he had not simply stumbled upon the fact that a 17-sided polygon is constructible, he had shown that if a number n is of a specific form, then a regular n-gon is constructible. The form depended on the prime factorization of n: if n is a power of 2, then the n-gon is certainly constructible. Otherwise, for n to be constructible, the only other primes allowed to occur in the factorization of n are **Fermat primes:** those of the form $2^{2^m} + 1$, such as 3 $(= 2^{2^0} + 1)$ and 5 $(= 2^{2^1} + 1)$, and of course 17 $(= 2^{2^2} + 1)$. What's more, each of these could only appear once in the factorization of n.

So Gauss' criterion guaranteeing constructability is that $n = 2^k \times p \times q \dots \times s$, where k is any natural number, and p, q, …, s are distinct Fermat primes. Gauss conjectured that this condition was also necessary: an n-gon is *only* constructible if this condition holds. This was proved in 1836 by Pierre Wantzel. Whether more constructible polygons will be found depends only on whether there are any more Fermat primes.

Constructing a square root

In 2.14 of Euclid's *Elements*, he showed how to construct a square root, using just a ruler and compass. Suppose we are given two lengths 1 and x. (For convenience we'll suppose that $x > 1$, but the method can easily be adapted if $x < 1$.) The challenge is to construct a new line, of length $\sqrt{x}$.

1 First put these two lengths end to end, to give a line AC of length $x + 1$. This can be done with the compass. Mark the point B where the two segments meet.
2 Bisect AC, to give a point D.
3 Set the compass to the length AD, and draw a circle centred at D.
4 Next, draw a straight line L through B, perpendicular to AC.
5 Mark a point E where L crosses the circle.
6 The line EB has length $\sqrt{x}$.

The reason this works is that DE has the same length as DC, namely half that of AC, so $\frac{x+1}{2}$. Also DB has length $\frac{x+1}{2} - 1 = \frac{x-1}{2}$. So, by Pythagoras' theorem in DBE, if y is the length of EB, then $y^2 + \left(\frac{x-1}{2}\right)^2 = \left(\frac{x+1}{2}\right)^2$. A little algebra completes the argument.

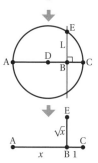

Squaring the circle

The challenge is as follows: we are given a circle, and required to construct a square of the same area, using only a ruler and compass. Also known as the *quadrature of the circle*, this problem has baffled mathematicians since the time of the ancient Greeks. It is intimately tied to an even older problem: the value of π.

Suppose the circle has radius of 1. Then its area is $\pi \times 1^2 = \pi$. So the sides of our square need to have length $\sqrt{\pi}$. Once we have a line of this length, building the square is straightforward. So the nub of the problem is to construct this length.

Since square roots are constructible, it is enough to construct a line of length π. The key to the problem then is whether or not π is constructible. In 1836, Pierre Wantzel showed that if π is a **transcendental number**, then it is not a constructible number. The final piece of the puzzle arrived in 1882, when the Lindemann–Weierstrass theorem implied that π is indeed transcendental, and so squaring the circle is impossible.

Ramanujan's approximate circle-squaring

In 1914, Srinivasa Ramanujan found a very accurate approximate method for squaring the circle. Although π is not constructible, he found an extraordinary constructible approximation to it, namely $\sqrt[4]{\frac{2143}{22}}$, accurate to eight decimal places. The number $\frac{2143}{22}$ can be constructed as a rational length, and then the procedure for constructing a square root can be applied twice to give $\sqrt{\frac{2143}{22}}$ and then $\sqrt[4]{\frac{2143}{22}}$. If the original circle has a radius of 1 metre, the resulting square will have sides accurate to the nearest nanometre.

Doubling the cube

In around 430 BC, Athens suffered a terrible plague. The Athenians consulted the Oracle of Apollo, on the island of Delos. To cure the plague, the Oracle said they must construct a new altar, double the size of the present one. When they consulted Plato, his response was that the Oracle intended to shame the Greeks for their neglect of geometry. So goes the tale of the origin of the problem of doubling the cube. Whatever the truth of the story, this problem did indeed preoccupy ancient mathematicians. Essentially it is a problem of a ruler and compass construction (albeit in three dimensions): we are given a cube and challenged to construct a new one of double the volume.

Suppose that the original cube has sides of 1 unit. Then it has volume 1, and so the new cube should have volume 2. This means that it must have sides of length $\sqrt[3]{2}$. Constructing this length is core of the problem. Plato's friend Menaechmus managed to solve it, with extra tools beyond simple ruler and compass: essentially, he realized that the parabolas $y^2 = 2x$ and $y = x^2$ would intersect at the point whose x-coordinate was $\sqrt[3]{2}$ (an astonishing insight since **Cartesian coordinates** were still thousands of years away). The problem was eventually proved impossible in 1836 by Pierre Wantzel, who demonstrated that $\sqrt[3]{2}$ is not a constructible number.

Constructible numbers

Many problems of ruler and compass construction boil down to this: given a line of length 1, and a number r, is it possible to construct a line of length r? If so, we say that r is *constructible*. Obviously integers, such as 4, are constructible: draw a long line, and then use the compass to measure off four successive segments of length 1. Numbers of the form $\frac{m}{n}$ are also constructible: the method of trisecting a line easily extends to give lines of rational length.

So far, this shows that every rational number is constructible. This is not all however: taking square roots is also a constructible procedure. Pierre Waltzel proved that this is now everything we can do: the only constructible numbers are those obtainable from the rational numbers by adding, subtracting, multiplying, dividing, and taking square roots. As this suggests, the constructible numbers form a **field**.

In particular, all constructible numbers are algebraic, meaning that no **transcendental number**, such as π, is constructible. But not every algebraic number is constructible: $\sqrt[3]{2}$ is not, (although $\sqrt[4]{2}$ is constructible, since it equals $\sqrt{\sqrt{2}}$). This insight handed Wantzel the solutions to several problems which had been outstanding for thousands of years: the trisection of the angle, and doubling the cube. He completed Gauss' work on constructible polygons, and made major inroads into the question of squaring the circle.

DIOPHANTINE EQUATIONS

Number theory

The term *number theory* might seem a good description of the whole of mathematics, but the focus of this subject is the system of ordinary whole numbers (or **integers**) rather than the more rarefied **real numbers** or **complex numbers**. The whole numbers are the most ancient and fundamental of all mathematical structures. But, below their surface, lie some of the deepest questions in mathematics, including the **Riemann hypothesis** and **Fermat's last theorem**.

The ancient Babylonians were interested in number theory, as is evidenced by tablets such as *Plimpton* 322, dating from around 1800 BC. An important development in the classical period was Diophantus of Alexandria's 13-volume *Arithmetica*, around AD 250. Modern number theory began in the works of the 17th-century French magistrate Pierre de Fermat.

Algebraic and analytic number theory

The two major concerns of contemporary number thoery are the behaviour of the **prime numbers** and the study of **Diophantine equations**: the formulas which describe relationships between whole numbers. These two topics roughly correspond to the two principal branches of the subject: *algebraic* number theory and *analytic* number theory.

The tools of the two subjects are different. The algebraic approach studies numbers via objects such as **groups** and **elliptic curves**, while analytic number theory uses techniques from **complex analysis**, such as **L-functions**. **Langland's program** provides the tantalizing suggestion that these two great subjects may be different perspectives on the same underlying objects.

Modular arithmetic

Modular arithmetic is the arithmetic of remainders. Saying '11 is congruent to 1 modulo 5', written '$11 \equiv 1 \pmod 5$', means that 11 leaves a remainder of 1 when divided by 5, because $11 = 2 \times 5 + 1$. It is only the remainder (1) which matters here, not the number of times that 5 goes into 11 (in this case 2). Similarly, we could write '$6 + 6 \equiv 0$ (mod 4)' or '$8 \times 3 \equiv 2$ (mod 11)'.

Modular arithmetic is widespread not just within mathematics, but in daily life. The 12-hour clock relies on our ability to do arithmetic modulo 12, and if you work out what day of the week it will be in nine days' time, you are doing arithmetic mod 7. Modular arithmetic is useful in number theory for providing information on numbers whose exact values are unknown, through powerful results such as Fermat's little theorem, and Gauss' quadratic reciprocity law.

Chinese remainder theorem
Some time between the third and fifth centuries AD, the Chinese mathematician Sun Zi wrote: 'Suppose we have an unknown number of objects. When counted in threes, two are left over. When counted in fives, three are left over. When counted in sevens, two are left over. How many objects are there?' In modern terms, this is a problem in modular arithmetic: what is needed is a number n, where $n \equiv 2$ (mod 3), $n \equiv 3$ (mod 5), and $n \equiv 2$ (mod 7).

The *Chinese remainder theorem* states that this type of problem always has a solution. The simplest case involves just two congruences: if a, b, r and s are any numbers, then there is always a number n where $n \equiv a$ (mod r), and $n \equiv b$ (mod s). The caveat is that r and s must be *coprime*, meaning they have no common factors. This immediately extends to solving any number of congruences (again providing the moduli are all coprime). The solution n is not quite unique: 23 and 128 both solve Sun Zi's original problem. In general there will be exactly one solution which is at most the product of all the moduli, in this case 105 ($3 \times 5 \times 7$).

Fermat's little theorem
A cornerstone of elementary number theory, Fermat's little theorem comes from noticing that $15^2 - 15$ is divisible by 2, and $101^7 - 101$ is divisible by 7. In 1640 Pierre de Fermat wrote to Bernard Frénicle de Bessy with the statement of his little theorem: if p is any prime, and n is any whole number, then $n^p - n$ must be divisible by p.

Writing this using modular arithmetic, we get:

$$n^p - n \equiv 0 \pmod{p} \quad \text{or} \quad n^p \equiv n \pmod{p}$$

If n itself is not divisible by p, this is equivalent to:

$$n^{p-1} \equiv 1 \pmod{p}$$

Fermat added the characteristic comment 'I would send you the proof, if I did not fear its being too long'. The first known proofs are due to Gottfried Leibniz, in unpublished work around 1683, and Leonhard Euler in 1736.

Quadratic reciprocity law

The great German mathematician Carl Friedrich Gauss loved this result, which he called the 'Golden Theorem'. First stated by Leonhard Euler in 1783, Gauss published its first complete proof in 1796.

For two odd primes p and q, this law describes an elegant symmetry between two questions: whether p is a square modulo q, and whether q is a square modulo p. It asserts that these questions always have the same answer, except when $p \equiv q \equiv 3 \pmod 4$ when they have opposite answers. Take 5 and 11, for example. Firstly $11 \equiv 1 \pmod 5$, and 1 is a square. So the theorem predicts that 5 mod 11 should also be a square. This is not immediately obvious, but on closer inspection, $4^2 \equiv 16 \equiv 5 \pmod{11}$.

Along with **Pythagoras' theorem**, the quadratic reciprocity law is one of the most profusely proved results in mathematics. Gauss alone produced eight proofs during his life, and over 200 more now exist, employing a wide variety of techniques.

Diophantus' *Arithmetica*

Known as 'the Father of Algebra', Diophantus of Alexandria lived around AD 250. Although the ancient Babylonians had begun probing integer solutions to quadratic equations, it was in Diophantus' thirteen volume *Arithmetica* that the study of Diophantine equations, named in his honour, was begun in earnest. This work marked a milestone in the history of number theory, but was believed to have been lost in the destruction of the great library at Alexandria. However in 1464 six of the books resurfaced and became a major focus for European mathematicians, most notably Pierre de Fermat.

Diophantine equations

A Diophantine equation is a **polynomial** much like any other. The difference is that we are only interested in whole numbers, or integers. Only integers can appear in the polynomial (although it makes no difference if we also allow fractions). Most importantly, we are only interested in integer solutions to the equation.

So, instead of analysing the real or complex numbers x, y, z which satisfy $x^3 + y^3 = z^3$, we ask whether there are any whole numbers which satisfy it. (In this case, the answer is no, as follows from the most famous of all Diophantine problems, Fermat's Last Theorem.) The reason for their enduring interest is that polynomials are the right way to express possible (or impossible) relationships between integers. For instance, **Catalan's conjecture** says that 8 and 9 are the only pair of integer powers next to each other.

Egyptian fractions

A *unit fraction* is one whose numerator is 1: such as $\frac{1}{2}, \frac{1}{3}$, or $\frac{1}{4}$ (but not $\frac{3}{4}$). Of course, any rational number can be written as unit fractions added together: for example $\frac{3}{4} = \frac{1}{4} + \frac{1}{4} + \frac{1}{4}$. A more interesting question is which rational numbers can be arrived at by adding unit fractions together, using only ones which are different.

For instance, $\frac{1}{2} + \frac{1}{4}$ is a representation of $\frac{3}{4}$ as an *Egyptian fraction*, so-called because this problem intrigued ancient Egyptian mathematicians. The *Rhind papyrus* dating from around 1650 BC includes a list of fractions of the form $\frac{2}{n}$ written in this form.

In 1202 Leonardo of Pisa (better known as Fibonacci) wrote his book *Liber Abaci*, in which he proved that every fraction can be split up in this way. He also provided an **algorithm** for finding this representation. But this has not completely settled the matter: questions still remain about how many unit fractions are needed to represent a particular number, which include the Erdős-Straus conjecture.

Erdős–Straus conjecture

Every known fraction of the form $\frac{4}{n}$ (with n at least 2) can be written as the sum of three unit fractions (that is, fractions whose numerator is 1). So, for every such n, there are three whole numbers x, y, z where $\frac{4}{n} = \frac{1}{x} + \frac{1}{y} + \frac{1}{z}$. For example, for $n = 5$, a solution is $\frac{4}{5} = \frac{1}{2} + \frac{1}{5} + \frac{1}{10}$. The claim that this is always true, formulated by Paul Erdős and Ernst Straus in 1948, has so far resisted all attempts at proof and counterexample.

Bézout's lemma

The highest common factor (highest common denominator) of 36 and 60 is 12. Named after Étienne Bézout, Bézout's lemma says that there are integers x and y where $36x + 60y = 12$. On closer inspection, one solution is $x = 2$ and $y = -1$. But there will be infinitely many others too, such as $x = 7$ and $y = -4$. This can be rephrased as a statement about Diophantine equations: if a and b have highest common factor d, then Bézout's lemma guarantees that the linear equation $ax + by = d$ has infinitely many integer solutions. Bézout's lemma is the key to understanding all linear Diophantine equations.

Linear Diophantine equations

The simplest Diophantine equations are ones which just involve plain x and y (no x^2 or xy or higher powers): *linear* equations, such as $6x + 8y = 11$. Any such equation defines a straight line on the plane. So asking whether it has integer solutions is the equivalent to looking for points on this line which have integer co-ordinates.

Bézout's lemma deals with the most important case: it says $ax + by = d$ has a solution when d is the highest common factor (highest common denominator) of a and b (and in this case it has infinitely many solutions). In fact, the equation $ax + by = d$ only has solutions when d is a multiple of the hcf of a and b. So the equation $6x + 8y = 11$ has no integer solutions, since the hcf of 6 and 8 is 2, and 11 is not a multiple of 2.

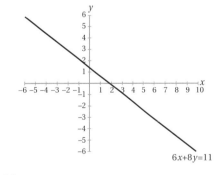

$6x + 8y = 11$

If a linear equation has any integer solutions, then it will have infinitely many. This result extends to equations in more variables, such as $3x + 4y + 5z = 8$. Here, the highest common factor of 3, 4 and 5 is 1. Since 8 is a multiple of 1, this equation will have infinitely many integer solutions.

Archimedes' cattle

Around 250 BC, Archimedes sent a letter to his friend Eratosthenes, containing a challenge for the mathematicians of Alexandria. It concerned the Sicilian 'cattle of the Sun', a herd consisting of cows and bulls of various colours. We will write W for the number of white bulls and w for the number of white cows, and similarly X and x for the black ones, Y and y for the dappled, and Z and z for the brown.

Archimedes described the herd through a system of linear Diophantine equations:

$$W = \left(\tfrac{1}{2} + \tfrac{1}{3}\right) X + Z \qquad X = \left(\tfrac{1}{4} + \tfrac{1}{5}\right) Y + Z \qquad Y = \left(\tfrac{1}{6} + \tfrac{1}{7}\right) W + Z$$

$$w = \left(\tfrac{1}{3} + \tfrac{1}{4}\right)(X + x) \qquad x = \left(\tfrac{1}{4} + \tfrac{1}{5}\right)(Y + y) \qquad y = \left(\tfrac{1}{5} + \tfrac{1}{6}\right)(Z + z) \qquad z = \left(\tfrac{1}{6} + \tfrac{1}{7}\right)(W + w)$$

The challenge was to find the make-up of the herd by solving the equations. We do not know how the Alexandrians fared. The first known solution is due to the European mathematicians who rediscovered it in the 18th century: $W = 10{,}366{,}482$, $X = 7{,}460{,}514$, $Y = 7{,}358{,}060$, $Z = 4{,}149{,}387$, $w = 7{,}206{,}360$, $x = 4{,}893{,}246$, $y = 3{,}515{,}820$, $z = 5{,}439{,}213$, a total of 50,389,082 head of cattle.

However, Archimedes warned that anyone who solved this problem 'would not be called unskilled or ignorant of numbers, but nor shall you yet be numbered among the wise'. In order to attain perfect wisdom, the problem had to be solved with two extra conditions included: $W + X$ should be a **square number**, and $Y + Z$ should be a **triangular number**. This removes the problem from the sphere of linear equations, and makes it substantially harder. In 1880 A. Amthor described a solution, based on reducing the problem to the Pell equation, $a^2 - 4729494b^2 = 1$. With the dawn of the computer age this was fleshed out to give the full 206,545-digit answer: incomparably more cattle than there are atoms in the universe.

Pell equations

The easiest Diophantine equations to manage are the linear ones, where there is a straightforward procedure to determine whether or not there are any integer solutions. The picture becomes much murkier once squares are introduced, as the unknown status of **perfect cuboids** demonstrates.

Around 800 BC the Hindu scholar Baudhayana gave $\frac{577}{408}$ as an approximation to $\sqrt{2}$. It is likely that this came from studying the equation $x^2 - 2y^2 = 1$, which has $x = 577$ and $y = 408$ as a solution (as well as $x = 17$, $y = 12$). This, $x^2 - 2y^2 = 1$, is the first example of a *Pell equation*. Another is $x^2 - 3y^2 = 1$, and generally $x^2 - ny^2 = 1$, where n is any non-square natural number. (In fact they have little to do with John Pell, but Leonhard Euler confused him with William Brouncker, and the name stuck.)

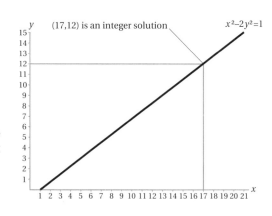

Pell equations were studied much earlier in India, notably by Brahmagupta in AD 628. The *chakravala method*, attributed to Bhāskara II in the 12th century, is a procedure for solving Pell equations by **continued fractions**. It was spectacularly deployed to solve the awkward case $x^2 - 61y^2 = 1$, to find minimum solutions of $x = 1,766,319,049$ and $y = 226,153,980$. Hermann Hankel called chakravala 'the finest thing achieved in the theory of numbers before Lagrange'. It was Joseph Louis Lagrange who gave the first rigorous proof that $x^2 - ny^2 = 1$ must have infinitely many integer solutions for any non-square number n.

Euler bricks

Pythagorean triples allow us to construct a rectangle whose sides and diagonals are all whole numbers. For instance, a rectangle with sides of 3 and 4 units will, by Pythagoras' theorem, have diagonals of length 5. An *Euler brick* generalizes this to three-dimensions: it is a cuboid (see **irregular polyhedra**) all of whose lengths are whole numbers, as are the diagonals of each face. The smallest Euler brick has sides of lengths 44, 117 and 240 units, and was discovered in 1719 by Paul Halcke. In 1740, the blind mathematician Nicholas Saunderson discovered a method to produce infinitely many Euler bricks. This approach was later augmented by Leonhard Euler himself. However, no way of listing every possible Euler brick is yet known.

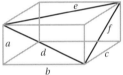

We can translate the geometry into algebra using Pythagoras' theorem. An Euler brick with sides a, b, c and face diagonals d, e, f must satisfy the Diophantine equations:

$$a^2 + b^2 = d^2, \qquad b^2 + c^2 = e^2 \qquad \text{and} \qquad c^2 + a^2 = f^2$$

The most sought after Euler bricks are the elusive perfect cuboids.

Perfect cuboids

Euler bricks have intrigued mathematicians since the 18th century: they are cuboids where the edges and face-diagonals all have integer lengths. A natural extension is to require that the cuboid's body-diagonal should also be a whole number. This describes a *perfect cuboid*. The problem is that no-one has ever seen one: whether or not they exist is a significant open problem.

In terms of Diophantine equations, what is required are integers a, b, c (representing the cuboid's edges), d, e, f (representing its face-diagonals), and g (representing its body-diagonal) which satisfy

$$a^2 + b^2 = d^2, \qquad b^2 + c^2 = e^2 \qquad \text{and} \qquad c^2 + a^2 = f^2,$$

as well as

$$a^2 + b^2 + c^2 = g^2$$

No perfect cuboid has ever been found. It is known that if one does exist one of its sides must be at least 9 billion units long. However in 2009, Jorge Sawyer and Clifford Reiter discovered the existence of perfect parallelepipeds (see **irregular polyhedra**). The smallest has edges of 271, 106, 103, parallelogram faces with diagonals of 101 and 183, 266 and 312, 255 and 323, and body-diagonals of 374, 300, 278 and 272.

Sums of two squares

An ancient mathematical question asks: which natural numbers can be written as two square numbers added together? **Fermat's two square theorem** answered this for **prime numbers**: the only ones which can are those of the form $4m + 1$ (as well as $2 = 1^2 + 1^2$). But what happens for composite numbers? 6 and 14 cannot, but 45 can: $45 = 3^2 + 6^2$. For large numbers like 6615, finding the answer could require a lot of experimentation.

Fermat's result suggests that prime numbers are especially important, and the main obstacle are the $(4m + 3)$-primes: 3, 7, 11, 19, 23, 31, etc. The solution (which can be deduced from Fermat's theorem) first breaks a whole number n into primes using the **Fundamental theorem of arithmetic**. Then the theorem states that n can be written as the sum of two squares if each $(4m + 3)$-prime appears in this break-down an *even number* of times. So $6 = 2 \times 3$ fails, since 3 appears once (an odd number of times). But $45 = 3^2 \times 5$ can be written as a sum of two squares, since 3 appears twice (an even number of times). With no further need for experiment, we now know that 6615 can never be written as a sum of two squares, because $6615 = 3^3 \times 5 \times 7^2$, where 3 appears three times.

Lagrange's four square theorem

In 1621, Claude Bachet translated the six surviving books of Diophantus' *Arithmetica* (dating from around AD 250) from ancient Greek into Latin. These volumes would play an important role in the development of modern number theory, most notably in the hands of Pierre de Fermat. Bachet was also a mathematician in his own right. He noticed that implicit in Diophantus' work was a remarkable claim: that every whole number could be written as the sum of four square numbers. For example, $11 = 3^2 + 1^2 + 1^2 + 0^2$ and $1001 = 30^2 + 8^2 + 6^2 + 1^2$.

Fermat later made the same observation, known as *Bachet's conjecture*. But the first published proof for all natural numbers was by Joseph-Louis Lagrange in 1770. This important theorem is generalized further by **Fermat's polygonal number theorem** and **Waring's problem**.

Legendre's three square theorem

Leonhard Euler published the first proof of Fermat's two square theorem in 1749, essentially settling the question of which numbers could be written as sums of two squares. In 1770, Joseph-Louis Lagrange proved his four square theorem, showing that every natural number can be written as the sum of four squares.

A puzzle remained: which numbers can be written as the sum of three squares? Most numbers can. The first which cannot are 7, 15, 23, 28, 31, 39, 47, 55, 60, …

In 1798, Adrien-Marie Legendre unravelled this sequence. It consists of numbers one less than a multiple of 8: 7, 15, 23, 31… along with this sequence repeatedly multiplied by four: 28, 60, 92, 124,… and 112, 240, 368, 496,… and so on. In short, Legendre's result is that every number can be written as the sum of three squares, except for those of the form $4^n(8k - 1)$.

Triangular numbers

In cue-sports such as billiards and pool, the *pack* comprises 15 balls arranged as an equilateral triangle at the start of the game. It is not fanciful to imagine that the number 15 was chosen because it is a *triangular number*: 14 or 16 balls cannot form an equilateral triangle. The smallest triangular number is 1, then 3, then 6. Looking at the corresponding packs of balls, in each case the first row contains 1 ball, the next 2, then 3 and so on. So triangular numbers are those of the form $1 + 2 + 3 + 4 + \cdots + n$, for some n. The formula for summing the numbers 1 to n (see **adding up the numbers 1 to 100**) gives a formula for the nth triangular number: $\frac{n(n + 1)}{2}$.

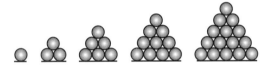

This is also the formula for the **binomial coefficient** $\binom{n}{2}$. In this guise, triangular numbers give the solution to the *handshaking problem*: how many handshakes take place when n people all shake hands with each other? Triangular numbers are the first of the polygonal numbers.

Polygonal numbers

Triangular numbers are those which count the number of balls which can be arranged in an equilateral triangle. Square numbers are the corresponding numbers for squares. Can this be extended to regular pentagons and other polygons? The answer is that it can, though care is needed since it is not immediately obvious how to arrange balls into pentagonal arrays. The convention is that the first pentagonal number is 1, the second is 5, and subsequent numbers are found by picking one corner, extending its two sides each by one ball, and completing the pentagon (enclosing all previous balls). See diagrams. This gives the next pentagonal numbers as 12, 22, 35, 51, etc.

A similar process works for *hexagonal numbers* (the first few of which are 1, 6, 15, 28, 45, 66), *heptagonal numbers*, and indeed *n-agonal numbers* for any n.

The formula for the mth triangular number is $\frac{m(m + 1)}{2}$ or to put it another way, $\frac{1}{2}m^2 + \frac{1}{2}m$ The formula for the mth square number is, of course, m^2. For the mth pentagonal number, it is $\frac{3}{2}m^2 - \frac{1}{2}m$. The general pattern might be visible now: the formula for the mth n-agonal number is $(\frac{n}{2} - 1)m^2 - (\frac{n}{2} - 2)m$. The fact that these numbers have genuine number theoretic significance is proved in Fermat's polygonal number theorem.

Fermat's polygonal number theorem

In 1796 Carl Friedrich Gauss proved that every number can be written as the sum of three triangular numbers. Echoing Archimedes, he wrote:

$$\text{EYPHKA} \qquad num = \triangle + \triangle + \triangle$$

In 1770 Lagrange had already proved his four square theorem: that every number can be written as the sum of four square numbers (or, as Gauss might have put it, '$num = \square + \square + \square + \square$').

Together these two results formed the first two cases of Fermat's polygonal number theorem. The next says that any number should be expressible as the sum of five pentagonal numbers. In general, the theorem asserts that any number can be obtained by adding together n n-agonal numbers. In his customary fashion, Pierre de Fermat claimed to have a proof, but did not apparently communicate it to anyone. The first known complete proof is due to Augustin Louis Cauchy in 1813.

Fermat's last theorem

The numbers 3, 4 and 5 form a **Pythagorean triple** because $3^2 + 4^2 = 5^2$ (that is $9 + 16 = 25$). Around 300 BC, Euclid realized that there must be infinitely many triples of whole numbers like this which satisfy $x^2 + y^2 = z^2$. In 1637, Pierre de Fermat contemplated what would happen if these squares were replaced with higher powers. In his copy of Diophantus' *Arithmetica*, he wrote 'It is impossible to separate a cube into two cubes, or a fourth power into two fourth powers, or any power higher than the second into two like powers.' Fermat was claiming that, for any value of n larger than 2, it would never be possible to find whole numbers x, y and z satisfying $x^n + y^n = z^n$. Infamously, he went on 'I have discovered a truly marvellous proof of this which this margin is too narrow to contain'.

Wiles' theorem

Fermat's famous claim became known as his *last theorem*, not because it was the last he wrote, but because it was the last to be proved. Despite his claim, most experts today do not believe that Fermat could possibly have had a complete proof (although he did prove the particular case of $n = 4$). Nor did he or anyone else find a counterexample.

It would more accurately have been called *Fermat's conjecture*, and remained unresolved over the centuries, despite the attentions of many of mathematics' greatest thinkers. In 1995, Andrew Wiles and his former research student Richard Taylor completed a proof which established Fermat's last theorem as a consequence of the **modularity theorem** for elliptic curves.

Beal's conjecture

Best known as a self-made billionaire, and for playing poker for the highest stakes in history, the Texas-based businessman Andrew Beal is also an enthusiastic amateur number theorist. In 1993, he was investigating Fermat's last theorem, which says that $x^n + y^n = z^n$ has no integer solutions when $n > 2$. Beal's idea was to alter the formula by allowing the exponents to differ: so, instead of $x^n + y^n = z^n$, he considered $x^r + y^s = z^t$, where r, s and t are allowed to be different (but all must be bigger than 2). Similar situations had been studied by Viggo Brun at the beginning of the 20th century.

The new formula does have solutions: for example, $3^3 + 6^3 = 3^5$, and $7^6 + 7^7 = 98^3$. However Beal noticed that, in every solution he found, the base numbers x, y and z all had a common factor (in the first example 3, and in the second 7). *Beal's conjecture* is the assertion that this must always be true. In 1997 the American Mathematical Society announced that Beal was offering a prize of \$50,000, now raised to \$100,000, for a proof or counterexample to his conjecture.

Catalan's conjecture (Mihăilescu's theorem)

The numbers 8 and 9 make strange neighbours: both are powers of other whole numbers: $8 = 2^3$ and $9 = 3^2$. The Belgian mathematician Eugène Catalan realized that there were no other examples of consecutive powers among the integers (0 and 1 excepted). *Catalan's conjecture* of 1844 was that this was indeed the only occurrence, but he wrote that he 'could not prove it completely so far'. In fact the problem predated him: around 1320 the rabbi and scholar Levi ben Gerson had already proved that there were no other instances of a power of 2 sitting next to a power of 3. The full conjecture stood open until 2002 when Preda Mihăilescu provided the long-sought proof. More formally, the statement is that the only solution to the Diophantine equation $x^a - y^b = 1$ (where a and b are both greater than 1) is $x = 3$, $y = 2$, $a = 2$, $b = 3$.

Waring's problem

By Lagrange's four square theorem, we know that every positive whole number can be written as the sum of four squares. What about writing numbers as sums of higher powers? In 1909, Arthur Wieferich showed that every number can be written as the sum of nine cubes. In 1986, Balasubramanian, Deshouillers, and Dress showed that 19 fourth powers are also enough. These results had been conjectured by Edward Waring in 1770. Waring had further suggested that this problem must have a solution for every positive power. That is, for every whole number n, there must be some other number g, so that any number can be written as the sum of at most g nth powers.

This important result was proved by David Hilbert in 1909, and is known as the *Hilbert–Waring theorem*. The first few numbers in the sequence are:

n	1	2	3	4	5	6	7	8	9	10
g	1	4	9	19	37	73	143	279	548	1079

The exact nature of this sequence, and related sequences, remains an active research topic.

The abc conjecture

To calculate the *radical* of a number, first break it down into primes, and then multiply together the distinct primes, ignoring any repetitions. So, $300 = 2 \times 2 \times 3 \times 5 \times 5$, and its radical, Rad(300) equals $2 \times 3 \times 5 = 30$. Similarly, $16 = 2 \times 2 \times 2 \times 2$ and Rad(16) = 2. (The radical is also called the *square-free part* of x, since it is the largest factor of x not divisible by a square. Note that the term 'radical' is separately used to refer to roots, as in *square root*.)

In 1985, Joseph Oesterlé and David Masser formulated a conjecture which would generalize Fermat's last theorem, Catalan's conjecture (Mihăilescu's theorem) and a great many other number theoretical problems. Dorian Goldfeld called it 'the most important unsolved problem in Diophantine analysis'. It concerns situations where three numbers a, b and c are coprime and satisfy $a + b = c$. The conjecture compares c and Rad($a \times b \times c$). In most cases $c <$ Rad($a \times b \times c$). For instance, if $a = 3$, $b = 4$, $c = 7$, then Rad($3 \times 4 \times 7$) = 42, which is more than 7. But occasionally this does not happen, for instance when $a = 1$, $b = 8$, $c = 9$, then Rad($1 \times 8 \times 9$) = 6, which is less than 9.

Masser proved that there are infinitely many of these exceptions, where $c >$ Rad($a \times b \times c$). The *abc conjecture* says that these are *only just* in violation of the rule: even a very minor tweak will get rid of most of them. For any number d bigger than 1 (even $d = 1.0000000001$), almost all of the exceptions should disappear, leaving just finitely many triples where $c >$ Rad($a \times b \times c$)^d.

PRIME NUMBERS

Prime numbers

The definition of a *prime number* is among the most profound and ancient there is: a positive whole number which cannot be divided into smaller whole numbers. The first examples are 2, 3, 5 and 7. On the other hand, 4 is not prime, as it equals 2×2. Up until the early 20th century, 1 was classed as prime, but no longer. Today we say that a prime must have exactly two factors: itself and 1. Non-prime numbers (apart from 1) are known as *composite*. The **fundamental theorem of arithmetic** tells us that prime numbers are to mathematicians as atoms are to chemists: the basic building blocks from which every other natural number is built. Despite the simplicity of their definition, the prime numbers still hold many mysteries. Major open questions about primes include **Landau's problems** and the **Riemann hypothesis**. There is no simple formula for generating prime numbers, and no immediate way to tell whether or not a large number is prime, making **primality testing** an important area of research.

Euclid's proof of the infinity of primes

The list of prime numbers starts with 2, 3, 5, 7, 11, 13, 17, … But where does it end? In fact, it goes on forever, a fact demonstrated by Euclid in Proposition 9.20 of his *Elements*. His method is a classical **proof by contradiction**: Euclid started by imagining that what he wanted to prove was false, and there really was a biggest prime number. Call this P. According to this assumption, he now had a list of all the prime numbers: 2, 3, 5, 7, …, P. Euclid then constructed a new number (Q, say), by multiplying all the primes together, and then adding 1. So:

$$Q = (2 \times 3 \times 5 \times 7 \times \ldots \times P) + 1.$$

Now, 2 cannot divide Q, because it leaves remainder 1. Similarly each of 3, 5, 7, …, P leaves remainder 1 when divided into Q. So none of the primes divide Q. Q itself is not necessarily prime (though it may be). Either way, according to the Fundamental Theorem of Arithmetic, it must have at least one prime factor, R. This, therefore, must have been missed from our list, and so be bigger than P. So Euclid's starting assumption, that there is a biggest prime number P, resulted in him finding a bigger one, R. Since this is a contradiction, it must be that the original assumption was false: there is no biggest prime.

The sieve of Eratosthenes

Dating from around 250 BC, Eratosthenes is famous for his remarkably accurate calculations of the circumference of the earth, as well as for his 'sieve' for generating lists of primes. The procedure is straightforward: first write out all the natural numbers up to some chosen point (such as 100, as in the illustration). First we remove 1 (which is not classed as a prime today, though Eratosthenes would have thought it one). Now the first number on the list is 2, which is prime, and which we circle. Next we delete from the list all its multiples: 4, 6, 8, 10,… Then we repeat: the first number on the list is now 3, which we identify as a prime, before removing all its multiples: 6, 9, 12,… . At each stage the first number on the list is identified as a prime, and all its multiples are deleted. This reveals the next prime.

This technique of *sieving* the natural numbers, much evolved since Eratosthenes' day, plays a major role in modern number theory, and forms the basis for the proof of **Chen's Theorems 1−3**.

1	2	3	4	5	6	7	8	9	10
11	12	13	14	15	16	17	18	19	20
21	22	23	24	25	26	27	28	29	30
31	32	33	34	35	36	37	38	39	40
41	42	43	44	45	46	47	48	49	50
51	52	53	54	55	56	57	58	59	60
61	62	63	64	65	66	67	68	69	70
71	72	73	74	75	76	77	78	79	80
81	82	83	84	85	86	87	88	89	90
91	92	93	94	95	96	97	98	99	100

1̶	②	3	4̶	5	6̶	7	8̶	9	1̶0̶
11	1̶2̶	13	1̶4̶	15	1̶6̶	17	1̶8̶	19	2̶0̶
21	2̶2̶	23	2̶4̶	25	2̶6̶	27	2̶8̶	29	3̶0̶
31	3̶2̶	33	3̶4̶	35	3̶6̶	37	3̶8̶	39	4̶0̶
41	4̶2̶	43	4̶4̶	45	4̶6̶	47	4̶8̶	49	5̶0̶
51	5̶2̶	53	5̶4̶	55	5̶6̶	57	5̶8̶	59	6̶0̶
61	6̶2̶	63	6̶4̶	65	6̶6̶	67	6̶8̶	69	7̶0̶
71	7̶2̶	73	7̶4̶	75	7̶6̶	77	7̶8̶	79	8̶0̶
81	8̶2̶	83	8̶4̶	85	8̶6̶	87	8̶8̶	89	9̶0̶
91	9̶2̶	93	9̶4̶	95	9̶6̶	97	9̶8̶	99	1̶0̶0̶

1̶	②	③	4̶	⑤	6̶	⑦	8̶	9̶	1̶0̶
⑪	1̶2̶	⑬	1̶4̶	1̶5̶	1̶6̶	⑰	1̶8̶	⑲	2̶0̶
2̶1̶	2̶2̶	㉓	2̶4̶	2̶5̶	2̶6̶	2̶7̶	2̶8̶	㉙	3̶0̶
㉛	3̶2̶	3̶3̶	3̶4̶	3̶5̶	3̶6̶	㊲	3̶8̶	3̶9̶	4̶0̶
㊶	4̶2̶	㊸	4̶4̶	4̶5̶	4̶6̶	㊼	4̶8̶	4̶9̶	5̶0̶
5̶1̶	5̶2̶	㊾	5̶4̶	5̶5̶	5̶6̶	5̶7̶	5̶8̶	㊾	6̶0̶
㊽	6̶2̶	6̶3̶	6̶4̶	6̶5̶	6̶6̶	㊿	6̶8̶	6̶9̶	7̶0̶
㋓	7̶2̶	㋒	7̶4̶	7̶5̶	7̶6̶	7̶1̶	7̶8̶	㋔	8̶0̶
8̶1̶	8̶2̶	8̶3̶	8̶4̶	8̶5̶	8̶6̶	8̶7̶	8̶6̶	8̶9̶	9̶0̶
9̶1̶	9̶2̶	9̶3̶	9̶4̶	9̶5̶	9̶6̶	9̶7̶	9̶8̶	9̶9̶	1̶0̶0̶

Goldbach's conjecture

Originating in a 1742 correspondence between Christian Goldbach and Leonhard Euler, *Goldbach's conjecture* asserts that every even number from 4 onwards is the sum of two prime numbers. So 4 = 2 + 2, 6 = 3 + 3, 8 = 3 + 5, 10 = 5 + 5, 12 = 5 + 7, and so on. Though checked by computer for even numbers up to 10^{18} by Tomás Oliveira e Silva in 2008, no full proof is known, making this one of the great open problems of number theory. The best progress so far is Chen's Theorem 1.

A related statement is the *weak Goldbach conjecture*, which says that every odd number from 9 onwards is the sum of three primes. In 1937, Ivan Vinogradov found a large number N and proved that the conjecture is true for all odd numbers bigger than N. So, theoretically, only finitely many numbers remain to be checked to prove the result. However, this threshold is still so huge (around 10^{43000}) that the weak conjecture also remains open today.

Landau's problems and the n^2+1 conjecture

At an international conference in Cambridge in 1912, Edmund Landau highlighted four problems about prime numbers, which he characterized as 'unattackable in the present state of science', and which still stand defiantly open nearly 100 years later:

1 Goldbach's conjecture

2 The twin primes conjecture

3 Legendre's conjecture

4 The n^2+1 conjecture

This last concerns primes which are 1 more than a square number. For example, $5 = 2^2 + 1$, and $17 = 4^2 + 1$. The conjecture is that there should be infinitely many other primes like these. It is known, as a consequence of Fermat's two square theorem, that there are infinitely many primes of the form $n^2 + m^2$. As with Chen's theorems, the most promising result to date comes from relaxing the criterion from primes to *semiprimes*: numbers which are the product of exactly two primes (e.g $9 = 3 \times 3$ or $15 = 3 \times 5$). In 1978, Henryk Iwaniec proved that there are infinitely many of these of the form $n^2 + 1$. A positive answer to the $n^2 + 1$ conjecture would be implied by **Hypothesis H**.

Fermat's two square theorem

With the exception of 2, all prime numbers are odd. So, when divided by 4, every odd prime must leave remainder either 1 or 3. This divides the odd primes into two families, and in the 17th century Pierre de Fermat noticed a surprising difference between them. The primes of the form $4n + 1$ could all be written as the sum of two squares: $5 = 2^2 + 1^2$, and $13 = 3^2 + 2^2$, for example, and furthermore this could be done in only one way. But none of the primes of the form $4n + 3$ (such as 7 or 11) could be expressed in this way.

This is also known as *Fermat's Christmas theorem* because it was first written in a letter to Marin Mersenne, on 25 December 1640. Typically, Fermat gave just the barest sketch of an argument that this should always be true. The first published full proof was by Leonhard Euler, in 1749. This is intimately related to the quadratic reciprocity theorem and, as G.H.Hardy remarked, 'is ranked, very justly, as one of the finest in arithmetic'. A consequence is that if a number can be written as the sum of two squares in two different ways (such as $50 = 7^2 + 1^2 = 5^2 + 5^2$), then it cannot be prime.

Fermat's result extends to a complete analysis of which numbers can be written as the **sums of two squares**.

Prime gaps

The study of prime numbers can be turned on its head by looking instead at the gaps between them: that is, the differences between successive prime numbers. So the first prime gap is that between 2 and 3, namely 1. Then between 3 and 5 we find 2, and so on. Can prime gaps be arbitrarily large? It turns out that they can; so, like the primes, the set of prime gaps is infinite. But where very large gaps first appear, and the relation of the possible size of a gap to that of the primes either side of it, remains a topic of research. The twin primes and de Polignac's conjectures are long-standing open problems about prime gaps. Bertrand's postulate implies that a prime number can't ever be followed by a gap bigger than it. Andrica's conjecture, if proved, would substantially strengthen this result.

Bertrand's postulate

How far apart can consecutive primes be? That was the question on Joseph Bertrand's mind in 1845, when he postulated that, between any natural number n and its double $(2n)$, you must always be able to find at least one prime number. Formally, for any number n (greater than 1), there is a prime number p, where $n < p < 2n$. Five years later, his postulate was proved by Pafnuty Chebyshev. Its subsequent reproof by Paul Erdős was immortalized in a verse by Nathan Fine: '*Chebyshev said it, but I'll say it again; there's always a prime between n and 2n*'.

Andrica's conjecture

In 1986, the Romanian mathematician Andrica Dorin formulated what he describes as 'a very hard open problem in number theory'. It concerns the differences between the square roots of successive primes. His conjecture is that these are always less than 1. More formally, if p and q are consecutive primes then $\sqrt{q} - \sqrt{p} < 1$.

If true, squaring this inequality translates to a rule about ordinary prime gaps: if p is a prime number; then the gap following p is less than $2\sqrt{p} + 1$. This would be a significant tightening of Betrand's postulate. Although no proof has been provided, experimental evidence suggests that *Andrica's conjecture* is likely to be true.

Legendre's conjecture

Adrien-Marie Legendre was active in several areas of mathematics. His work in number theory saw him clash with Gauss, whom he accused of 'excessive impudence', in appropriating without credit his discovery of the **quadratic reciprocity law**. Legendre's later work in geometry was marred by his doomed attempts to prove Euclid's **parallel postulate**.

His investigations of prime numbers led him to conjecture that between any two square numbers there will always be a prime. That is, for any natural number n, there is a prime number p where $n^2 < p < (n + 1)^2$. The format of the conjecture is similar to Bertrand's postulate. But while that quickly succumbed to mathematical ingenuity, Legendre's conjecture remains stubbornly resistant. The most promising progress on Legendre's conjecture to date is Chen's Theorem 3.

Twin prime conjecture

Pairs of prime numbers which are 2 apart, such as 5 and 7, or 17 and 19, are known as *twin primes*. Although the infinity of the primes has been known since the time of Euclid, whether there are infinitely many pairs of twin primes is still an open question.

At time of writing, the largest known twin primes are the two numbers either side of $2{,}003{,}663{,}613 \times 2^{195{,}000}$, discovered in 2007 by Eric Vautier as part of *Primegrid* and the internet *Twin Prime Search*. The twin primes conjecture is generalized by de Polignac's conjecture and the first Hardy–Littlewood conjecture. It is the first of many attempts to analyse the different constellations which can be found among the primes.

de Polignac's conjecture

Proposed in 1849 by Alphonse de Polignac, this asserts that every even number appears as the gap between two consecutive primes, infinitely often. So there should be infinitely many pairs of consecutive primes which are 4 apart, infinitely many which are 6 apart, and so on. This generalizes the *twin primes conjecture*, which corresponds to the particular case of gaps of size 2, and is a particular case of the first Hardy–Littlewood conjecture. de Polignac's conjecture is still open, though Chen's theorem 2 marks a significant step towards proving it.

Chen's theorem 1

Between 1966 and 1975, Chen Jingrun proved three theorems which marked major progress towards several outstanding open problems about prime numbers. At the centre of Chen's work was the notion of a *semiprime*: a number which is the product of exactly two primes (e.g. $4 = 2 \times 2$ or $6 = 2 \times 3$). Chen's first theorem was aimed in the direction of the Goldbach conjecture. It says that, after a certain threshold, every even number can be written either as the sum of two primes (as Goldbach would have it), or of a prime and a semiprime.

Chen's theorem 2

Chen's second theorem took a stride towards de Polignac's conjecture. It states that every even number appears infinitely often either as the difference of two prime numbers (which would follow from de Polignac's conjecture), or of a prime and a semiprime. A consequence of Chen's theorem 2 is that there are infinitely many prime numbers p where $p + 2$ is either prime (as the twin prime conjecture states) or semiprime.

Chen's theorem 3

In 1975, Chen proved a theorem aimed at Legendre's conjecture. *Chen's Theorem 3* says that between any two square numbers there exists either a prime, or a semiprime.

Constellations of primes

The twin primes and de Polignac's conjecture both concern pairs of primes, a fixed distance apart. But what of longer patterns: triples, quadruples of primes, in a prescribed pattern? Care is needed: not all possible arrangements of primes are possible. For instance, why does the twin primes conjecture ask for pairs of prime numbers of the form n and $n + 2$, instead of n and $n + 1$? The answer is that one of n and $n + 1$ must always be divisible by 2, and so they cannot both be prime (the sole exception being the pair 2 and 3).

The same phenomenon occurs for longer patterns. It is no use searching for triples of primes of the form n, $n + 2$ and $n + 4$: one of these will always be divisible by 3, and so not prime (except for the triple 3, 5, 7). These forbidden patterns can be described neatly using modular arithmetic.

The remaining permissible patterns such as n, $n + 2$, $n + 6$ are known as *constellations*. Linear constellations such as this are the subject of the first Hardy–Littlewood conjecture. Even more general constellations such as $n^2 + 1$, $2n^2 - 1$ and $n^3 + 3$ are the subject of Hypothesis H.

The first Hardy–Littlewood conjecture

In 1923, Hardy and Littlewood addressed the problem of longer constellations of primes. Just as the prime number theorem gives a rule for the average number of individual primes over a large range, their idea was to predict how often a particular constellation should appear, on average. They calculated a precise estimate. A consequence, if true, would be that every permitted linear constellation does appear among the primes, infinitely often. So, since n, $n + 2$, $n + 6$ is not forbidden, there should be infinitely many triples of primes of this form (starting with $11, 13, 17$ then $17, 19, 23$ and $41, 43, 47$). If proved, the first Hardy–Littlewood conjecture would be a broad generalization of both Dirichlet's theorem and the Green–Tao theorem. It would be generalized even further by Hypothesis H.

Hypothesis H

From the twin prime conjecture (still unproved!) to the first Hardy–Littlewood conjecture is already a long journey of generalization. But in 1959 Andrzej Schinzel and Wacław Sierpiński suggested yet another step. The first Hardy–Littlewood conjecture is concerned only with constellations of primes obtained through addition, such as n, $n + 2$, $n + 6$. Each such pattern requires finding a suitable n, and then adding specified constants to it.

This does not encompass the $n^2 + 1$ conjecture. There the condition on the prime also involves multiplication. To come up with one broad conjecture which included this too, Schinzel and Sierpiński needed the language of **polynomials**. The idea was that any suitable collection, such as $n^2 + 1$, $2n^2 - 1$, and $n^3 + 3$, should all give simultaneously prime outputs, for infinitely many values of n. Only *irreducible* polynomials need be considered, which cannot be split up (n^2 for example, is reducible to $n \times n$, and can never produce a prime result).

Mimicking the prime number theorem, Schinzel and Sierpiński provided a detailed estimate of how often they expected their systems to be simultaneously prime.

Dirichlet's theorem

Undoubtedly, the primes form the most enigmatic of all sequences of natural numbers: 2, 3, 5, 7, 11, 13,… More prosaic are *arithmetic progressions*: start with a natural number (3 for example), and then repeatedly add on another number (such as 2), to get a sequence: 3, 5, 7, 9, 11, 13,… . Understanding the relationship between these simple sequences and the infinitely subtler sequence of primes is an ongoing area of research.

A fundamental question to ask is: does the sequence above contain infinitely many primes?

Not every arithmetic progression can contain primes: the sequence with initial term 4 and common difference 6 is 4, 10, 16, 22,… But this can never reach a prime number: every number in the sequence is even, and so divisible by 2. To have a hope of finding infinitely many primes, the initial term of the sequence and its common difference should be *coprime* (that is, have highest common factor equal to 1). This sequence fails because 4 and 6 have 2 as a common factor.

In 1837, Johann Dirichlet proved his most celebrated and technically impressive result: he showed that this is indeed the only obstacle. Every arithmetic progression does contain infinitely many primes, as long as its initial term and common difference are coprime. That is, if a and b are coprime, then there are infinitely many prime numbers of the form $a + bn$. Dirichlet's proof was the first to deploy **L-functions**, and as such is often considered as marking the birth of analytic number theory.

Green-Tao theorem

Dirichlet's theorem demonstrates that most arithmetic progressions contain infinitely many primes. Turning this question around leads to the search for arithmetic sequences embedded in the primes. For example 3, 5, 7 is an arithmetic progression of primes of length 3 (and common difference 2), and 41, 47, 53, 59 is one of length 4 (and common difference 6).

An old conjecture, whose roots are lost in mathematical folklore, but likely to date back at least to 1770, says that there is an arithmetic progression of primes of any length you care to name. At time of writing, the longest known arithmetic progression of primes was found by Benoāt Perichon, as part of Primegrid. It has length 26, and starts at 43142746595714191, with common difference 5283234035979900. Progress towards the general conjecture was slow over the 20th century, but it was dramatically proved in 2004, by Terence Tao and Ben Green, which contributed to Tao winning the Fields medal in 2006. This theorem would be generalized by the first Hardy–Littlewood conjecture and Hypothesis H.

Fermat primes

Pierre Fermat was interested in primes of a particularly simple sort, those which are one more than a power of two: $2^n + 1$. For example, $3 = 2^1 + 1$, and $17 = 2^4 + 1$. Not every number of this form is prime: $2^3 + 1 = 9$ for example; and nor does every prime fit this form (7 does not). Fermat noticed that in all cases where $2^n + 1$ is prime,

n itself was a power of 2. So he began studying numbers of the form $2^{2^m} + 1$, now called *Fermat numbers*. Fermat conjectured that every such number is prime.

These numbers grow large very quickly, so it was difficult to test this conjecture for many values. In 1953, however, J.L.Selfridge showed that $2^{2^{16}} + 1$ is not prime, refuting Fermat's conjecture.

In fact, most Fermat numbers are not prime. To date, only five *Fermat primes* are known: 3, 5, 17, 257 ($= 2^{2^3} + 1$), and 65537 ($= 2^{2^4} + 1$). After that the sequence seems to stop yielding primes ($2^{2^5} + 1 = 4,294,967,297 = 641 \times 6,700,417$ for example). Whether there are any more Fermat primes remains an open question. One consequence would be for the theory of **constructible polygons**.

Mersenne primes

Mersenne numbers are those which are 1 less than a power of 2: that is, numbers of the form $2^n - 1$. Not all such numbers are prime, for example $15 = 2^4 - 1$. Also, not all primes are Mersenne (5 and 11 are not, for instance). But those which are, such as $3 = 2^2 - 1$ and $7 = 2^3 - 1$, comprise the important class of *Mersenne primes*, named after the French monk Marin Mersenne who began listing them in 1644.

As it happens $2^n - 1$ can only be prime if n is itself prime. However, this is still no guarantee of primality, for example $2^{11} - 1 = 2047 = 23 \times 89$. But this provides the best starting point in the ongoing search for large primes. At time of writing, 47 Mersenne primes are known. It is an open question whether there are really infinitely many.

The study of Mersenne primes is intimately connected to that of **perfect numbers** since even perfect numbers are all the numbers of the form $\frac{M \times (M + 1)}{2}$ where M is a Mersenne prime. The first two examples are $6 = \frac{3 \times 4}{2}$ and $28 = \frac{7 \times 8}{2}$.

Large primes

The infinity of the primes has been known since Euclid's time, proving beyond doubt that there is no biggest prime number. Nevertheless the quest to pin down ever larger primes dates back for hundreds of years. With a handful of exceptions, the largest known prime numbers have always been Mersenne primes. In 1588, Pietro Cataldi showed that the Mersenne number $2^{19} - 1 = 524,287$ is prime, and this record stood for almost 200 years.

In the 20th century the search accelerated not only through improved techniques of primality testing, such as the discovery of the **Lucas–Lehmer test**, but through sheer computing power. Since 1951, the hunt for large primes has relied on ever faster computers and, since 1996, it has been dominated by the *Great Internet Mersenne Prime Search* (*GIMPS*), which exploits idle time on the computers of thousands of volunteers worldwide. The largest known prime at the time of writing is a Mersenne prime discovered in August 2008 through GIMPS (by Smith, Woltman, Kurowski and colleagues): $2^{43,112,609} - 1$. When written out in full, this is 12,978,189 digits long.

Primality testing

The simplest method to tell whether a number n is prime is just to try to divide it by every smaller number. (In fact it is enough just to try prime numbers up to $\sqrt{n}$.) However, for really large numbers this is impractically slow: to test a number 100 digits long would take longer than the lifetime of the universe.

The sieve of Eratosthenes is the oldest known method for generating lists of primes, and remains in use today. Other primality tests such as the Fermat Primality Test, and the *Miller–Rabin test* are probabilistic: no true primes fail them, but the numbers which pass are only 'probably prime'. (These tests might more accurately be termed *compositeness tests*.) All modern tests have to be grounded in modular arithmetic to bring the numbers involved down to a manageable size. Today, the search for large prime numbers hinges on the non-probabilistic *Lucas–Lehmer test*. In 2002, the *AKS primality test* demonstrated for the first time that testing the primality of general numbers can be done comparatively efficiently.

Fermat's primality test

The *Chinese hypothesis* was an incorrect method for identifying prime numbers. It said that the number q is prime if and only if q divides $2^q - 2$. It is understandable that people believed this: 3 divides $2^3 - 2(= 6)$, 4 does not divide $2^4 - 2(= 14)$, 5 divides $2^5 - 2(= 30)$, and this continues to work for some time. The first counterexample is the non-prime $341 = 11 \times 31$, which does divide $2^{341} - 2$.

Despite being false, this conjecture contains the seeds of a primality test based on **Fermat's little theorem**. A consequence of this is that if q is possible prime, and we find some number n where $n^q - n$ is not divisible by q, then q cannot be prime after all. The chinese hypothesis corresponds to the case $n = 2$: so it forms a necessary condition for primality, but not sufficient.

Fermat's primality test tries a putative prime q against base numbers n. If the conclusion to the theorem fails (so q does not divide $n^q - n$), it follows that q is not a true prime after all.

However, this test is not perfect, as there are *pseudoprimes* which can pass it for particular bases: 341 is a 2-pseudoprime. The worst cases are non-primes q which can pass this test for every base n which is coprime to q. These are called *Carmichael numbers*, after their discoverer in 1910, Robert Carmichael. The first is 561. In 1994, Alford, Granville and Pomerance proved that there are infinitely many Carmichael numbers.

Lucas–Lehmer test

The Lucas–Lehmer test dates from 1930 and is used today by the *Great Internet Mersenne Prime Search* to test for Mersenne primes. If p is a large prime number, the question to answer is whether the enormous number $2^p - 1$ is also prime. Call this number k. The basis of the test is the Lucas–Lehmer sequence which is defined recursively: $S_0 = 4$ and $S_{n+1} = S_n^2 - 2$. *Lehmer's theorem* says that k is prime if and only if it divides S_{p-2}.

The difficulty is that the Lucas–Lehmer sequence grows very fast indeed: the first few terms are $S_0 = 4$, $S_1 = 14$, $S_2 = 194$, $S_3 = 37,634$ and $S_4 = 1,416,317,954$. By S_8 the sequence has grown bigger than the number of atoms in the universe, making any further direct calculations impossible. An approach using modular arithmetic is needed, in which successive values of S_n are calculated modulo k instead: this is where all the computing time is used. The final step is to calculate $S_{p-2} \bmod k$. If this is zero, then k does divides S_{p-2} and so is prime.

The prime counting function

There are of course infinitely many prime numbers. So what can it mean to 'count' them? The answer is given by the *prime counting function* π (not to be confused with the number π). For a natural number n, $\pi(n)$ is defined to be the number of primes up to and including n. So $\pi(7) = 4$ because there are exactly four prime numbers up to 7 (namely 2, 3, 5 and 7). Similarly $\pi(8) = 4$, $\pi(11) = 5$, $\pi(100) = 25$, and $\pi(1000) = 168$.

Understanding this function is a major aim of number theory. By examining it over a large range (see graph below), the unpredictability of the individual primes is smoothed out, to reveal general trends. The general rate of growth of $\pi(n)$ is described in the Prime number theorem, with more precise information coming from the **Riemann hypothesis** (if true).

The second Hardy–Littlewood conjecture

The partnership between G.H.Hardy and J.E.Littlewood is one of mathematics' greatest collaborations. Their second conjecture, made in 1923, is that the prime counting function is *subadditive*: for any two numbers x and y, $\pi(x + y) \leq \pi(x) + \pi(y)$.

A consequence of this would be that no block of n consecutive numbers can contain more primes than the first n numbers does. In a surprising development in 1974, Ian Richards proved that the first and second Hardy–Littlewood conjectures are incompatible, and concluded that the second is 'probably false'.

The prime number theorem 1

At the age of 14, Carl Friedrich Gauss made a remarkable observation about the prime counting function. He noticed that the ratio $n : \pi(n)$ was roughly the **natural logarithm** of n, that is, $\ln n$. This means that if you pick a number between 1 and n at random, the probability of getting a prime is approximately $\frac{1}{\ln n}$. So the number of primes between 1 and n is approximately $\frac{n}{\ln n}$. The young Gauss conjectured that, even for big values of n, this would continue to be true. That is, the two functions $\pi(n)$ and $\frac{n}{\ln n}$ should be *asymptotically equal*. This is written as $\pi(n) \sim \frac{n}{\ln n}$. More precisely, this means that one divided by the other will get closer and closer to 1, as n gets larger.

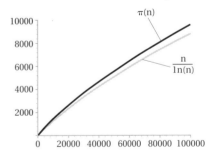

The prime number theorem 2

Later, Gauss refined his estimate for the number of primes from $\frac{n}{\ln n}$ to the more accurate *logarithmic integral* function, $\mathrm{Li}\, n$. Technically, $\mathrm{Li}\, n = \int_2^n \frac{\mathrm{d}x}{\ln(x)}$. Again, Gauss believed, but did not prove, that $\pi(n) \sim \mathrm{Li}\, n$. This assertion is the *prime number theorem*, and it also implies the weaker result that $\pi(n) \sim \frac{n}{\ln n}$.

In 1896, by studying the Riemann zeta function, Jacques Hadamard and Charles de la Vallée Poussin each independently proved the prime number theorem. This theorem describes the general pattern, but not the exact values, of $\pi(n)$. It predicts that the number of primes up to 10,000,000,000 should be approximately 455,055,614. In fact there are 455,052,511, an error of around 0.0007%. The **Riemann hypothesis**, if proved, would fill in the details of these errors, and so turn the prime number theorem into a method for describing $\pi(n)$ at the best possible level of accuracy.

The Riemann zeta function
Bernhard Riemann only published one paper on number theory: 'On the Number of Prime Numbers less than a Given Quantity' of 1859. But it has a strong claim to be the most influential ever written. The paper concerned a certain function defined by a **power series**:

$$\zeta(s) = \sum_{n=1}^{\infty} \frac{1}{n^s} = 1 + \frac{1}{2^s} + \frac{1}{3^s} + \frac{1}{4^s} + \frac{1}{5^s} + \cdots$$

The function was not new: Leonhard Euler had contemplated it in 1737, and had realized its connection with the prime numbers by spotting a different way to write it, now known as Euler's product formula:

$$\zeta(s) = \prod_{p \text{ prime}} \frac{p^s}{p^s - 1} = \frac{2^s}{2^s - 1} \times \frac{3^s}{3^s - 1} \times \frac{5^s}{5^s - 1} \times \frac{7^s}{7^s - 1} \times \cdots$$

Here, the product is over the prime numbers p, where the sum above was over all natural numbers n.

Riemann's first insight was that the zeta function made sense as a complex function: he could feed in complex values of s and get out complex values of $\zeta(s)$. In fact the formulas above are only valid in half the complex plane. (They only *converge* when the real part of s is greater than 1.) His second stroke of genius was to find a third formula for ζ, valid everywhere (except at the single point $s = 1$), which tallies exactly with Euler's versions.

Extracting explicit values from the zeta function is extremely difficult: studying Riemann's new presentation shows that $\zeta(0) = -\frac{1}{2}$, and that $\zeta(2n)$ is *transcendental* for every positive integer n. But it was only in 1978 that Roger Apéry was able to show that $\zeta(3)$ is *irrational*. The nature of $\zeta(s)$ for other odd values of n remains mysterious. The further behaviour of this function is the focus of Riemann's famous hypothesis.

Euler's product formula
Applying the **generalized binomial theorem** to $(1 - \frac{1}{2})^{-1}$ gives:

$$\left(1 - \frac{1}{2}\right)^{-1} = 1 + \frac{1}{2} + \frac{1}{2^2} + \frac{1}{2^3} + \cdots$$

Similarly:

$$\left(1 - \tfrac{1}{3}\right)^{-1} = 1 + \tfrac{1}{3} + \tfrac{1}{3^2} + \tfrac{1}{3^3} + \cdots$$

Multiplying these two together gives:

$$\left(1 - \tfrac{1}{2}\right)^{-1} \times \left(1 - \tfrac{1}{3}\right)^{-1} = 1\left(1 + \tfrac{1}{3} + \tfrac{1}{9} + \cdots\right) + \tfrac{1}{2}\left(1 + \tfrac{1}{3} + \tfrac{1}{9} + \cdots\right) + \tfrac{1}{4}\left(1 + \tfrac{1}{3} + \tfrac{1}{9} + \cdots\right) + \cdots$$

By expanding the brackets and re-ordering, the right-hand side becomes:

$$1 + \tfrac{1}{2} + \tfrac{1}{3} + \tfrac{1}{4} + \tfrac{1}{6} + \tfrac{1}{8} + \tfrac{1}{9} + \tfrac{1}{12} + \cdots$$

Every number which has 2 and 3 as its prime factors will appear in this expression exactly once. Similarly:

$$\left(1 - \tfrac{1}{2}\right)^{-1} \times \left(1 - \tfrac{1}{3}\right)^{-1} \times \left(1 - \tfrac{1}{5}\right)^{-1} = 1 + \tfrac{1}{2} + \tfrac{1}{3} + \tfrac{1}{4} + \tfrac{1}{5} + \tfrac{1}{6} + \tfrac{1}{8} + \tfrac{1}{9} + \tfrac{1}{10} + \tfrac{1}{12} + \cdots$$

In order for every natural number to appear on the right-hand side, every prime number would have to appear on the left. This suggests taking the infinite product:

$$\prod_{p \text{ prime}} \left(1 - \tfrac{1}{p}\right)^{-1}$$

The problem is that the corresponding series $1 + \tfrac{1}{2} + \tfrac{1}{3} + \tfrac{1}{4} + \tfrac{1}{5} + \tfrac{1}{6} + \tfrac{1}{7} + \tfrac{1}{8} + \tfrac{1}{9} + \tfrac{1}{10} + \cdots$ does not *converge*: it just gets bigger and bigger. But if $s > 1$, then the series given by the Riemann zeta function $\zeta(s)$ $1 + \tfrac{1}{2^s} + \tfrac{1}{3^s} + \tfrac{1}{4^s} + \tfrac{1}{5^s} + \tfrac{1}{6^s} + \tfrac{1}{7^s} + \cdots$ does converge, and the same argument suggests that this should equal:

$$\prod_{p \text{ prime}} \left(1 - \tfrac{1}{p^s}\right)^{-1}$$

This is the core idea behind *Euler's product formula*, the first hint that the Riemann zeta function is intimately linked to the prime numbers

$$\zeta(s) = \prod_{p \text{ prime}} \left(1 - \tfrac{1}{p^s}\right)^{-1}$$

or equivalently

$$\zeta(s) = \prod_{p \text{ prime}} \left(\tfrac{p^s}{p^s - 1}\right)$$

The Riemann hypothesis

'If I were to awaken after having slept for a thousand years, my first question would be: Has the Riemann hypothesis been proven?', said the great mathematician David Hilbert. Using the zeta function and some spectacular insights into **complex analysis**, Riemann was able to concoct an exact formula for the *prime counting function*: arguably the holy grail of number theory. However, his formula depended on knowing where the zeta function disappeared: the values of *s* for which $\zeta(s) = 0$. It is comparatively easy to check that $\zeta(-2)$, $\zeta(-4)$, $\zeta(-6)$,… are all zero: these are called the *trivial zeroes*. But there are infinitely many non-trivial zeroes too. Riemann showed that these all lie in a *critical strip* between $\text{Re}(s) = 0$ and $\text{Re}(s) = 1$ (see diagram), and that they must lie symmetrically across the *critical line* $\text{Re}(s) = \frac{1}{2}$.

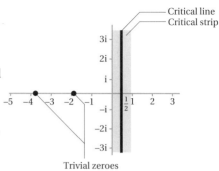

The *Riemann hypothesis* is the claim that all the non-trivial zeroes of the zeta function lie on the critical line. Riemann wrote that he thought this 'very probable' but had 'temporarily put aside the search for this after some fleeting futile attempts'; 150 years later, after the concentrated efforts of hundreds of our greatest minds, a proof remains elusive.

Riemann's zeroes

Riemann's manuscript immediately caused excitement in the mathematical world, and has continued to inspire professional and amateur mathematicians ever since. In 1896, Hadamard and de la Vallée Poussin independently managed to prove that none of Riemann's zeroes lie on the very edge of the strip, on the lines $\text{Re}(s) = 0$ or $\text{Re}(s) = 1$. Shaving off this infinitesimal sliver was enough to deduce the prime number theorem. In 1914, G.H. Hardy proved that there are infinitely many zeroes on the critical line (though this does not rule out there being others off it). In 1974 Norman Levinson proved that at least 34.74% of the non-trivial zeroes do lie on the critical line, and this was improved to 40.1% by Brian Conrey in 1989.

Of course, the Riemann hypothesis implies that that the true proportion is 100%, and so far the experimental evidence backs this up: in 2004, Xavier Gourdon and Patrick Demichel harnessed the power of **distributed computing** to verify that the first 10 trillion non-trivial zeroes do indeed lie on the critical line.

L-functions

Euler's original definition of the zeta function was:

$$\zeta(s) = \sum_{n \geq 1} \frac{1}{n^s} = 1 + \frac{1}{2^s} + \frac{1}{3^s} + \frac{1}{4^s} + \frac{1}{5^s} + \cdots$$

Notice that all the fractions have 1 as their numerator. Other functions can be formed by replacing these 1s with other (carefully chosen) complex numbers. This is the profoundly important family of *L-functions*. These behave somewhat similarly to the zeta function (including in their reluctance to give up their secrets). Each comes with its own version of Euler's product formula, its own extension to the complex numbers, and its own variant of the Riemann hypothesis, which are collected together in the generalized Riemann hypothesis.

L-functions first arose in the original proof of Dirichlet's theorem. Since then, they have become central to the subject of analytic number theory, despite the formidable technical difficulties they pose. An analogue of the Riemann hypothesis for closely related functions was the centrepiece of the **Weil conjectures**, proved in 1980 by Pierre Deligne. Then in 1994 L-functions were exploited by Andrew Wiles to prove Fermat's last theorem. Other families of L-functions hold the keys to many further treasures of the mathematical world, including the **Birch and Swinnerton–Dyer conjecture** and **Langlands' program**.

The generalized Riemann hypothesis

A proof of the Riemann hypothesis would be momentous for the light it shed on the prime numbers. In particular, if it is true, the behaviour of the prime counting function can be captured far more precisely than the prime number theorem so far allows. But the implications go even wider: the zeta function is just the first in the infinite family of L-functions which pervade modern number theory. Techniques which allowed it to be understood would have deep implications for the rest.

The generalized *Riemann hypothesis* says that the non-trivial zeroes of every L-function lie on the critical line $\text{Re}(s) = \frac{1}{2}$. Individually, no instance of this has yet been proved or falsified.

GEOMETR

MANY OF THE THEOREMS OF GEOMETRY that are taught in school today date back to ancient Greece. In fact most of them are found in one book: Euclid's *Elements*. In it, Euclid combined his own research with the amassed knowledge of the ancient world to study the interactions of points, straight lines and circles. The *Elements* was as significant for its axiomatic approach as it was for the theorems it contained. However, in the 19th century new forms of non-Euclidean geometry were discovered, in which Euclid's axioms were not valid.

From here, geometry diverged into several strands, which can be illustrated by one of the simplest mathematical figures: the circle. In *differential geometry*, a circle is the result

TES • THE PARALLEL POSTULATE • THE
ATE • ANGLES • PARALLEL LINES • RIGHT
SIAN COORDINATES • PLOTTING GRAPHS
GHT LINE • THE REAL LINE • EUCLIDEAN
LTI-DIMENSIONAL SPHERES • TRIANGLES
D TRIANGLES • PYTHAGORAS' THEOREM
THAGOREAN TRIPLES • THE AREA OF A
AND TANGENT • COSECANT, SECANT, AND
• TRIGONOMETRIC VALUES • THE LAW OF
TRIANGLES • THE EULER LINE • CIRCLES
AL TANGENT THEOREM • THE INSCRIBED
ANGLES IN THE SAME SEGMENT • CYCLIC
THEOREM • INTERSECTING CHORDS
CONVEX QUADRILATERALS • NON-CONVEX
THE PLATONIC SOLIDS • IRREGULAR

Y

of patching together several segments of line, smoothly curved in a precise way. In topology, circles are allowed to be bent out of shape, making a square an example of a circle. Between these two lies *differential topology*. This allows deformed circles such as ovals or rounded oblongs, but only ones which are *smooth,* without the sharp corners of a square. *Knot theory* is the study of the different ways that smooth circles can get tangled in 3-dimensional space. A different view comes from the algebraic description of a circle, by the equation $x^2 + y^2 = 1$. *Algebraic geometry* is the study of objects described by polynomials in this way. This abstract approach allows geometrical methods to apply in settings far beyond anything we can draw on a piece of paper.

EUCLIDEAN GEOMETRY

Euclid's *Elements*

Euclid of Alexandria's 13-volume treatise *Elements* marks a milestone in mathematical writing. Dating from 300BC it ranks as the most successful textbook of all time.

Euclid did not just include his own work, but collected that of his contemporaries and forebears into an impressive body of knowledge. For around 2000 years, it was held in such high regard that it remained a standard text around the world. The first printed edition was produced in 1482. Over one thousand subsequent editions have been published, a number exceeded only by the Bible.

The proof of the **infinity of the primes** is a highlight of the section on number theory, and Euclid also developed the theory of **ruler and compass constructions**. But it is for Euclid's geometrical work that the *Elements* is chiefly famous, celebrated as much for his modern approach as for the facts proved. For the first time, a mathematician wrote down in detail his starting assumptions (or *postulates* or *axioms*), and deduced all his results with precision, directly from these axioms.

Euclid's postulates

People had been studying points and lines for thousands of years before Euclid. However, in his hands this subject (plane geometry) became the first area of mathematics to be *axiomatized*. Euclid showed that his work rested on five fundamental assumptions:

1 Between any two points you can draw a straight line segment.
2 Any straight line segment can be extended indefinitely in both directions.
3 A circle can be drawn with radius of any length, centred at any point.
4 Any two right angles are equal.
5 If two straight lines are crossed by a third, so that the interior angles on one side are less than two right angles ($x + y < 180°$), then the two lines, when extended indefinitely, will eventually cross.

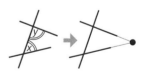

It is likely that Euclid was not satisfied by the long-winded fifth axiom, known as the *parallel postulate*. The first 28 propositions of the *Elements* use only postulates 1–4. But in the 29th, on parallel lines, he had no choice but to rely on postulate 5.

The parallel postulate

For most of the 2300 years since Euclid's *Elements*, the term 'Euclidean' geometry was redundant; there was no other sort. Only geometrical systems obeying Euclid's five postulates were known. Nevertheless, most geometers shared Euclid's view that the fifth postulate was somewhat troublesome, being less obviously true than the others. Easier alternative formulations were found, including one known as *Playfair's axiom* (although John Playfair was not the first to discover it):

> *Given a straight line, and a point not on that line, there is at most one straight line passing through the point, parallel to the given line.*

'Parallel' means that the lines, extended indefinitely, will never cross. Euclid had already shown (using just the first four postulates) that there is *at least one* such line, which explains the strange expression 'at most one'.

The independence of the parallel postulate

For centuries, European and Islamic mathematicians debated this axiom. Could it be derived from the first four of Euclid's postulates? A huge number of fallacious proofs were written and debunked, and several logically equivalent conditions were discovered (including the assertion that the angles in a triangle always sum to 180°). Recently, Scott Brodie observed that **Pythagoras' theorem** is logically equivalent to the parallel postulate.

It was not until the 19th century when Carl Friedrich Gauss, Nikolai Lobachevsky and János Bolyai conceived of a system of **hyperbolic geometry** which obeyed postulates 1−4, but not the parallel postulate, that the parallel postulate was finally shown to be independent of the first four postulates (see **independence results**).

Angles

An *angle* is an amount of turn. But how can we quantify this? Perhaps the easiest way is in terms of *revolutions*. So one complete rotation counts as 1, and a right angle counts as $\frac{1}{4}$. This is convenient for measuring large amounts of turn, which is why rotational speed is usually measured in *revolutions per minute*. For smaller amounts of turn, a finer scale is helpful.

The traditional measure of angle is the *degree* (°), where 360° constitutes a complete rotation. This has its roots in the ancient Babylonian year, which had 360 days. Traditionally a degree is divided into 60 *arc-minutes*, and each minute into 60 *arc-seconds*. Another system, no longer much used, is the *grade* (or *gradian*), invented in France in an attempt to incorporate angles into the metric system. 100 grades constitute a right angle, making a complete revolution 400 gradians. Though historically interesting, these systems are mathematically arbitrary. Scientists generally prefer to use **radians**, a system based on the circle, where one full revolution equals 2π radians.

Parallel lines

Two straight lines are *parallel* if they lie in the same plane and never meet, even when extended indefinitely in either direction, like an infinitely long set of rail-tracks. (The condition that the lines lie in a plane is necessary: in three dimensions two

lines can be *skew*, neither parallel nor crossing each other.) If we
assume the parallel postulate, an equivalent condition is that
the perpendicular distance between the two lines is always the
same, wherever it is measured.

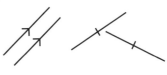

Parallel lines are usually represented with matching arrows
on the lines, while line segments of equal length are drawn with
matching dashes across them.

Two important facts concern what happens when a pair of
parallel lines are crossed by a third line, and were proved by Euclid as
Proposition 1.29 of *Elements*, his first result depending on the parallel
postulate:

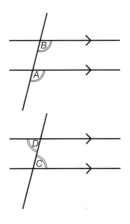

> **1** *Corresponding angles are equal*: in the illustration A and B are
> *corresponding* angles (sometimes called 'F angles') because they
> occupy corresponding positions on different parallel lines. Such
> angles are always equal.
>
> **2** *Alternate angles are equal*: C and D are *alternate* angles
> (or 'Z angles') and again are always equal.

Right angles

90°, $\frac{\pi}{2}$ radians, or a quarter of a revolution: however you measure
it, a right angle is a crucial ingredient in Euclidean geometry. Perhaps the most elementary
definition is to take a straight line. On one side this creates an angle of π (or 180°). If we divide
this into two equal angles, we get a right angle. The fourth of Euclid's postulates says that,
whenever you do this, the result is always the same. A right angle is indicated by two lines
marking off a small square in the corner.

The opposite notion to parallel: two lines are *perpendicular* if they are at right angles to
each other. (In some settings, such as vector geometry, the word *orthogonal* is used.)

Right angles also provide a notion of measurement. It is obvious how to measure the distance
between two points on a plane. But between two lines (or a point and a line, or a line and a
plane) we need the perpendicular distance: the length of a new line segment which crosses the
original ones at right angles.

Cartesian geometry

René Descartes (or *Cartesius* when he wrote in
Latin) is principally famous for his epistemological breakthrough 'I think therefore I am'. In the
17th century, the dividing lines between philosophy, science and mathematics were less marked
than today. In his work *Discours de la Méthode* (*Discourse on Method*) Descartes first wrote
'*je pense, donc je suis*'. But it was in an appendix to that work, entitled *La Géométrie*, that
Descartes introduced his *coordinates*, which had an equally profound impact on modern
thought. Descartes described the Euclidean plane (an unbounded flat 2-dimensional surface)
by drawing two lines across it. These cross at right angles at the *origin*, and divide the plane into
four *quadrants*. The horizontal line is usually called the *x*-axis, and the vertical one the *y*-axis.
The position of any point on the plane can be given using *x* and *y*-coordinates.

Cartesian coordinates

In Cartesian geometry, every point on the plane can be identified by saying which quadrant it is in, and how far it is from each axis. This is much like reading a map; indeed the two numbers which capture this information are called the point's *coordinates*.

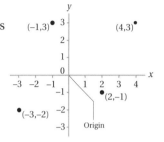

The point $(4, 3)$ is in the top right quadrant (because both numbers are positive), four units to the right (along the x-axis), and three units up. (Some people use a mnemonic such as 'Along the corridor and up the stairs' to remember the order of the x and y coordinates.) Similarly, $(-3, -2)$ is in the bottom left quadrant, three units to the left and two units down. The origin has coordinates $(0, 0)$.

Plotting graphs

As far back as the ancient Babylonians, algebraic methods have been used in geometry. But Cartesian coordinates opened up geometry to more sophisticated techniques. With Cartesian coordinates, points are precisely defined by numbers. The relationships between those numbers can be used to describe geometric figures.

For instance we could look at all the points whose two coordinates are the same: $(0, 0)$, $(1, 1)$, $(-10, -10)$, etc. Plotting these on the plane, we find that they all lie on a straight line. Looking at the numbers, it is obvious that a point (x, y) lies on this line precisely if $y = x$. So we call $y = x$ the *equation* of the line.

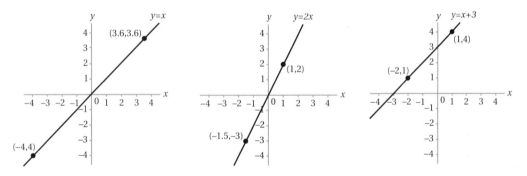

Another example is those points where the second coordinate is double the first: $(1, 2)$, $(-6, -12)$, $(0, 0)$, etc. Again they lie on a straight line, and in this case its equation is $y = 2x$. Similarly, starting with the equation $y = x + 3$, substituting values for x (say 0, -4 and 8) gives points $(0, 3)$, $(-4, -1)$ and $(8, 11)$ on the line. All of these points have y-coordinate 3 bigger than their x-coordinate.

Gradients

Gradients are marked on road-signs to indicate the steepness of hills. A gradient of 25% means that for every 1 metre travelled horizontally you climb $\frac{1}{4}$ of a metre. Another way to say this is that, over any range, the increase in height divided by the horizontal distance covered is always $\frac{1}{4}$. If you were going downhill, the sign might say -25%.

Of course, physical hills have beginnings and ends, and the steepness in between is irregular; this is not so with straight lines in the plane. To calculate the gradient of a straight line, pick any two points on the line, measure the height gained vertically and divide this by the distance covered horizontally. (If left to right is a downwards slope, then the gradient is negative.)

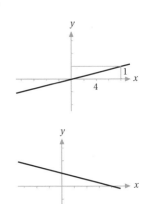

For a smooth curve the gradient is not fixed, but changing. It makes sense to pick a point on the curve and ask about the gradient of the *tangent* to the curve there. This can be calculated by **differentiation**.

The equation of a straight line

A straight line in the plane can be identified by two pieces of information: firstly, its gradient, and secondly the coordinates of any point the line passes through. A convenient point to pick is the *y-intercept*: the place where the line crosses the vertical axis (or y-axis). Since it lies on the axis, the first coordinate of this point will be 0. Call it $(0, c)$.

If the gradient of the line is m, then the equation of the line is $y = mx + c$. So the line $y = 4x + 2$ crosses the y-axis at $(0, 2)$ and has gradient 4.

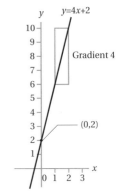

An exceptional class is that of the vertical lines. Since they are parallel to the y-axis, they do not cross it anywhere, and there is no meaningful value of c to calculate. Similarly the gradient of these lines is not defined, since they do not cover any horizontal distances. (Some would say that they have 'infinite gradient'.) Despite this, the equations of these lines are straightforward, since they are defined by the first coordinate staying fixed: $x = 3$, for example.

Horizontal lines have a gradient of 0 and are defined by their fixed second coordinate: $y = 4$, for example.

The real line

Euclid's *Elements* set up geometry for the millennia to come. His fundamentals – parallel lines, right-angled triangles, circles – are as important now as ever. Nevertheless, by the 19th century, mathematicians had become concerned not to take too much for granted. The *Elements* opens with some definitions: '1. A point is that which has no part. 2. A line is a breadthless length … 4. A straight line is a line which lies evenly with the points on itself. 5. A surface is that which has length and breadth only'. Any modern mathematician will have no trouble grasping Euclid's meaning. At the same time, what really is 'that which has no part'? It was necessary to translate Euclid into the precise terminology of modern mathematics.

These ideas can be formalized through the system of **real numbers** (**R**). This comes ready wrapped as a geometric object, the *real line*. In modern terms this is the first Euclidean space.

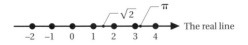

Euclidean plane

The set (**R**) of real numbers is also known as the *real line*: it provides the perfect model of a 1-dimensional line. *Points* on this line are simply numbers, and the *distance* between two points is the larger number minus the smaller. So two points are either a positive distance apart (with infinitely many points in between), or are actually the same point; no two points are true neighbours. Also, points come in a natural order, given by their size.

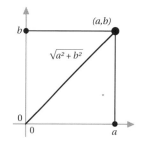

A *plane* can be modelled by replacing the individual numbers with pairs of real numbers (a, b). This is the system of Cartesian coordinates. Formally, these pairs of numbers are not merely *directions* to a point; they *are* the points. The distance between a point (a, b) and the origin $(0, 0)$ is given by Pythagoras' theorem: $\sqrt{a^2 + b^2}$. (This is easily extended to give the distance between any two points.)

This collection of all pairs of real numbers is denoted **R²** and called *Euclidean 2-space*, or simply *the plane*. Repeating this process gives higher-dimensional spaces.

Higher-dimensional spaces

We can repeat the trick of Euclidean space in three dimensions: *Euclidean 3-space* (**R³**) is the set of all triples of real numbers (a, b, c), with the distance from (a, b, c) to $(0, 0, 0)$ given by $\sqrt{a^2 + b^2 + c^2}$. There's no need to stop here. It may not be possible to visualize 4-dimensional space but mathematically it is clear what needs to happen: *Euclidean 4-space* (**R⁴**) is the set of all quadruples of real numbers (a, b, c, d). This can continue to any dimension we name. In general, *Euclidean n-space* (**Rⁿ**) is the set of all n-tuples of real numbers $(a, b, ..., z)$, with the distance from the origin given as $\sqrt{a^2 + b^2 + ... + z^2}$.

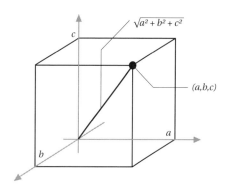

Multi-dimensional spheres

On a plane, a circle is defined to be the set of points a fixed distance (the *radius*) from a given point (the *centre*). For simplicity, take the radius to be 1, and the centre to be the origin $(0, 0)$. Then Pythagoras' theorem produces a formula for the circle as the set of points (x, y) where $\sqrt{x^2 + y^2} = 1$, that is, $x^2 + y^2 = 1$.

The same idea in three dimensions produces a sphere: the set of points (x, y, z) which are distance 1 away from $(0, 0, 0)$. So, $x^2 + y^2 + z^2 = 1$. It is obvious how to extend this to higher dimensions: in Euclidean n-space, the n-sphere is the set of points $(x, y, ..., z)$ a distance of 1 unit away from $(0, 0, ..., 0)$, that is, the points for which $x^2 + y^2 + \cdots + z^2 = 1$.

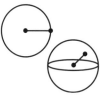

Despite the human impediments to visualizing shapes in higher dimensions, they can often be accessed fairly painlessly, by generalizing from spaces which are more familiar.

TRIANGLES

Triangles

The world of human affairs is best described by the system of **Euclidean geometry**. Here, there are no **biangles**, shapes built from just two straight lines. So the humble triangle is among the most elementary figures, and therefore one of the most significant.

Triangles come in many forms: the first **regular polygon**, an *equilateral triangle*, is one whose sides are all the same length. An *isosceles* triangle is one with two sides of the same length. Triangles whose sides are all different lengths are known as *scalene*.

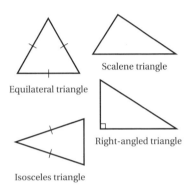

Scalene triangle

Equilateral triangle

Right-angled triangle

Isosceles triangle

A mainstay of Euclidean geometry, a **right-angled triangle**, is simply one which has a right angle at one corner. If instead all its angles are less than 90°, then we have an *acute* triangle. If a triangle contains an *obtuse* angle (one greater than 90°), then it is called an *obtuse* triangle.

Angles in a triangle

The angles in a triangle sum to 180°.

Proved by Euclid in Proposition 1.32 of *Elements*, this is the first important result of triangular geometry. The theorem follows from the alternate angles theorem on parallel lines. Starting with a triangle ABC, draw a new line through C parallel to AB. Then the three angles at C lie on a straight line, and so sum to 180°. But the two new angles at C equal the angles at A and B, by the alternate angles theorem. Among many consequences, this fact implies that the angles in an equilateral triangle are each 60°.

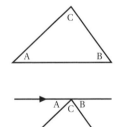

Right-angled triangles

A *right-angled triangle* is a triangle, one of whose angles is a right angle. The *hypotenuse* of a right-angled triangle is its longest side: the one opposite the right angle.

Right-angled triangles are everywhere: they are fundamental objects of Euclidean geometry. It was over 3000 years ago that Pythagoras' theorem first described the relationship between the three sides of a right-angled triangle. In the system of Cartesian coordinates, every point (x, y) comes with its own right-angled triangle, given by the points $(0, 0)$, $(x, 0)$ and (x, y). So Pythagoras' theorem provides the principal method for calculating distances on the plane. Right-angled triangles also provide the setting for **trigonometry**.

Pythagoras' theorem

Perhaps the most famous theorem of all, Pythagoras' theorem is also one of the oldest. Though attributed to the Greek mathematician and mystic Pythagoras (circa 569−475 BC), there is strong evidence that the ancient Babylonians knew this result, over a thousand years earlier. Euclid included it as Proposition 1.47 of his *Elements*.

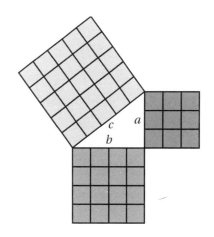

The theorem concerns a right-angled triangle. It says: the square of the hypotenuse (c) is equal to the sum of the squares of the other two sides (a and b). This can be thought of geometrically, as relating the areas of squares built on the sides of the triangle, or purely algebraically: $a^2 + b^2 = c^2$.

Proof of Pythagoras' theorem

Pythagoras' theorem is perhaps the most proved of all mathematical theorems. In the 1907 book *The Pythagorean Proposition*, Elisha Loomis collected together 367 different proofs.

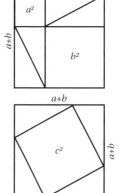

A geometric proof due to the 12th-century Indian mathematician Bhāskara is illustrated here. A square with side $a + b$ is divided up in two different ways: once into four copies of the triangle with sides a, b, c along with a square of area a^2 and a square of area b^2; once into four copies of the triangle and a square of area c^2. Since the total area in both cases is the same, it must be that $a^2 + b^2 = c^2$.

Pythagorean triples

The easiest lengths to work with (especially in the days before pocket calculators) are those with whole-number values. Once right-angled triangles had assumed their rightful place as the gatekeepers to geometry, there was a need to find some which had integer-valued lengths. Unfortunately, most do not. For instance, if you make the two shorter sides each 1 unit long, then Pythagoras' theorem shows that the length of the hypotenuse (c) must satisfy, $c^2 = 1^2 + 1^2 = 1+1= 2$. So $c = \sqrt{2}$, which is not only not a whole number, but worse, it is an **irrational** number. Inconveniently, this is what usually happens.

The first right-angled triangle you can build from whole numbers has sides of length 3, 4 and 5. This satisfies Pythagoras' theorem: $3^2 + 4^2 = 9 + 16 = 25 = 5^2$. So $(3, 4, 5)$ is called a *Pythagorean triple*. Multiples of this, such as $(6, 8, 10)$ and $(9, 12, 15)$ are also Pythagorean triples. *Primitive* Pythagorean triples are those which are not multiples of a smaller one. The next few are $(5, 12, 13)$, $(7, 24, 25)$, $(8, 15, 17)$, and $(9, 40, 41)$.

Euclid, in Proposition 10.29 of *Elements*, produced a general formula for generating these. It follows that the list of primitive Pythagorean triples goes on for ever. The solution to the problem of finding Pythagorean triples was an early result in the study of **Diophantine equations**. Fermat's last theorem states that if we replace the squares with a higher power ($n \geq 3$), then we will never find any integer solutions to $a^n + b^n = c^n$.

The area of a triangle

Over the years, geometers have discovered a phenomenal number of formulas for calculating the area of a triangle. The commonest is 'half the base times the height': $\frac{1}{2} \times b \times h$, where b is the length of one of the edges (the 'base') and h is the perpendicular distance from the base to the third point of the triangle. (This is really three formulas in one, depending on which edge you choose as the 'base'.)

A little trigonometry transforms this formula into another: $\frac{1}{2} \times a \times b \times \sin C$ (here C is the angle opposite the side c).

If your triangle happens to have integer coordinates, the easiest method may be to use **Pick's theorem**.

A more sophisticated formula is $r \times s$, where r is the radius of the triangle's *incircle* (see **centres of triangles**) and s is its *semiperimeter*, $s = \frac{a+b+c}{2}$. The semiperimeter is also involved in **Heron's formula**, due to Heron of Alexandria (circa AD 50), and perhaps the most spectacular formula for the area of a triangle: $\sqrt{s(s-a)(s-b)(s-c)}$.

Trigonometry

In a right-angled triangle, Pythagoras' theorem relates the lengths of the sides. Meanwhile, the fact that angles in a triangle sum to 180° relates the angles: if we know that one angle is 60°, then the final remaining angle must be 30°. This means that all right-angled triangles containing a 60° angle are the same basic shape, with just their sizes differing. (Technically, this makes them *similar*.) Coming from the Greek words *trigōnon* ('triangle') and *metron* ('measure'), trigonometry is about uniting the two: the relationships between the lengths and the angles.

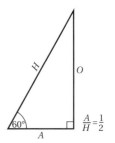

Suppose we know that one angle of our right-angled triangle is 60°. What are the possible lengths of the sides? Focusing on the two edges on either side of the 60° angle, the *adjacent* side (A) and the *hypotenuse* (H), possible lengths are $A = 1\,$cm, $H = 2\,$cm; $A = 5\,$km, $H = 10\,$km; or $A = 8\,\mu$m, $H = 16\,\mu$m. These solutions are different, but they have something in common. In each case the hypotenuse is twice the length of the adjacent side.

Knowing that one angle is 60° is not enough to determine the lengths of the triangle's sides, but it *is* enough to tell us the value of $\frac{A}{H}$, in this case $\frac{1}{2}$. Then, if we were told that $A = 4\,$m, we would immediately know that $H = 8\,$m.

If we had started with a 45° angle we would have found that $\frac{A}{H} = \frac{1}{\sqrt{2}}$, around 0.707. The function which takes the angle and tells us the value of $\frac{A}{H}$ is called the *cosine function*.

Sine, cosine and tangent

If x is an angle in a right-angled triangle, we can call the three lengths H (hypotenuse), O (opposite the angle) and A (adjacent to the angle). Although H, A and O can change without altering x (by enlarging or shrinking the triangle), these lengths always remain in the same proportions to each other. So the values of $\frac{O}{H}, \frac{A}{H}$ and $\frac{O}{A}$ are fixed, and totally determined by the angle x. These ratios are given by the *sine*, *cosine* and *tangent* functions respectively (or *sin*, *cos* and *tan* for short): $\sin x = \frac{O}{H}$, $\cos x = \frac{A}{H}$ and $\tan x = \frac{O}{A}$. For example, in a $(3,4,5)$-triangle (see **Pythagorean triples**), if x is opposite the side of length 3, then $\sin x = \frac{3}{5}$, $\cos x = \frac{4}{5}$ and $\tan x = \frac{3}{4}$.

Evaluating these functions for a particular angle, say 34.2°, is decidedly tricky; happily most pocket calculators have buttons dedicated to the task. In the past, people had to wade through trigonometric tables or construct accurate scale drawings of triangles. These important functions have long since outgrown their humble geometric origins. In new guises as **power series**, they play pivotal roles in complex analysis and **Fourier analysis** (the study of waveforms), among other areas.

Cosecant, secant and cotangent

In the definitions of sine, cosine and tangent, you might wonder why $\frac{O}{H}, \frac{A}{H}$ and $\frac{O}{A}$ were chosen instead of $\frac{H}{O}, \frac{H}{A}$ and $\frac{A}{O}$. After all, these are also fixed for any right-angled triangle containing the given angle x. These are the definitions of the three less well-known trigonometric functions: *cosecant*, *secant* and *cotangent*, respectively (or *cosec*, *sec* and *cot* for short).

It immediately follows from the definitions that $\csc x = \frac{1}{\sin x}$, $\sec x = \frac{1}{\cos x}$ and $\cot x = \frac{1}{\tan x}$. So translating information about *cosec*, *sec* and *cot* into terms of *sin*, *cos* and *tan* (or vice versa) is never difficult.

Trigonometric identities

1 The three main trigonometric functions are connected by the formula $\tan x = \frac{\sin x}{\cos x}$. This is immediate from the definitions, since $\frac{O}{A}$ is equal to $\frac{O}{H}$ divided by $\frac{A}{H}$

2 If x is an angle in a right-angled triangle, applying Pythagoras' theorem to the hypotenuse (H), opposite (O) and adjacent (A) sides gives $O^2 + A^2 = H^2$. Dividing this equation by H^2 produces $\frac{O^2}{H^2} + \frac{A^2}{H^2} = 1$, which says that:

$$(\sin x)^2 + (\cos x)^2 = 1$$

Usually rewritten as $\sin^2 x + \cos^2 x = 1$, this formula often makes an appearance when sin and cos are in use.

3 If we know the value of $\sin x$ and $\sin y$, what can we say about $\sin(x + y)$? An elementary mistake is to believe that $\sin(x + y) = \sin x + \sin y$. The situation is a little more complicated, but there are still manageable formulas:

$$\sin(x + y) = \sin x \cos y + \cos x \sin y$$

Similarly:

$$\cos(x + y) = \cos x \cos y - \sin x \sin y$$

4 Applying the formulas above to the case where $x = y$, gives the so-called *double-angle formulas* for $\sin 2x$ and $\cos 2x$

$$\sin 2x = 2 \sin x \cos x$$

and

$$\cos 2x = \cos^2 x - \sin^2 x$$

These can be deduced from **de Moivre's theorem**.

Trigonometric values

1 In most cases, the values of $\sin x$, $\cos x$ and $\tan x$ are best left to an electronic calculator. But some values are suitable for human consumption, notably 0°, 90° and 180°. Translating these into radians:

$$\sin 0 = 0 \qquad \sin \tfrac{\pi}{2} = 1 \qquad \sin \pi = 0$$

Similarly

$$\cos 0 = 1 \qquad \cos \tfrac{\pi}{2} = 0 \qquad \cos \pi = 1$$

and

$$\tan 0 = \tan \pi = 0$$

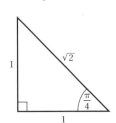

Other important values are for 30°, 60° and 45°. In radians, they are:

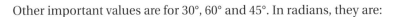

$$\sin \tfrac{\pi}{6} = \tfrac{1}{2} \qquad \cos \tfrac{\pi}{6} = \tfrac{\sqrt{3}}{2} \qquad \tan \tfrac{\pi}{6} = \tfrac{1}{\sqrt{3}}$$

$$\sin \tfrac{\pi}{3} = \tfrac{\sqrt{3}}{2} \qquad \cos \tfrac{\pi}{3} = \tfrac{1}{2} \qquad \tan \tfrac{\pi}{3} = \sqrt{3}$$

$$\sin \tfrac{\pi}{4} = \tfrac{1}{\sqrt{2}} \qquad \cos \tfrac{\pi}{4} = \tfrac{1}{\sqrt{2}} \qquad \tan \tfrac{\pi}{4} = 1$$

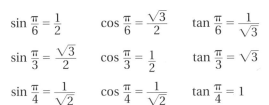

2 In a right-angled triangle, the hypotenuse is always the longest side: so $H \geq O$ and $H \geq A$. Since $\sin x = \tfrac{O}{H}$ and $\cos x = \tfrac{A}{H}$, it must be that $\sin x \leq 1$ and $\cos x \leq 1$. In a slightly extended form, sine and cosine can assume negative values too, but only ever between -1 and 1 (until **complex** values of x are permitted).

3 The tangent function, meanwhile, can give any value as output. But it cannot accept any value for x as input: no triangle can have two 90° angles in it (no Euclidean triangle at any rate, but see **elliptic geometry**). So it is not clear what $\tan \tfrac{\pi}{2}$ might mean. Worse, as x gets closer and closer to 90°, the ratio of the side O to A gets ever larger: $\tan(89°) = 57$, $\tan(89.9°) = 573$, and $\tan(89.99999°)$ is almost 6 million. So there is no sensible value which can be assigned as $\tan(90°)$ or $\tan \tfrac{\pi}{2}$, and the function is left undefined at this point.

The law of sines
The definitions of sin, cos and tan all involve working in a right-angled triangle, as does Pythagoras' theorem. What can be said about non-right-angled triangles? The *sine rule*, or *law of sines*, says that if you take one side of a triangle, and

divide its length by the sine of the opposite angle, then you will always get the same answer (d), irrespective of which side you picked. Suppose a triangle has sides of length a, b and c, with opposite angles A, B and C respectively. The sine rule is the statement:

$$\frac{a}{\sin A} = \frac{b}{\sin B} = \frac{c}{\sin C} = d$$

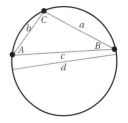

The number d has a nice geometric interpretation: it is the diameter of the triangle's *circumscribing circle* (the unique circle which passes through the three corners of the triangle; see also **centres of triangles**).

The sine rule has its roots in results 1.18 and 1.19 of Euclid's *Elements*, but was first written down explicitly by the 13th-century Persian mathematician and astronomer Nasīr al-Dīn al-Tūsī.

The law of cosines

The *law of cosines* is an extension of Pythagoras' theorem to non-right-angled triangles. Although trigonometric functions such as cosine were not developed until later, a geometric version of this result was proved in 2.12 and 2.13 of Euclid's *Elements*. Also known as the *cosine rule*, it says that if a triangle has sides of length a, b and c, with opposite angles A, B and C respectively, then:

$$a^2 = b^2 + c^2 - 2bc\cos A$$

If the triangle happens to be right-angled, this collapses back down to the ordinary statement of Pythagoras' theorem, since $\cos(90°) = 0$.

Any triangle can be divided into two right-angled-triangles: the cosine rule comes from piecing together the ordinary trigonometry of these two.

Centres of triangles

The *centre* of a circle or rectangle has a clear and unambiguous meaning. But where is the centre of a triangle? According to the *Encyclopedia of Triangle Centers*, maintained by the mathematician Clark Kimberling at the University of Evansville, there are 3587 different answers to this question (at time of writing)! Worse, the potential number of *triangle centre functions* is infinite.

Here are the most common centres:

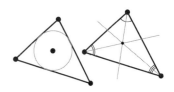

1 Inside any triangle you can draw a unique *inscribed circle*, which has all three sides of the triangle as tangents. The *incentre* is this circle's centre. So this point is the same perpendicular distance from each side. The incentre is also where the bisectors of the three angles meet.

2 If you join each corner of the triangle to the midpoint of the opposite side, these three lines meet at the triangle's *centroid*. If you cut the triangle out of sheet metal, this would be its centre of gravity.

3 Every triangle also has a unique *circumscribing circle*, which passes through all three of the triangle's corners. The *circumcentre* is the centre of this circle. It is also the point where the perpendicular bisectors of the three sides meet. (It's not much of a 'centre' perhaps, since it will lie outside the triangle if one of the angles is obtuse.)

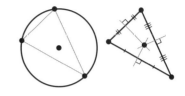

4 If you join each corner of the triangle to its opposite side so that the lines meet at a right-angle, then the three lines (or *altitudes*) meet at the triangle's *orthocentre*. (For obtuse triangles you need to extend the sides, and again this centre can lie outside the triangle.)

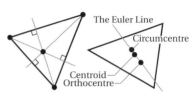

5 The *first Brocard point* is defined as the point P so that the angles PAB, PBC and PCA are all equal. It is not quite a triangle centre. There is a *second Brocard point* Q where QBA, QAC and QCB are equal. The *third Brocard point* (R) is defined as the midpoint of the first and second. It is a triangle centre.

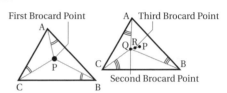

The Euler line

Only for an equilateral triangle do all the different centres of the triangle coincide. In 1765, however, the great Swiss mathematician Leonhard Euler proved that the orthocentre, centroid and circumcentre always lie on a straight line, now known as the *Euler line*. He also demonstrated that the distance between the orthocentre and centroid is twice that between the centroid and circumcentre.

CIRCLES

Circles

Mark a spot on the ground, ask 10 people each to stand 1 metre away from it, and an approximation to a circle should appear. This is, more or less, Euclid's definition of a circle, which has survived till today essentially unchanged. Two pieces of data are required: a point (O) to be its *centre*, and a length (*r*) for its *radius*. Formally then, the circle is the set of all points in the plane *r* units away from O. (The same definition in three dimensions will give a sphere.)

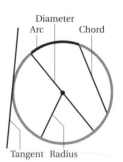

A *disc* is the set of points *at most r* units away from O: a filled-in circle, in other words. In practice, this distinction is often abandoned, as in discussion of the area of a circle (which to an irrepressible pedant is 0, though the area of the corresponding disc is given by πr^2).

Circles come with their own lexicon of terms: the *circumference* is the perimeter of the circle; an *arc* is a portion of the circumference between two points; a *radius* is a line segment from the centre to the circumference; a *chord* is a line segment from one point on the circle to another, dividing it in two; a *diameter* is a chord which passes through the centre (so its length is twice the radius); a *tangent* is a straight line outside the circle which touches it at exactly one point.

π

At approximately 3.14159265358979323846264338832795…, the number π (pi) is defined as the ratio of a circle's circumference to its diameter. Around 1500 BC, the authors of the Egyptian Rhind papyrus posited that a circular field of diameter 9 units has the same area as a square field whose sides are 8 units. If correct, this would give π a value of $\frac{256}{81}$. In fact, π is an **irrational number**, meaning that it can never be written exactly as a fraction or a recurring decimal. So its decimal representation continues for ever, without repeating. More than this, it is a **transcendental number**, even further removed from the familiar territory of the whole numbers. Perhaps this fact, combined with its ancient history and succinct definition, explains the incredible superstardom of this number, well beyond mathematical circles. The eponymous hero of many books, π also stars in several films and songs. March 14th (3.14) is celebrated as international π day.

π has also become a test-bed for various feats of human endeavour. The current record for computing digits of π belongs to a team of computer scientists led by Fabrice Bellard, who established the first 2699999990000 digits as an ordinary desk-top computer, in 2009. The mathematician John Conway tells of romantic walks with his wife, when they would recite the digits of π, taking alternate groups of twenty. Conway has memorized the first thousand decimal places, some way short of the current Guinness world record of 67,890, set by Chao Lu in 2005. However, verification is continuing on Akira Haraguchi's 2006 attempt at the 100,000 digit mark.

Circle formulas

As ancient as the Pyramids, the original definition of the number π is as the ratio of a circle's circumference (*c*) to its diameter (*d*). So the formula for the circumference of a circle is a true mathematical antiquity: $c = \pi d$. Equivalently, $c = 2\pi r$, where *r* is the circle's radius.

The formula for the area (*A*) of a disc was first explicitly derived by Archimedes around 225 BC: $A = \pi r^2$. This states that a square built on the radius of the circle fits into the circle exactly π times. In situations with Cartesian coordinates, the *unit circle* is the gold standard. It has centre at the origin, and radius 1. So it is formed by all the points (*x*, *y*) which are 1 unit away from (0, 0), which (by Pythagoras' theorem) amounts to the equation $x^2 + y^2 = 1$. (In the **complex numbers**, this is neatly expressed as $|z| = 1$.) More generally, a circle with centre at (*a*, *b*) and radius *r* is described by the formula $(x - a)^2 + (y - b)^2 = r^2$.

Radians

As an **angle** is a quantity of turn, the obvious geometric figure to consider is the circle. Starting at a point on the circle, we can measure the amount of turn through the corresponding distance around the circle. Since the total circumference of the unit circle has length 2π, this is the number of *radians* which constitute a complete revolution. So one radian is $\frac{1}{2\pi}$ (around 15.9%) of a complete revolution. Equivalently, it is the angle which gives an arc of length 1.

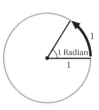

An advantage of this approach is that measuring arc lengths becomes immediate. If an arc takes us an angle θ (measured in radians) around a circle of radius r, then its length is just $r\theta$. This is the first example of radians chiming well with mathematical functions. Trigonometry similarly works far more smoothly with radians than with any other measurements of angle.

The equal tangent theorem

A basic fact about circles, proved by Euclid in Proposition 3.18 of his *Elements*, is that the tangent at a point on the circle is at right angles to the radius at that point. A consequence is the *equal tangent theorem*: take any point X outside a circle, and there are exactly two tangents to the circle you can draw, which pass through X. The theorem says that the distance from X to the circle is the same along both of these lines. To see this, call the points where the tangents touch the circle A and B, and the centre of the circle O. Then OA and OB are both radii, and so the same length. Also, OA and AX are perpendicular, as are OB and BX. So we have two right-angled triangles: OAX and OBX. OX is the hypotenuse of both, and OA and OB have the same length, so Pythagoras' theorem gives the lengths for AX and BX as equal.

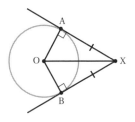

The inscribed angle theorem

An angle at the centre of a circle is twice that on the circumference.

Proposition 3.20 of Euclid's *Elements*, this fundamental theorem about angles in circles incorporates several other useful results. More formally, the theorem says that if you take two points (A and B) on a circle's circumference, and join each to the circle's centre (O), the angle is double that which you find from joining A and B to a third point (C) on the circumference. (A little care is needed, because there are of course two angles at O, one large and one smaller. The theorem always concerns the one in the corresponding position to that at C, not the one opposite it.)

This result has several important consequences, including the theorem of Thales, the theorem on angles in the same segment, and the characterization of cyclic quadrilaterals.

The theorem of Thales

Angles in a semicircle are right angles.

To unpack this result a little, if you draw a diameter across a circle, and connect its two endpoints A and B to any point C on the circumference, then

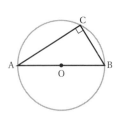

the resulting angle ACB is always a right angle. This follows from the inscribed angle theorem: since A and B form an angle of 180° at the centre, the angle on the circumference must be half that, 90°. Euclid included this fact as Proposition 3.31 of his *Elements*, but it was first proved by the 'Father of Geometry', the Egyptian-influenced philosopher Thales of Miletus, around 600 BC.

Angles in the same segment

If you take two points on the circumference of a circle, and join them with a chord, you have divided the circle into two *segments*.
An *angle in a segment* is the angle created when the two endpoints (A and B) of the chord are connected to a third point (C) on the circle's circumference. In Proposition 3.21 of *Elements*, Euclid proves that any two angles in the same segment are equal. So, in the illustration, the angles ACB and ADB are equal. In fact, this is a consequence of the inscribed angle theorem, since each of the angles ACB and ADB must be half that at the centre (AOB), and so must be equal.

Cyclic quadrilaterals

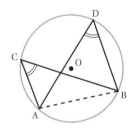

Any triangle may be *inscribed* in a circle. That is, you can draw a circle so that all three of the triangle's corners sit on the circumference. However, the same thing is not true for *quadrilaterals* (four-sided shapes). A rhombus, for example, cannot be inscribed in a circle (except when the rhombus is actually a square). *Cyclic quadrilaterals* are those which can be inscribed in a circle, and in Proposition 3.22 of *Elements*, Euclid proved their defining characteristic: pairs of opposite angles add up to 180°.

Really this is two results in one. The first says that cyclic quadrilaterals have this property. This is a consequence of the inscribed angle theorem: join two opposite corners of the quadrilateral to the centre (O). Then the two angles at the centre obviously add up to 360° ($2x + 2z = 360°$), and so the angles at the two remaining corners must add up to half this: $x + z = 180°$. The second half of the theorem says that every quadrilateral with this property has an inscribing circle.

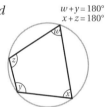

The alternate segment theorem

Suppose we have a circle with three points, A, B and C, on its circumference, and a tangent to the circle at A. Take D to be any point on the tangent line which is on the opposite side of the line AB from C. In Proposition 3.32 of *Elements*, Euclid shows that the angles ACB and BAD are equal. (Angle ACB is the angle produced at C by the lines from A and B.) Also known as the *Tangent–Chord theorem*, psychologically, this theorem says that the angle ACB is equal to the angle between the chord AB and the arc from A to B. But curved lines and angles don't necessarily mix well, so the straight tangent is used for precision.

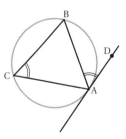

95

Intersecting chords theorem

If A, B, C and D are any four points on a circle, and X is the point where the lines AC and BD meet, then the triangles ABX and DCX are similar. This theorem still applies even if X is outside the triangle.

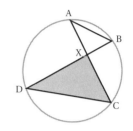

Eyeball theorem

The analysis of circles did not stop with Euclid. Since he laid the foundations, a multitude of exotic and beautiful facts have been discovered. One such is the *eyeball theorem*: Take two circles C_1 and C_2 (not necessarily the same size), and draw two tangents from C_1 which meet at the centre of C_2. Say these cross C_2 at A and B. Similarly draw two tangents from C_2 meeting at the centre of C_1, crossing C_1 at points D and E. Then the distance from A to B (along a straight line) is the same as that from D to E.

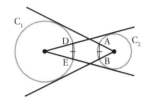

POLYGONS AND POLYHEDRA

Polygons

Triangles, quadrilaterals and pentagons all fall under the broader category of *polygons*: 2-dimensional shapes bounded by straight lines, meeting at vertices (corners).

Polygons have been studied since antiquity. The easiest to analyse are convex polygons. (Stars are examples of non-convex polygons.)

A *regular* polygon is one where the edges all have the same length, and all the angles are equal. Equilateral triangles and squares are the first examples of regular polygons, followed by regular pentagons, and so on.

Starting at three, a regular convex polygon can exist with any number of sides. As the number of sides increases, the polygon gets ever closer to a circle. In 1796, at the age of 19, Carl Friedrich Gauss proved that not all regular polygons can be drawn using elementary **ruler and compass constructions**.

Convex

Non-convex

Convexity

An object *X* is *convex* if, whenever you mark two points inside it, and join them together with a straight line, the whole line segment between them lies entirely inside the shape.

A regular pentagon is convex, but a star shape is not, as you can find two points where the line crosses outside the shape. The same definition holds in higher dimensions. So a spherical ball is convex, but a curvy banana is not.

Convexity is a fundamental criterion for classifying geometrical objects, in different dimensions. Generally convex shapes are well-behaved and classifiable, while non-convex ones can be more awkward. Regular **polygons** are usually assumed convex, but there are non-convex possibilities too, in the form of star-polygons, such as the pentagram.

The **Platonic solids** are convex, but they have non-convex, self-intersecting analogues in the form of the four regular **Kepler–Poinsot polyhedra**.

Convex quadrilaterals

Constructed from four sides of equal lengths, and four right angles, a *square* is the simplest *quadrilateral*: 2-dimensional shapes built from four straight edges. Squares are the only quadrilaterals classed as regular polygons. Relaxing these conditions, other types of quadrilaterals emerge:

- A *rectangle* contains four right angles (consequently its sides come as two pairs of equal length).
- A *rhombus* has four equal sides, which come in two parallel pairs. The angles will not be right angles (except when it happens also to be a square).
- A *parallelogram* comprises two pairs of parallel sides (each pair is necessarily of equal length).
- A *trapezium* (or *trapezoid*) has one pair of parallel sides. Subspecies include *isosceles trapezia* (where the remaining pair of sides are of equal length) and *right trapezia* (which also contain two right angles).
- A *kite* (or *deltoid*) has two pairs of sides of equal length, like a parallelogram. But in this case the equal sides meet, rather than being opposite each other.

If a quadrilateral has no equal lengths, no parallel sides, and no right-angles, it is deemed *irregular:* most **cyclic quadrilaterals** are irregular, for example.

Non-convex quadrilaterals

The definition of a *kite* can be satisfied by both convex and non-convex shapes. A *chevron* is the name usually given to a non-convex kite, the most symmetric of the *reflex quadrilaterals* (which contain one angle of more than 90°).

On the fringes of acceptability are formations of four straight lines where two crash through each other: the *self-intersecting* quadrilaterals, of which the most symmetric are the *bow-ties*.

Rectangle

Rhombus

Parallelogram

Trapezium

Kite

Chevron

Bow-tie

Polyhedra

The definition of a polyhedron has changed over time. But essentially it is a surface which is built from flat 2-dimensional *faces* meeting at straight *edges* and *vertices* (corners). Polyhedra are the 3-dimensional analogues of polygons, and are divided into convex *solids* and the more intricate non-convex polyhedra.

The story of polyhedra is a succession of mathematical classifications based on ever more encompassing notions of symmetry. It starts with the most regular of all: the Platonic solids.

Next are the Archimedean solids, prisms and antiprisms, which are also highly symmetric in that their vertices are all identical, and faces are all regular convex polygons. Unlike the Platonic solids, however, their faces may be different shapes.

Other attractive polyhedra are the *Catalan solids* which make fair dice (since their faces are all identical, though not regular polygons), and the **Johnson solids**: all the convex polyhedra whose faces are regular polygons.

Non-convex polyhedra can be classified too: the most symmetrical of these are the regular **Kepler–Poinsot polyhedra**. Polyhedra whose vertices are all identical are known as *isogonal*.

The Platonic solids

The *Platonic solids* comprise five beautiful and important polyhedra:

- the *tetrahedron*, built from four equilateral triangles
- the *cube* with its six square faces meeting at right angles
- the *octahedron* constructed from eight equilateral triangles
- the *dodecahedron* with its twelve regular pentagonal faces
- the *icosahedron* where twenty equilateral triangular faces meet at each corner in fives.

The philosopher Plato held these five highly symmetrical shapes in the highest regard. Around 350 BC, he wrote that the tetrahedron, cube, octahedron and icosahedron correspond to the four elements: fire, earth, air and water, respectively. The dodecahedron he considered 'God used for arranging the constellation of the whole universe'.

Mathematically, these five are convex and *regular*: each face is a regular polygon, identical to every other, and similarly the edges are all indistinguishable, as are the vertices (corners). In one of the world's first classification theorems, Plato presented a proof that these are the *only* convex regular polyhedra; no sixth would ever be found. The final book of Euclid's *Elements* is also devoted to these shapes.

Irregular polyhedra

The world is full of polyhedra and solids, and most do not have the high levels of symmetry that mathematicians enjoy. A brick, for example, is not a Platonic cube, but

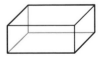

a *cuboid*, with sides of three different rectangular shapes. A cuboid is a special case of a *parallelpiped*, built from three pairs of parallel parallelograms. Pyramids are another important family of irregular polyhedra: the square- and pentagon-based pyramids can have sides with equilateral triangles (as can the triangle-based pyramid, or tetrahedron). Beyond this, all pyramids must have irregular triangular sides.

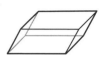

Once we allow polyhedra with irregular polygons, the most symmetrical are those which make fair dice, where every face is identical. Beyond this, there is no limit to the list of possible irregular polyhedra. The **Johnson solids** at least provide a complete lexicon of those convex polyhedra which can be constructed from regular polygons. The **Stewart toroids** extend this list to non-convex shapes.

Nets

The German artist Albrecht Dürer was also a mathematician, with a particular interest in polyhedra. In his 1538 work *Instruction on Measurement*, Dürer introduced an invaluable tool for understanding polyhedra. A *net* is a flat arrangement of polygons, some joined along their edges. By folding and gluing this pattern, you can create a model of the polyhedron.

Every polyhedron can be described by a net. Indeed, the cube has 11 different nets.

Dürer's fascination with polyhedra led him to rediscover two of the **Archimedean solids** (the *truncated cuboctahedron* and the *snub cube*), and to design a shape of his own: the melancholy octahedron.

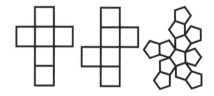

Polyhedral duality

A cube has six faces and eight vertices, while an octahedron has eight faces and six vertices. Both have twelve edges. This symmetry comes from the fact that the cube and octahedron are *dual polyhedra*.

To obtain the dual of a polyhedron, mark a spot in the middle of each face, and join two spots with a line if the two corresponding faces meet. The resulting framework of spots and lines describe a new polyhedron: the dual of the original. Repeating the process, taking the dual of the dual, gives back the original shape. Among the Platonic solids, the tetrahedron, with its four faces and four vertices, is self-dual. The dodecahedron and icosahedron form a dual pair.

The Archimedean solids are dual to the Catalan solids.

The Archimedean solids

No other polyhedra can possess the perfect symmetry of the *Platonic solids*, but slightly loosening the requirements opens up an exciting range of new shapes. The fourth-century mathematician Pappus credits Archimedes with the discovery of 13 convex polyhedra with faces which are regular polygons (though not all the same), and which are symmetrical in their vertices: that is, the arrangement of faces and edges at every vertex is identical to every other, so moving any vertex to any other is a **symmetry** of the shape.

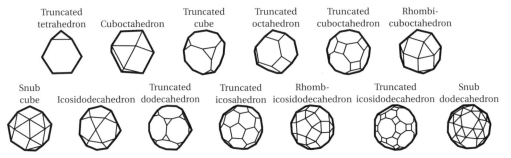

The *truncated icosahedron* is famous as a soccer ball, and as the structure of a Buckminsterfullerene (Carbon-60) molecule.

Prisms and antiprisms

The Archimedean solids are defined as convex polyhedra whose faces are regular and whose vertices are all indistinguishable. But these 13 shapes do not list every possibility. There are also two infinite families of polyhedra which satisfy these criteria.

The *prisms* are formed by taking two identical regular polygons (such as two hexagons of the same size), and joining their edges with squares. In general a prism is a solid formed by two regular *n*-gons joined by a ring of *n* squares. (Polyhedra with rectangles in place of squares are also sometimes known as prisms, but these do not have all regular polygonal faces.)

Hexagonal prism

In a hexagonal *antiprism*, the two hexagons are twisted out of sync, and then joined by equilateral triangles. In general an antiprism comprises two regular *n*-gons joined by a ring of 2*n* alternating equilateral triangles.

Hexagonal antiprism

Fair dice

Which shapes make fair dice? To make a fair die, a polyhedron should be convex, with all its faces identical. Certainly the Platonic solids satisfy these requirements, but they are not alone. In 1865, Eugène Catalan published a list of 13 beautiful new solids with this property. He did not discredit Plato's classification, since the faces are not regular polygons: they are rhombuses, non-equilateral triangles, kites or irregular pentagons.

Trapezohedron

Strombic hexecontahedron

The *Catalan solids* are obtained as the duals of the Archimedean solids. These elegant shapes go by clumsy names, such as the *rhombic dodecahedron* with 12 rhombus faces and the *strombic hexecontahedron* with its 60 kite-shaped faces.

Disphenoid

Bipyramid

There are also three infinite families satisfying the criteria:

1 *bipyramids*, where two pyramids with *n*-agonal bases are glued at their bases (obtainable as the duals of the prisms)
2 *trapezohedra* where two 'cones' built from kites are glued together (these are the duals of the antiprisms)
3 *disphenoids* which are tetrahedra, but whose four faces are identical non-equilateral acute triangles.

Kepler–Poinsot polyhedra

Small stellated dodecahedron

Great dodecahedron

Are the Platonic solids the only solids with identical faces of regular polygons? As so often in mathematics, the answer closely depends on definitions.

In 1619 Johannes Kepler noticed two non-convex polyhedra which also fit the bill. Artists such as Paolo Uccello had already exploited their beauty, but previous mathematicians had overlooked them, perhaps because their faces do not only meet along edges, they also pass through each other at *false edges*.

Great stellated dodecahedron

Great icosahedron

If we are really interested in 3-dimensional *solids*, these should probably be ruled out. But they are usually classed as legitimate *polyhedra*. Louis Poinsot identified two more in 1809, and the resulting four *Kepler–Poinsot polyhedra* arguably complete the list of regular polyhedra begun by Plato. They are the small stellated dodecahedron, great dodecahedron, great stellated dodecahedron, and the great icosahedron.

Star polygons and star polyhedra

Two of the Kepler-Poinsot polyhedra are *star polyhedra*: the great and small stellated dodecahedra. Both are obtained from the dodecahedron by a process of *stellation*: extending edges and faces until they meet. The basic idea can be seen with polygons: the pentagram is a stellation of the pentagon, for instance. Usually there are choices about which edges to connect: a heptagon can be stellated in two ways, to give two different heptagrams.

The possibilities for stellating an icosahedron were documented in the book *The Fifty-Nine Icosahedra*, by Coxeter, Du Val, Flather and Petrie, in 1938. Some of the Archimedean solids have hundreds of millions of distinct stellations.

Seventeen of the **uniform polyhedra** are star polyhedra deriving from Archimedean solids.

Because of the way their faces pass through each other, star polyhedra are typically not topologically spherical, and so will not satisfy **Euler's polyhedral formula.**

Compound polyhedra

As well as recognizing the first two of the Kepler–Poinsot polyhedra, Johannes Kepler also discovered the first *compound polyhedron*: the *stella octangula*, obtained by pushing two tetrahedra into each other, so that they have a common centre. (It can also be obtained as a stellation of an octahedron.) As with other star polyhedra, this shape is non-convex and self-intersecting, resulting in false edges and vertices. Shapes such as this, which can be pulled apart into separate polyhedra, are not usually classed as polyhedra themselves. Nevertheless, all its faces are identical, as are its edges and vertices, making it a regular compound polyhedron. There are four others, built similarly from five tetrahedra, ten tetrahedra, five cubes and five octahedra.

Uniform polyhedra

The discovery of star polyhedra opened the door for a new classification of *uniform polyhedra*: polyhedra with regular polygonal faces (including star polygons), and whose vertices are all identical. The convex examples were long known: the Platonic solids, the Archimean solids, prisms and antiprisms. The Kepler–Poinsot polyhedra are the first non-convex, self-intersecting examples, and in 1954 Coxeter, Longuet-Higgins and Miller produced a list of 53 more, starting with the tetrahemihexahedron, created from three intersecting squares and four equilateral triangles. The last, the *great dirhombicosidodecahedron* or 'Miller's monster' has 60 vertices, at each of which four squares, two triangles, and two pentagrams meet, making 124 faces in total.

Star prism

Star antiprism

The uniform polyhedra are completed by two more infinite families: *star prisms* (where two regular *n*-grams are joined by a ring of intersecting squares) and *star antiprisms* (where two regular *n*-grams are joined by a ring of intersecting equilateral triangles). See illustrations.

In 1970 S.P. Sopov proved that this list was complete, although in 1975 John Skilling made a curious discovery. If edges were allowed to coincide (that is, one edge could be shared between four faces), he found one further possibility: *Skilling's figure* (the *great disnub dirhombidodecahedron*, pictured) with 204 faces meeting at 60 vertices.

Skilling's figure

Isogonal polyhedra

The definition of a uniform polyhedron has two parts: the vertices are all the same (the arrangement of faces and edges at every vertex is identical to every other), and every face is a regular (possibly non-convex) polygon. If this second requirement is dropped, the number of qualifying shapes becomes infinite: these are the *isogonal* polyhedra.

This can be seen with the very first examples: the *disphenoid tetrahedra* (see **fair dice**). Take any cuboid, and pick four corners which do not share an edge. Joining these four creates a disphenoid. No complete classification of isogonal polyhedra is yet known.

Isogonal polyhedron

The Johnson solids

In 1966, Norman Johnson ignored all questions of symmetry and asked simply: what convex polyhedra can be built from regular polygons (not necessarily all the same)? He produced a catalogue of 92 convex polyhedra. In 1969 Victor Zalgaller proved that Johnson's list, along with the Platonic and Archimedean solids, and the prisms and antiprisms, are indeed all there are.

The first Johnson solid (J_1) is a pyramid with square base and equilateral triangular sides. Another is the gyrobifastigium (J_{26}), built from four squares and four equilateral triangles.

Gyrobifastigium

The Stewart toroids

Adventures among the *Toroids* was a little-known book by Bonnie Stewart. Hand-written, and published privately in 1970, it grew into a cult classic among people interested in polyhedra.

In it, Stewart considered polyhedra that can be built from regular polygons. But, departing from the Johnson solids, he did not confine himself to convex shapes. Others may have given this up as a lost cause, since there is no limit to the number of Johnson solids which can be glued together.

However, some of these shapes do have a high-level of symmetry. Eight octahedra, for example, can be glued at their faces to form a ring. Some of the most breathtaking of Stewart's shapes come not from gluing solids together, but from the opposite philosophy: he took large models of several Johnson and Archimedean solids, and analysed the possibilities for drilling through them, lining the tunnels with regular polygons.

Topologically these shapes are not spheres but *n*-tori (see **orientable surfaces**), with the number of holes *n* given by their *genus*. The arrangements must therefore satisfy the appropriate **polyhedral formula**.

Eight octagon ring

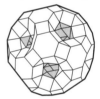

Cupola-drilled
truncated
icosidodecahedron

Polychora

With the study of polyhedra having yielded such spectacular fruits, a mathematician's immediate reaction is to seek to generalize it. Just as polyhedra are the 3-dimensional analogues of polygons, so *polychora* are the equivalent objects in four dimensions. These are built from 3-dimensional polyhedral *cells*, meeting at 2-dimensional polygonal *faces*, 1-dimensional straight *edges*, and zero-dimensional *vertices*.

Just as Plato had classified the regular convex polyhedra, so in 1852 the Swiss geometer Ludwig Schläfli classified the regular convex polychora. He found:

- the pentachoron (or 4-*simplex*), built from five tetrahedra, and analogous to the tetrahedron
- the tesseract (or 4-*hypercube*), built from eight cubes
- the hexadecachoron (4-*orthoplex* or 4-*cross-polytope*), built from 16 tetrahedra: the analogue of the octahedron

- the icositetrachoron (*octaplex*), built from 24 octahedra, a new shape, with no 3-dimensional analogue
- the hecatonicosachoron, built from 120 dodecahedra, analogous to the dodecahedron
- the hexacosichoron, built from 600 tetrahedra, analogous to the icosahedron.

Schläfli and Edmund Hess also listed the 10 non-convex regular polychora: these Schläfli–Hess polychora are the equivalents of the Kepler–Poinsot polyhedra.

Hypercube

The four vertices of a square can be described neatly by **Cartesian coordinates**: $(0,0)$, $(0,1)$, $(1,0)$, and $(1,1)$. Similarly, the eight vertices of a cube are given by: $(0,0,0)$, $(0,0,1)$, $(0,1,0)$, $(1,0,0)$, $(0,1,1)$, $(1,0,1)$, $(1,1,0)$, and $(1,1,1)$. It does not need much insight to see where the sixteen vertices of the 4-dimensional hypercube should be: $(0,0,0,0)$, $(0,0,0,1)$, $(0,0,1,0)$, $(0,1,0,0)$, $(1,0,0,0)$, $(0,0,1,1)$, $(0,1,0,1)$, $(0,1,1,0)$, $(1,0,0,1)$, $(1,0,1,0)$, $(1,1,0,0)$, $(0,1,1,1)$, $(1,0,1,1)$, $(1,1,0,1)$, $(1,1,1,0)$, and $(1,1,1,1)$. This is a good start for mathematical analysis, but can this shape actually be visualized? One tactic is to use a net. Just as a cube is built from six square faces folded together, so a 4-hypercube is built from eight cubic cells 'folded together'. This net was depicted by Salvador Dali in his painting *Corpus Hypercubus*, and was contemplated in Robert Heinlein's 1941 short story '–And He Built A Crooked House–', in which an architect builds a house in the shape of this net, which an earthquake then 'folds up' to form a hypercubic home.

An alternative is to give a *projection:* a 3-dimensional cube can be drawn on a 2-dimensional page as one square within another, with their corners joined by edges. Similarly one possible 3-dimensional projection of the 4-hypercube consists of one cube inside another, with their corners joined by edges. The eight cubic cells are then the outer cube, the inner cube, and the remaining six (distorted by perspective) joining the faces of the first two.

A third possibility is the *Petrie projection* of the hypercube: analogous to drawing a cube by joining the corners of one square to those of another of the same size, diagonally displaced from it.

Uniform polychora

Many questions which led to interesting insights about polyhedra can be asked again in the context of 4-dimensional polychora. In 1965 John Horton Conway and Michael Guy completed by computer the classification of the 4-dimensional equivalents of the Archimedean solids, begun in 1910 by the self-taught prodigy Alicia Boole Stott. These *Archimedean polychora* are convex, with all vertices identical, and all faces regular polygons. Consequently, their cells must be traditional Platonic or Archimedean solids, prisms or antiprisms. In total, there are 64 individual shapes listed, plus two infinite prismatic families.

Research is ongoing to extend this work to incorporate their non-convex equivalents, in a complete classification of uniform polychora.

Regular polytopes

A *polytope* is the general word for a polygon, polyhedron, polychoron, or the equivalent in some higher dimension. Of particular interest are the most symmetric of these objects: the *regular* polytopes. In two dimensions, there are infinitely many regular polygons: the equilateral triangle, square, regular pentagon, and so on. In three dimensions, there are five Platonic solids. In four dimensions, there are six Platonic polychora.

Ludwig Schläfli proved that a remarkable thing occurs when we look in higher dimensions than four. There are only ever three regular polytopes: the *simplex*, *hypercube* and *orthoplex* (the equivalents of the tetrahedron, cube and octahedron respectively).

There are also the troublesome self-intersecting, non-convex polytopes to account for. In two dimensions these are the star polygons, beginning with the pentagram. In three dimensions, we find the four Kepler–Poinsot polyhedra, and in four dimensions, ten *Schläfli–Hess polychora*. Again higher dimensions turn out to be far simpler and, from five-dimensions onwards, there are none at all. This is an example of a phenomenon well-known to geometers: that life in three and four dimensions is in many ways more complicated than in higher-dimensional spaces.

TRANSFORMATIONS

Isometries of the plane

Having drawn a picture on the plane, there are various ways to move it into a new position, without it becoming twisted or distorted. These are called *isometries* of the plane; technically defined this means that a line will have the same length before and after the move.

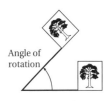

Angle of rotation

- A *rotation* is given by two pieces of information: a point (the centre of rotation), and an angle describing the amount of rotation. (As always in mathematics, a positive angle corresponds to an anticlockwise rotation, and a negative one to a clockwise rotation.)

- A *translation* slides a figure around and is expressed by a **vector**, with the top row corresponding to movement right (or left, if negative), and the bottom row corresponding to shifting the figure up (or down). So $\begin{pmatrix} 4 \\ -3 \end{pmatrix}$ corresponds to moving 4 units right, and 3 down, for example.

- A *reflection* is defined by a straight line which acts as a mirror, moving a point to a position on the opposite side, the same distance away.

Line of reflection

- A *glide* is a reflection followed by a translation along the same line. (See **glide symmetry**.)

Rotations and reflections can also be captured as **transformation matrices**.

Symmetry

For an object drawn on the plane, a *symmetry* is an isometry which leaves the shape looking the same. A square, for example, has both rotational and reflectional symmetry. With the centre of the square as the centre of rotation, rotating by 90° leaves it looking the same. Repeating this manoeuvre produces two other symmetries, rotations by 180° and 270°, before bringing the square back to its starting position. So we say that a square has rotational symmetry 'of order 4'.

A square also has four different lines of reflectional symmetry: the two diagonals, the horizontal and the vertical. Altogether this produces eight symmetries (including the trivial one: just leaving the square as it is). This information is encapsulated in the **symmetry group** of the square.

Shapes can have just rotational symmetry, just reflectional symmetry, or both. Infinite patterns, such as **tessellations**, may also have **translational symmetry** and **glide symmetry**.

Symmetry groups

For mathematicians, a *symmetry* is an action which results in an object looking the same as it did before. This active approach has a nice consequence. If A and B are symmetries they can be combined to produce a third: $A \circ B$. In the case of a square, if A is 'rotate 90° anticlockwise about the centre', and B is 'reflect in the horizontal line', then $A \circ B$ is the result of performing B first, and then A. This turns out to be 'reflect in the diagonal line $y = x$'.

Of course there is one symmetry which has no effect when combined with any other: the trivial symmetry ('1') which leaves the square as it is.

Every symmetry has an inverse: if A is 'rotate 90° anticlockwise', then its inverse (denoted A^{-1}) is 'rotate 90° clockwise'. These facts suggest that the collection of symmetries of the square forms a **group**. This turns out to be true for the symmetries of any object.

For 2-dimensional shapes, there are two families of groups, depending on whether the object has any reflectional symmetry. If it does not, as is the case with a swastika, the group is *cyclic*, meaning that it resembles addition modulo some number (in this case 4). If there is also reflectional symmetry, as in the case of the square, the group is *dihedral*.

In three dimensions, a cube has a symmetry group of size 48. Other polyhedra and higher dimensional polytopes come with more complicated groups.

In the case of the most symmetrical of all shapes, circles and spheres, these groups are infinite **Lie groups**.

Tessellations also have infinite symmetry groups, namely **frieze groups** or **wallpaper groups**.

Similarity

Two triangles are *similar* if their angles match. So, triangles *A*, *B* and *C* both have angles of 30°, 60° and 90° for example, then they are similar. (It is not required that the lengths of their sides should be equal.) Similarity also allows the triangle to be reflected.

This is a quick way to capture a broader phenomenon. In general, two shapes are *similar* if they are the same shape but not necessarily the same size or in the same position. (If they are the same shape and size, then they are *congruent*, and there will be an isometry taking one to the other.) As it stands, this definition is unsatisfactorily imprecise, but can be made exact through the idea of an **enlargement**.

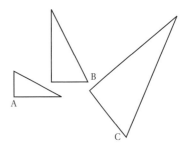

Enlargement

An *enlargement* is specified by two pieces of information: a point (the centre of the enlargement), and a number (the **scale factor**). Any shape in the plane can then be transformed as follows: pick a point on it, and draw a line through that and the centre of the enlargement. If the scale factor is 2, the corresponding point on the enlarged shape will lie on this line, twice as far from the centre as the original point. If the scale factor is 3, it will be three times as far, and so on. Repeating this process for several points will reveal the position of the enlarged shape.

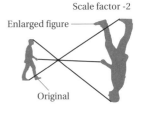

It is an unfortunate consequence of terminology that an 'enlargement' can actually make things smaller. If the scale factor is between 0 and 1, the 'enlarged' shape will be a shrunken version of the original. If the scale factor is negative, then the constructing lines take the shape out of the other side of the centre, flipping it round in the process. (This is sometimes called 'reflection through a point'.)

The same procedure works for objects in higher dimensions. Objects which are enlargements of each other are said to be *similar*.

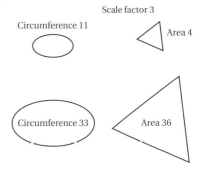

Scale factors

Two objects which are *similar* look the same, have the same proportions, but are different sizes. This difference in size is measured by the scale factor. If a shape is 3 units long, and the scale factor is 2, then the length of the enlarged shape is 3 × 2 = 6 units. This does not only work for straight lines. If an ellipse with a circumference of 11 units is enlarged by a scale factor of 3, then the new ellipse has a circumference of 33 units.

However, the same process does not apply to area. If a triangle of area 4 is enlarged by scale factor 3, the area of the new triangle is not 4×3, but is calculated by multiplying the original area by the *square* of the scale factor: $4 \times 3^2 = 36$. Again this procedure works for any shape, and is useful when direct methods for calculating area are not available.

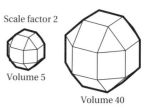

Scale factor 2

Volume 5

Volume 40

In three dimensions, if a solid has a volume of 5, and is enlarged by a scale factor of 2, then to find the volume of the new shape, we multiply by the scale factor cubed: $5 \times 2^3 = 40$.

TESSELLATIONS

Tessellations

From the tombs of the Pharaohs to the etchings of M.C. Escher, humans have always been fascinated by patterns built from the repetition of simple shapes. The art of tiling was widespread in the Islamic world, where a religious injunction against figurative art spurred artists to explore the aesthetic possibilities of abstract design, such as those within the Alhambra Palace in Spain.

It took longer for the mathematics behind these patterns to be revealed. The first idea is that a 2-dimensional figure *tessellates* if it can function as a tile: copies of it can be placed side by side, so that as large an area as you like can be covered, with no overlaps or gaps.

The simplest tilings are the **regular tessellations**, though **irregular** and **semiregular** ones are equally common in design. Once a larger variety of tiles become involved, a more careful account of the possible symmetries is provided by the **wallpaper** and **frieze groups**. Of course, mathematicians also explore the same phenomenon in higher dimensions.

Regular tessellations

The most basic tilings are those involving just one regular polygon: these *regular tessellations* are the equivalents of the Platonic solids. A grid is the commonest example: tessellating squares with four meeting at every vertex. Equilateral triangles also tessellate when arranged so that six meet at every vertex.

Which other regular polygons tessellate? Pentagons do not: if you try it, you will end up with **pentaflakes**. (The interior angle of a pentagon is 108°, which does not divide 360°.) The regular hexagon is the only other regular polygon to tessellate, a fact exploited by bees. Heptagons (and *n*-gons for larger *n*) cannot. If you place two side by side, the remaining angle is too small to fit a third.

Heptagons

Not enough room

Irregular tessellations

It is not only regular polygons that can tessellate, every triangle does. To see this, draw and cut out any triangle. Using this as a template, draw around it once. Then move the template so that one side matches up with the corresponding side of the drawn triangle (be careful only to rotate it in the plane, do not flip it over) and draw around it again. Repeating this will create a pattern that tiles the plane.

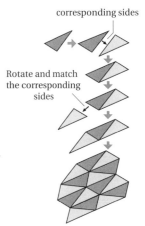
corresponding sides

Rotate and match the corresponding sides

The same procedure works for quadrilaterals: every figure with four straight sides tessellates. Pentagonal tilings are more complicated. Some pentagons do tessellate, though regular ones do not. Regular hexagons tessellate, and in 1918 Karl Reinhardt showed there are exactly three classes of irregular convex hexagon which do too. For $n \geq 7$, there are no convex n-gons which tessellate.

Pentagonal tilings

Regular pentagons do not tessellate, but there are some irregular convex pentagons which do. There are 14 essentially different known ways for this to happen. One of these is the *Cairo tessellation* (illustrated) which adorns the pavements of that city. Others include the four discovered in 1977 by the amateur mathematician Marjorie Rice, and the most recent found in 1985 by Rolf Stein.

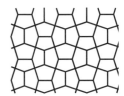

It is not yet established that these 14 list every possible tiling by convex pentagons.

Semiregular tessellations

Like regular tessellations, the *semiregular* ones use only regular polygons, but this time more than one type of tile is allowed, and it is required that every vertex is identical to every other. There are eight such tilings; each involves two or three shapes out of equilateral triangles, squares, regular hexagons, octagons and dodecagons. One of these tilings comes in a left-handed and a right-handed version.

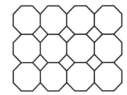

Sometimes called the *Archimedean tessellations* as they correspond to Archimedean solids, these eight tilings have formed the basis of decorative patterns in palaces and temples for thousands of years.

Translational symmetry

Finite objects such as polygons or polyhedra can have at most two types of symmetry: reflectional, and rotational. Infinite tilings admit a third possibility: *translational symmetry*. Shifting the rhombus tiling overleaf right by 1 unit leaves it looking the same, and so is a symmetry. Of course, moving it right by 2 units, or 3, 4, 5, … are also symmetries, meaning that a transitionally symmetrical pattern automatically has infinitely many symmetries.

The translational symmetry of a pattern is the first step towards classification. Many common patterns have two distinct types of translational symmetry: left–right and up–down. This allows 17 essentially different patterns, classified by the 17 **wallpaper groups**. Other tilings have only one type of translational symmetry: left–right (or up–down, but not both). These are classified by the seven **frieze groups**.

The **aperiodic tilings**, including the Penrose and Ammann tilings, are remarkable for having rotational and reflectional symmetry, but no translational symmetry at all.

Glide symmetry

As well as rotational, reflectional and translational symmetry, the final possible type of symmetry of 2-dimensional pattern is the *glide*: a combination of translation and a reflection. In the illustration, the picture can be reflected in the horizontal line, and then translated along that same line. The resulting symmetry is neither a reflection nor a translation, but a *glide* (or *glide-reflection*). In contrast, the combination of a rotation with a translation always gives another rotation (although locating its centre is not always straightforward). Many tilings and patterns have glide symmetries, and so they feature in the frieze and wallpaper groups.

Frieze groups

In architecture, a *frieze* is a narrow band along the top of a wall. Since classical times, friezes have often been decorated with repeating geometrical patterns. In this context there is translational symmetry, but only of the left–right variety. Patterns like this, with just one form of translational symmetry, come in seven types. Their names, given to them by John Horton Conway, describe trails of footsteps with the correct symmetries:

- The *hop* is the simplest of the groups, consisting only of translations.
- The *sidle* has translations together with reflections in vertical lines.
- The *jump* has translations and one horizontal reflection.
- The *step* has translations and glides.
- The *spinning hop* has translations with rotational symmetries of 180°.
- The *spinning sidle* has translations, glides, vertical reflections and rotational symmetries of 180°.
- The *spinning jump* is the largest group, with translations, vertical reflections, one horizontal reflection and rotational symmetries of 180°.

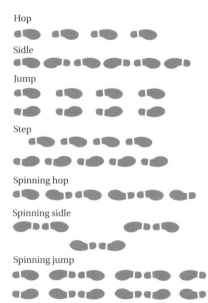

Hop

Sidle

Jump

Step

Spinning hop

Spinning sidle

Spinning jump

Conway's orbifolds

The frieze groups classify patterns with left–right or up–down translational symmetry. What of patterns with both? The simplest such pattern has only translational symmetry, without any reflectional, rotational or glide symmetry. In the *orbifold notation* devised by John Conway, this is denoted *o*.

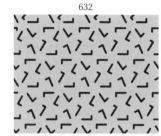

632

Unlike polygons, tilings can have more than one centre of rotation. One possibility has centres of order 6, other centres of order 3 and a final set of order 2. The illustrated example has these centres of rotation and no reflectional symmetries. This group is denoted 632.

Patterns which additionally have reflective symmetry are identified by *. Rotations come in two types: *kaleidoscopes* whose centres lie on a line of reflection and *gyrations* which do not. Gyrations are placed before the asterisk, and kaleidoscopes after it. A chessboard pattern has group *442, with three kaleidoscopes of orders 4, 4 and 2, and no gyrations. A pattern denoted 4*2 has a gyroscope of order 4 and a kaleidoscope of order 2.

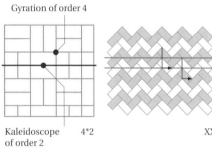

Gyration of order 4

Kaleidoscope of order 2   4*2   XX

The final type of symmetry to consider is the *glide*, denoted x. The possibilities are xx which has two types of glide (and no reflections or rotations), *x which has one glide and one reflection, and 22x which has no reflections, two rotations of order 2 and a glide.

Wallpaper groups

Wallpaper groups classify those patterns which contain two different translational symmetries. In 1891, Evgraf Fedorov proved that there are exactly 17 different possibilities. Here we will use the later orbifold notation of John Conway. The starting point of the classification is the following fact: patterns with two types of translation can only have rotational symmetry of order 1, 2, 3, 4 or 6. This is the *crystollagraphic restriction theorem*.

333

The simplest pattern has translational symmetry alone: this is denoted *o*. The possibilities for a pattern with rotations, but without reflection or glide symmetries, are 632, 442, 333 and 2222. With both rotational and reflectional symmetry, the possibilities are *632, *333, 3*3, *442, 4*2, 22*, *2222, 2*22 and ** (the last having two parallel lines of reflection, and no rotational symmetry). Finally, those with glides are xx, *x and 22x.

*2222

Heesch's tile

One of the questions posed by David Hilbert in his 18th problem was about shapes which can tessellate on their own, but only in a rather strange way. Although every tile is identical, nevertheless they appear in positions which are non-identical. This would mean that you can always find two tiles which cannot be matched by a symmetry, because the arrangements of tiles around them are different. It is not difficult to construct a tiling like this (any aperiodic tiling fits this description). Hilbert's question was whether there was any shape which can *only* tessellate this way.

Heesch's tile

In fact, he posed it in the context of 3-dimensional tilings (*space-filling polyhedra*), presumably believing that no such thing could exist in two dimensions. But the great mathematician had overlooked a possibility, and the first *anisohedral tile* was discovered by Heinrich Heesch in 1935. There is also a 3-dimensional tiling which satisfies Hilbert's criteria, discovered by Karl Reinhardt in 1928.

Aperiodic tilings

Tilings can have two types of translational symmetry (classified by the wallpaper groups), or just one, as described by the frieze groups. There are also tilings which have no translational symmetry at all. These are the *aperiodic* tilings, which never repeat themselves: even if you were to tile a square mile, there would be no way to slide the pattern around so that it fits back over itself. Having only rotational and reflectional symmetry, their possible

Radial tiling

symmetry groups are the same as for a polygon. Common examples are radial tilings (with dihedral symmetry groups), and the beautiful spiral tilings (with cyclic symmetry groups), such as the *Voderberg tiling*. More exotic possibilities are the Penrose and Ammann tilings.

Uncomputable tilings

Suppose I present you with a collection of shapes and challenge you to tile the plane with them. I am not concerned about symmetry; all that is required is that you can cover as large an area as I ask, with no gaps or overlaps. You can have as many of each shape as you like. If I give you squares and equilateral triangles, you will have no trouble. If I pick regular pentagons, heptagons and decagons, you will be unable to do it. But if I present you with a collection of 100 intricate and irregular polygons, then you will have to stop and think.

The question Hao Wang addressed was whether there is some definite procedure that you could follow, to decide whether or not my selection of shapes does tile the plane. In other words, he was searching for an **algorithm**. In 1961, he believed he had found one. But to prove it would work, Wang had to make an assumption, that no set of tiles should only tile the plane aperiodically. The discovery of Penrose and Ammann tilings demolished this hypothesis, and with it Wang's algorithm. In fact, there can be no algorithm to solve this problem: it is *uncomputable*.

Penrose and Ammann tilings

Aperiodic tilings were known to the mosaic-makers of ancient Rome. But every example seemed to have a periodic cousin. If a set of tiles could tile the plane, then it could be rearranged to do so in a way with translational symmetry. In 1961 Hao Wang conjectured that this must always be the case: no finite set of tiles should *only* give rise to aperiodic patterns. Wang's conjecture was refuted by his student Robert Berger in 1964, who concocted a set of 20,426 notched square tiles which tile the plane only aperiodically.

During the 1970s the mathematical physicist Roger Penrose and the amateur mathematician Robert Ammann independently discovered beautiful simplifications of this result. In 1974 Penrose found an aperiodic set of just two tiles: both rhombuses and with equal sides, one fat and one skinny. These tiled the plane, but by notching the edges (or colouring the edges and insisting that touching edges match) Penrose ensured that they could never do so in a periodic fashion. Penrose dubbed these tiles the 'rhombs' and he found two other purely aperiodic sets of tiles: the 'kites and darts', which also use two different tiles, and the 'pentacles', which use four. Ammann discovered further examples including, with Frans Beenker, the *Ammann–Beenker* tiling involving squares and rhombuses.

An outstanding question in tiling theory is: is there a single tile which can tile the plane, but only aperiodically?

Honeycombs and crystals

A honeycomb is a tessellation in more than two dimensions. Instead of filling the plane with polygonal tiles, a 3-honeycomb fills 3-space with 3-dimensional *cells*. A solid which tessellates is called a *space-filling polyhedron*. Only one of the Platonic solids qualifies: the cube. (Aristotle mistakenly believed that the tetrahedron also tessellates, perhaps because a mix of tetrahedra and octahedra together can fill space.)

There are other space-fillers, however. One Archimedean solid, the truncated octahedron, tessellates, as do triangular and hexagonal prisms and one Johnson solid (the gyrobifastigium). For polyhedra with irregular faces, one Catalan solid, the rhombic dodecahedron, fills space, as do over 300 assorted other irregular polyhedra.

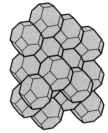
Tessellating truncated octahedra

For patterns involving more than one solid, a key result is the *crystallographic restriction theorem*, which says that a tessellation of 3-space which has translational symmetry can only have rotation of order 2, 3, 4 or 6. The analogues of the wallpaper groups are the 230 crystallographic groups. This is a fundamental result in materials science, as it restricts the possible arrangements of molecules in crystalline solids.

n-dimensional honeycombs

In four dimensions, three of the regular polychora tessellate: the hypercube, the hexadecachoron and the icositetrachoron. In five and more dimensions, the hypercube is the only regular polytope to tessellate.

However more complicated patterns can be formed with the inclusion of more than one type of tile. To analyse these, *space groups* play the role of the 17 wallpaper groups and 230 crystallographic groups.

In his 18th problem, David Hilbert asked a key question: are there only finitely many space groups in each dimension? In 1911, Ludwig Bieberbach was able to answer this question in the positive. (Bieberbach made other notable contributions to mathematics, but was disgraced for his Nazi politics.) In 1978, Harold Brown, Rolf Bülow and Joachim Neubüser showed that there are 4,895 space groups in four dimensions. In 2001, Wilhelm Plesken and Tilman Schulz used a computer to list the 222,097 space groups in five dimensions, and the 28,934,974 in six dimensions.

Quasicrystals

On the molecular scale, solids come in two basic forms: *amorphous*, where molecules are arranged haphazardly as in a liquid (an example is glass), and *crystalline* where they are arranged in fixed geometric patterns (such as diamond).

In 1982, Dan Schechtman discovered an alloy of aluminium and manganese with an unexpected property: it had rotational symmetry of order 5. The scientific establishment were dismissive; this seemed to contradict the fundamental *crystallographic restriction theorem*, which states that the only permissible rotations for a crystalline solid are of orders 2, 3, 4 and 6. On closer inspection however, the substance seemed not to be crystalline after all, having no translational symmetry. But amorphous solids are not organized enough to exhibit rotational symmetry, so what was it?

Mathematicians had encountered similar structures before, in the aperiodic tilings of Penrose and Ammann. Schechtman had discovered a 3-dimensional analogue of these, since dubbed a *quasicrystal*. Since then, numerous quasicrystals have been found in nature, occupying a niche between amorphous and crystalline solids. This development spurred on mathematical investigations. Many aperiodic tilings are now interpreted as slices through patterns in higher dimensions (rather as the conic sections are slices through a conical surface). The Penrose rhombs, for example, can be interpreted as a slice through a 5-dimensional hypercubic lattice.

CURVES AND SURFACES

Curves

A *curve* is a 1-dimensional geometric object. The simplest examples are those which are not 'curved' at all: straight lines. These are also the simplest when looked at algebraically. On a plane with Cartesian coordinates, straight lines are described by equations such as $x + y + 1 = 0$, or $3x - y - 7 = 0$, or generally $Ax + By + C = 0$ for some numbers A, B, C (and not with $A = B = 0$). These are the polynomial equations of *degree* 1.

Equations of degree 2 additionally contain x^2, y^2 or xy terms. These **quadratic curves** have an elegant description as the **conic sections**. Beyond these are the cubic curves, those of degree 3. Among these are the **elliptic curves**, which play a central role in modern number theory. Other curves, such as the **Archimedean sprials**, can most naturally be described in **polar coordinates**.

Conic sections

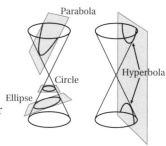

Which curves are described by equations of degree 2? Even 1800 years before Descartes invented his system of coordinates, Greek geometers, most notably Appolonius of Perga around 220 BC, were able to address this problem. It has a very elegant solution: first we consider a pair of infinite cones joined opposite at their tips. The *conic sections* are the curves formed by taking slices through this surface. These make up the family of curves given by equations of degree 2. Slicing the plane through horizontally will produce a circle (such as that given by $x^2 + y^2 - 1 = 0$). Slicing vertically, right through the centre gives a pair of intersecting straight lines (described by the equation $x^2 - y^2 = 0$, for example).

Quadratic curves

The three principle types of conic section are **ellipses**, **parabolas**, and **hyperbolas**. Any equation of degree 2 is of the form $Ax^2 + Bxy + Cy^2 + Dx + Ey + F = 0$, for some numbers A, B, C, D, E and F. The number $B^2 - 4AC$ largely determines what sort of curve it is: if this is zero, then we have a parabola (or in degenerate cases a straight line, or a pair of parallel lines). If it is negative we have an ellipse (or a single point, or nothing at all), and if it is positive we have a hyperbola (or pair of intersecting straight lines).

Focus and directrix

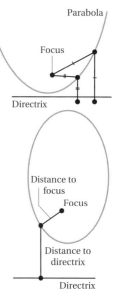

As well as arising when a cone intersects with a plane, the conic sections can be obtained by another construction. Take a point on the plane, and a straight line (not passing through the point). We call these the *focus* and the *directrix*, respectively. For any point on the plane we can ask what its distance is to the focus, and what its distance is to the directrix. (The 'distance' from a point to a line always means the shortest, or perpendicular, distance.) What pattern is formed by the collection of points for which these two distances are the same? The answer is a *parabola*.

By subtly changing the question we get different curves: if we want the distance to the focus to be *half* that to the directrix, the resulting curve is an *ellipse*. If we require it to be *double*, we get a *hyperbola*.

In these constructions, the crucial quantity is the ratio of a point's distance from the focus to its distance from the directrix. Call this number e. The defining characteristic of conic sections is that, whichever point on the curve you choose, you get the same value of e, called the curve's *eccentricity*. If $0 < e < 1$ the curve is an ellipse if $e = 1$ it is a parabola, and if $e > 1$ it is a hyperbola.

Ellipses

An *ellipse* can be defined as a slice through a cone, or by the focus/directrix method. In fact, an ellipse is symmetric, and has two foci on either side of its centre. These two foci provide another way to define an ellipse: suppose the distances from a point to the two foci are given by a and b. The ellipse is also defined by the condition that $a + b$ is always constant.

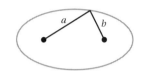

This provides a nice method for drawing an ellipse: push two pins into a piece of paper with a piece of slack string tied between them. Tracing out the places where the string is pulled taught will produce an ellipse.

The *major axis* of the ellipse is the longest straight line segment inside it, passing through both foci and the centre. The *minor axis* is perpendicular to this: the shortest straight line through the centre, connecting two points of the ellipse. Taking the distance between the foci and dividing by the length of the *major axis* gives the *eccentricity* of the ellipse. A circle is the special case where the two foci coincide: an ellipse of eccentricity 0.

In 1609, Johannes Kepler formulated his first law of planetary motion: that the orbit of a planet is an ellipse with the sun at one focus.

Parabolas

Unlike an ellipse, a parabola is a not a closed curve, but has infinite length. One of the conic sections, a parabola can be defined as a slice through a cone along a plane parallel to the edge of the cone. Alternatively it is the set of points whose distance from a given line (the *directrix*) is equal to that from a particular point (the *focus*).

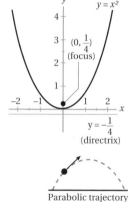

A common parabola is given by $y = x^2$ (or equivalently $x^2 - y = 0$). This has its focus at $\left(0, \frac{1}{4}\right)$, and its directrix is $y = -\frac{1}{4}$.

During the middle ages, the physics of motion was not well understood. People generally believed that when you fired a canon, the missile would fly in a straight line until it lost 'impetus' and fell to the ground. It was Galileo who, in the 17th century, combined mathematical knowledge with experimental skill to challenge this assumption. He devised a series of tests which demonstrated that projectiles actually travel in parabolic paths (ignoring the effect of air resistance). It was not until the work of Isaac Newton that scientists understood why this is true.

NASA distinguishes between *periodic comets*, which have elliptical orbits (and so reappear – in the case of Halley's comet, every 75 years, but for some long-period comets every ten million years), and *single-apparition* comets which travel on parabolic or hyperbolic paths, only passing through the solar system once.

Hyperbolas

A hyperbola is the only conic section with two separate branches, which come from slicing through both halves of the double cone. Like their cousins the ellipses, hyperbolas have two foci and two directrices. Again the two foci provide an alternative characterization. If the distances from a point on the curve to the two foci are a and b, then the numbers $a - b$ and $b - a$ are fixed for all points on the curve (swapping them over moves between the two branches). For this reason, hyperbolas occur as interference patterns in waves. If you drop two pebbles into a pond, the two sets of circular ripples will interfere in a family of hyperbolas.

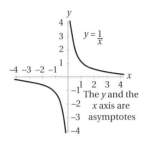
The y and the x axis are asymptotes

All hyperbolas have two asymptotes. In the case of the the most famous hyperbola, $yx = 1$ (or equivalently $y = \frac{1}{x}$), the asymptotes are the x- and y-axes. Of course these are at right angles to each other, making this a *rectangular hyperbola* (which always has eccentricity $\sqrt{2}$).

Hyperbolas of interference

Asymptotes

An *asymptote* to a curve is a straight line which the curve approaches arbitrarily closely but never reaches. A classic example is the hyperbola $yx = 1$. Here, as x gets bigger and bigger, y gets closer and closer to 0, without ever arriving there. So the line $y = 0$ is an asymptote to the curve. Similarly, as y gets larger, x gets closer to 0, making the line $x = 0$ an asymptote too.

Not every curve has any asymptotes; ellipses and parabolas do not, for example.

Newton's cubics

The three types of conic section are the curves given by quadratic equations. Curves of higher degree come in a far greater variety than three. Some of them can be more neatly expressed in polar coordinates.

In 1710, Isaac Newton considered curves defined by equations which involve x^3, x^2y, xy^2, or y^3: the cubic curves. Unlike the conic sections, these curves may cross over themselves. Some cubic curves come in two parts, and they may have *cusps*, places where they are not smooth, but have a sharp point. Newton discovered 72 distinct types of cubic curve; subsequent research revealed another 6. Cubic curves are fully classified by these 78 different families.

The most important cubics are the **elliptic curves**, which remain of central importance to mathematics today. Not every cubic curve is elliptic, but Newton showed that all cubic curves can be constructed by suitably squashing or stretching an elliptic curve.

Quadric surfaces

In 2-dimensional space, the simplest curves are straight lines (given by linear equations), followed by the conic sections: curves on the plane defined by quadratic equations. Looking at surfaces in 3-dimensional space, linear formulas define flat *planes*, and quadratic formulas produce the family of *quadric surfaces*. These are essentially obtained by lifting the conic sections into three dimensions, in different ways. The simplest versions are *cylinders* formed from ellipses, parabolas and hyperbolas, obtained by building straight walls on top of these curves. The equations for these are the same as for the original curves, with the z-coordinate free to take any value.

After rotating and centring the surface, many quadrics are given by an equation of the form $Ax^2 + By^2 + Cz^2 = 1$. If A, B and C are all positive, an **ellipsoid** is defined. If two are positive and one negative, a one-sheeted **hyperboloid** is defined, and if two are negative and one positive, a two-sheeted hyperboloid is defined.

The equation $z = Ax^2 + By^2$ defines a **paraboloid**: *elliptical* if A and B have the same sign, and *hyperbolic* if they have opposite signs.

The remaining cases are *elliptic cones* (cones with elliptical cross-sections) and pairs of planes.

Ellipsoid

An *ellipsoid* resembles a squashed and/or stretched sphere, whose cross-sections are ellipses. A special case is the *spheroid*, where the cross-sections in one direction are all circles. We have known that the earth is approximately spheroidal since Isaac Newton's work on universal gravitation. However, in the 18th century, there   was some debate about whether it forms a stretched sphere (a *prolate* spheroid, like a rugby or American football) as the French astronomers Geovanni and Jacques Cassini believed, or a squashed sphere (an *oblate* spheroid) as Newton himself held. Further measurements proved Newton correct, although smaller celestial bodies can form prolate spheroids. One such is the dwarf planet Haumea.

Ellipsoids are given by an equation of the form $Ax^2 + By^2 + Cz^2 = 1$ for some positive numbers A, B, C. If A, B, C are all different, the result is a *scalene* ellipsoid. If two are equal (say $A = B$) we have a spheroid (which is oblate if $C < A$ and prolate if $C > A$). The situation $A = B = C$ describes a sphere.

Paraboloids

Paraboloids come in two types. *Elliptical paraboloids* are shaped like parabolic cups with elliptical cross-sections. When these ellipses are circles, the result is a *circular paraboloid*, widely exploited in telecommunications, in the designs of satellite dishes and radiotelescopes. Its useful property is that all rays perpendicular to the base are reflected straight to the paraboloid's focus. Reversing this process, circular paraboloids are also used as reflectors in spotlights and commercial illumination. Placing a bulb at the focus, the paraboloid reflects out light as a beam of parallel rays.

The second form is the saddle-shaped *paraboloid*, famous as the shape of Pringles and also used for roofing in modern The archetypal hyperbolic paraboloid is by $z = xy$.

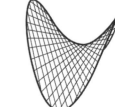

Hyperboloids

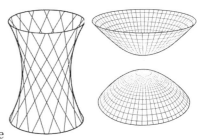

Hyperboloid is the name given to two different surfaces: one-sheeted and two-sheeted. The two-sheeted version resembles two elliptical paraboloids facing each other across a gap (though this is not an exact description, the curvature being slightly different). The one-sheeted hyperboloid is famous as the shape of a cooling tower. Sliced vertically these both produce hyperbolas, but their horizontal cross-sections are ellipses.

Most applications involve one-sheeted *circular* hyperboloids, where these ellipses are actually circles. (Often 'hyperboloid' is understood as short-hand for 'one-sheeted circular hyperboloid'.) As these are **ruled surfaces**, they are easily constructed. Take two identical circular rings, and join the corresponding points with wires. Pulling them apart will form a cylinder. Twisting turns this into a hyperboloid. Since this surface is *doubly ruled*, two sets of wires can simultaneously be straightened. Being constructible from straight beams, hyperboloids have been popular in art and architecture ever since Vladimir Shukhov built a hyperboloid water tower, in 1896.

Surfaces of revolution

A nice way to generate 2-dimensional surfaces from 1-dimensional curves is through rotation. Starting with a circle, draw a straight line passing through it as a diameter. As the circle spins around this axis in three dimensions, it sweeps out a surface, namely a sphere. The same trick works for any curve, although the resulting surface depends not only on the choice of curve, but also on the position of the axis. Rotating a circle around a line not passing through it, for example, produces a torus.

Rotating a straight line about a parallel axis produces a circular cylinder. If the two lines are not parallel on the plane, then the result is the double cone that encapsulates the conic sections. Starting with two skew lines in 3-dimensional space (that is, two lines which are not parallel, but do not cross), the surface swept out is a circular hyperboloid of one sheet. Rotating an ellipse about one of its axes produces a spheroid, while a parabola gives a circular paraboloid, and a hyperbola produces a circular hyperboloid. More intricate curves can produce very beautiful surfaces of revolution, a fact long exploited by potters and sculptors.

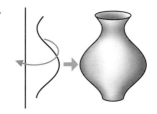

Ruled surfaces

A plane is completely built from straight lines: the straight lines on the surface together totally cover it.

More surprising is that there are other, apparently curvier, surfaces with the same property. Cylinders are just obtained by building straight walls along curved paths. Cones are also ruled surfaces. Another famous example is the *helicoid*, swept out by a straight line spiralling down a vertical axis, reminiscent of a ramp in a multi-storey carpark (garage).

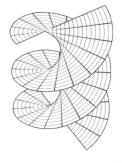

There are three *doubly ruled* surfaces, where every point lies on two straight lines. These are the plane, the circular single-sheeted hyperboloid, and the hyperbolic paraboloid.

Surfaces of higher degree

Quadric surfaces are given by polynomial equations of degree 2. If we raise this degree, a wealth of more complicated surfaces appears. An example is the *ding-dong surface*, a cubic surface obtained as a surface of revolution of a self-intersecting cubic curve.

An interesting class of surfaces are the *superquadrics*. These mimic the quadric surfaces, replacing the x^2 terms with higher degree. For example, a spheroid is a surface of revolution of an ellipse. An ellipse has equation $\left|\frac{x}{a}\right|^2 + \left|\frac{y}{b}\right|^2 = 1$. By replacing the squares with some higher power n, we obtain a *superellipse*, given by $\left|\frac{x}{a}\right|^n + \left|\frac{y}{a}\right|^n = 1$. (If $a = b$, as in the diagram, we get a *squircle*, unrelated to the ancient problem of **squaring the circle**. The illustrated example has equation $|x|^7 + |y|^7 = 1$.) If we now take the surface of revolution, we get a *superegg*. An interesting property of this shape is that it has zero curvature at its tips. So, unlike a spheroid, it can stand perfectly upright, a fact exploited by the mathematician and sculptor Piet Hein.

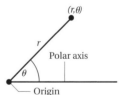

POLAR COORDINATES

Polar coordinates

The system of **Cartesian coordinates** identifies a point on the plane through its distances from a pair of perpendicular axes. An alternative is to use *polar coordinates*; these also involve two pieces of information: a distance and an angle, usually denoted r and θ respectively. The distance states how far the point is from the origin: so, in the diagram, the points A and B (with Cartesian coordinates (1,0) and (0,1) respectively) are each 1 unit away from the origin, as is C with Cartesian coordinates $\left(\frac{1}{\sqrt{2}}, \frac{1}{\sqrt{2}}\right)$. (This follows from Pythagoras' theorem: $\left(\frac{1}{\sqrt{2}}\right)^2 + \left(\frac{1}{\sqrt{2}}\right)^2 = \frac{1}{2} + \frac{1}{2} = 1$.)

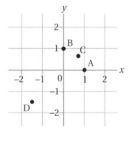

Distance alone cannot distinguish between these, so we also provide the angle that that point makes at the origin, against the *polar axis*. This is the horizontal line starting at the origin and going right (the positive *x*-axis in Cartesian coordinates). So the point A which sits on this line has angle 0°. In mathematics a positive angle is always anticlockwise, so the point C has angle 45° and B has 90°.

Usually, however, we measure the angles in **radians**: so, writing the distance before the angle, the polar coordinates of A, B and C are (1,0), $\left(1, \frac{\pi}{2}\right)$, and $\left(1, \frac{\pi}{4}\right)$.

The point D is 2 units away from the origin, at an angle of 270°, so it has polar coordinates $\left(2, \frac{3\pi}{2}\right)$.

Polar geometry

Polar and Cartesian coordinates are two different languages for talking about the same objects. Anything which can be said in one can be translated into the other. Nevertheless, polar coordinates efficiently describe certain geometrical shapes in the plane. For instance, the circle of radius 1 has a simple formula: $r = 1$. This describes the set of points of the form $(1, \theta)$, each of which is one unit away from the origin.

On the other hand, keeping the angle fixed, say at $\frac{\pi}{4}$, and letting r vary produces a straight line at that angle to the horizontal. This is described by the equation $\theta = \frac{\pi}{4}$.

Other examples of shapes that are well described by polar coordinates are Archimedean and logarithmic spirals, and cycloids.

Polar coordinates are ubiquitous in complex analysis: every complex number z comes equipped with a distance r (its *modulus*) and an angle θ (its *argument*). These are tied together in the formula $z = re^{i\theta}$.

Archimedean spirals

Polar coordinates are perfect for describing spirals, where a point's distance from the origin depends on some quantity of rotation. The simplest case is the *Archimedean spiral*, given by the equation $r = \theta$. This consists of all points whose first and second polar coordinates agree: those of the form (θ, θ). When $\theta = 0$, the length r is also 0, which happens only at the origin. When $\theta = \frac{\pi}{4}$ (that is 45°) the length is $\frac{\pi}{4}$ too (around 0.8). When $\theta = \frac{\pi}{2}$ (or 90°), $r = \frac{\pi}{2}$ (around 1.6), and so on. Once θ gets up to 2π, the spiral has performed one revolution, and crosses the polar axis. But we can continue plotting larger values of θ, until it crosses the axis again at 4π (720°), and again at 6π, 8π, 10π, and so on.

The spiral can be expanded or contracted by including a multiplying constant: $r = 2\theta$ defines one twice as sparse as that above, and $r = \frac{1}{2\pi}\theta$ describes a spiral which crosses the axis at values 0, 1, 2, 3, … The defining characteristic of these spirals (unlike hyperbolic or logarithmic spirals) is that consecutive turns are a fixed distance apart (like the grooves on a vinyl record).

The *Parker spiral* is an Archimedean spiral formed by the sun's magnetic field, as it permeates space.

Logarithmic spirals

The shape that Jakob Bernoulli dubbed *Spira Mirabilis* ('the miraculous spiral') has the polar equation $r = e^{\theta}$ (or equivalently $\theta = \ln r$). Starting at $\theta = 0$, $e^0 = 1$, so the curve crosses the polar axis at 1. It crosses again at $e^{2\pi}$ (around 535.5), $e^{4\pi}$ (around 286751.3), and so on. It also makes sense to spiral back the other way, by allowing θ to take negative values. So it also crosses the axis at $e^{-2\pi}$ (about 0.002), $e^{-4\pi}$ (around 0.000003), and infinitely often as it winds ever more tightly towards the origin (although, unlike the hyperbolic spiral, the distances decrease so quickly that the length along the curve from any point to the origin is finite).

Jakob Bernoulli was stunned by the fractal-like self-similarity of the logarithmic spirals: if you enlarge or shrink it by a factor of $e^{2\pi}$, the result is exactly the same curve. Even more, if you take the inverse of the spiral (given by $r = e^{-\theta}$) the result is again the same. Bernoulli was so besotted by this curve that he instructed that one be engraved on his tombstone. (Unfortunately the stonemason was no geometer, and carved an Archimedean spiral instead.)

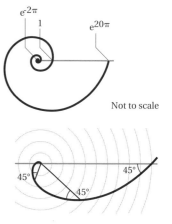

Not to scale

The logarithmic spiral is also known as the *equiangular spiral* because of another defining property: the angle between the tangent and the radius is constant, at $\frac{\pi}{4}$ (that is, 45°).

Logarithmic spirals with different angles are given by the equation $r = e^{c\theta}$, for different numbers c, producing an angle of $\tan^{-1}\left(\frac{1}{c}\right)$.

Logarithmic spirals, and approximations to them such as **Fibonacci spirals**, are common in nature, from spiral galaxies and cloud formations, to nautilus shells.

The problem of the four mice

In 1871, the astronomer and mathematician Robert Kalley Miller set a tricky problem in the notorious mathematical tripos exam at Cambridge University. It concerned four mice A, B, C and D, which start at four corners of a square room. They are released simultaneously and all run at the same speed, with A chasing B, B chasing C, C chasing D, and D chasing A. The problem was to predict the paths they will follow.

Initially each mouse runs along the wall. But, as its target also moves, it deviates from that path. The answer is that their paths produce four intertwined logarithmic spirals converging at the centre of the room. This generalizes to rooms of other polygonal shapes. In 1880 Pierre Brocard considered three mice in an irregular triangular room. The three spirals meet at the first or second Brocard points of the triangle, depending on their direction (see **centres of triangles**).

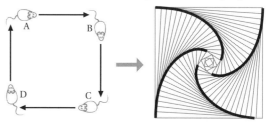

Roses

Why might a superstitious mathematician think the equation $r = \cos 2\theta$ lucky? In polar coordinates, it describes a four-leafed clover: a *quadrifolium*. This is one of the family of *rose curves*, given by the equations $r = \cos k\theta$ (or $r = \sin k\theta$) for different values of k. Roses were first studied by the Italian priest Luigi Guido Grandi in the early 18th century.

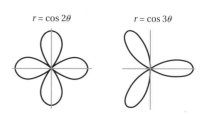

$r = \cos 2\theta$ $r = \cos 3\theta$

The curve depends on the number k. If k is odd, the rose has k petals. So the equation $r = \cos 3\theta$ describes a 3-petalled rose, or *trifolium*. When k is even, the rose has not k but $2k$ petals. It also makes sense to consider non-integer values for k. This time the petals will overlap. If k is a rational number, say $k = \frac{a}{b}$, with a and b coprime, again there are two cases: if a and b are both odd, the rose will have a petals, and a pattern which starts to repeat when θ reaches $b\pi$ (a period of $b\pi$). Otherwise it will have $2a$ petals and a period of $2b\pi$. When k is irrational (such as $k = \sqrt{2}$) the rose never repeats and has infinitely many petals.

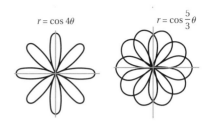

$r = \cos 4\theta$ $r = \cos\dfrac{5}{3}\theta$

The tautochrone problem

In 1659, Christiaan Huygens was considering beads sliding down slopes. Assuming zero friction, he discovered a remarkable curve which is *tautochronous* ('of the same time'). No matter how far up he put the bead at the start, it always took exactly the same time to slide to the bottom. Huygens curve was a cycloid. If you draw a spot on a bicycle tyre, this is the path the spot follows as you cycle along. However, the tautochrone cycloid is the other way up, as if you are cycling across the ceiling.

The brachistochrone problem

In 1696, Johann Bernoulli set the readers of *Acta Eruditorum* a challenge. Suppose we have two points A and B, with A higher than B (but not vertically above it). The idea is to draw a curve from A to B, and let a bead slide down. What curve should we use, if we want the bead to reach B in the shortest possible time? This is the *brachistochrone* ('shortest time') problem.

Several mathematicians were able to provide the answer, including Newton, Leibniz, Bernoulli himself, as well as his brother Jakob. The solution was the same as for the tautochrone problem: a cycloid.

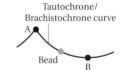

Tautochrone/
Brachistochrone curve

A

Bead   B

Cycloids

Draw a straight horizontal line, and roll a circle along it. If you mark one point on the circle, the curve traced out is a *cycloid*, famous as the solution of the tautochrone and brachistochrone problems. A cycloid is described parametrically in Cartesian coordinates by:

$$x = t - \sin t$$
$$y = 1 - \cos t$$

As t increases, the centre of the circle moves horizontally along the line $y = 1$. So, at time t, the centre is at $(t, 1)$, with the curve cycling around it.

Hypocycloids and epicycloids are similar constructions, with circles taking the place of the straight line.

Cycloid

Centre of the circle,
at point $(t, 1)$

Hypocycloids and epicycloids

If a small circle rolls around the inside of a larger circle, we can mark one point on the small circle and trace out its path. The resulting curve is a *hypocycloid*. If the outer circle has double the radius of the inner one, they are called a *Tusi couple*. As Nasir al-Din al-Tusi realized in the 13th century, this hypocycloid is just a straight line segment. If the larger circle is triple the length of the smaller, we obtain a three-pointed *deltoid*. If it is quadruple the length, we obtain a curve with four cusps (sharp corners), and so on.

Deltoid

Cardioid

A similar construction is an *epicycloid*, where one circle rolls around the outside of another. A notable epicycloid is where the two circles are the same size. In this case, the result is a *cardioid*. If the radius of the outer circle is half that of the inner, the result is a *nephroid* ('kidney shape').

Nephroid   Hypocycloid with $k = \frac{11}{2}$

For hypocycloids and epicycloids, the key is the ratio of the larger circle's radius to that of the smaller. Call this k. If k is a whole number, we get a curve with k cusps. If k is a rational number, say $k = \frac{a}{b}$, with a and b coprime, the curve will self-intersect and have a cusps. If k is irrational, the curve never closes up, and will produce an inky mess.

Epicycloid with $k = \frac{11}{2}$

Roulettes

The *spirograph* is a mathematical toy, invented by Denys Fisher in the 1960s for creating beautiful, intricate patterns. Probably it appeals as much to geometers as to children. The spirograph relies on a principle similar to that of cycloids, hypocycloids and epicycloids: a small plastic disc is rolled along a line, or around a fixed larger circle. The difference is where the pen is placed: rather than being on the perimeter of the smaller circle, it fits in a hole somewhere inside the disc. The resulting curves are called *trochoids*, *hypotrochoids* and *epitrochoids* (coming from the Greek *trochos* meaning 'wheel').

Epitrochoid

This idea can be extended to allow the pen-point to be outside the smaller disc, a fixed distance from the centre (as if on a matchstick glued to the disc).

Roulettes are the most general curves formed in this way, obtained by attaching a point to a curve (not necessarily on it) and 'rolling' that curve along another, tracing out the path of the point. For example, rolling a parabola along a straight line and tracing out the path of its focus produces a catenary.

Hypotrochoid

Catenary

Fix the two ends of a chain to a wall, and let the chain hang down in between. What curve will be formed? That was the question posed by Jakob Bernoulli in *Acta Eruditorum* in 1690. (We assume the chain is infinitely flexible and of even density.) Galileo Galilei had already considered this in 1638.

Catenary

He said the curve was a parabola. But Galileo was wrong, as Joachim Jungius demonstrated in 1669.

Bernoulli received three correct solutions to his challenge: from Gottfried Leibniz, Christiaan Huygens and his brother Johann Bernoulli. Their result was one of the early triumphs of differential calculus. The curve which answers the question is a *catenary* (chain curve), given by the equation $y = \frac{e^x + e^{-x}}{2}$, or more concisely as $y = \cosh x$, where cosh is the hyperbolic cosine function.

DISCRETE GEOMETRY

Pick's theorem

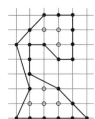

There are many ways to calculate the area of a shape on the plane. Generally, as the figure becomes increasingly complicated, so does the formula for its area. An elegant way around this was found by the Austrian mathematician Georg Pick, in 1899.

Create a grid of dots by placing one at every point on the plane whose coordinates are both integers. Pick's method applies to any shape which can be formed by joining these dots with straight lines. There are only two ingredients: the number of points on the boundary of the shape (call this A), and the number of points enclosed within it (B). Pick's theorem states that the area is $\frac{A}{2} + B - 1$. In the pictured example, $A = 22$ and $B = 7$, so the area is $\frac{22}{2} + 7 - 1 = 17$ square units. This provides a fast method to calculate the areas of complicated shapes which would otherwise involve dividing the shape into triangles, in a rather longwinded way.

Thue's circle packing

Suppose you have a table and a bag of coins, and are challenged to fit as many coins on the table as you can. All the coins are the same size and may not be piled up, just laid flat on the table-top. What is the best strategy?

There are two obvious candidates: *square packing*, where the coins lie in vertical and horizontal rows, each coin touching four others; and *hexagonal packing*, where each coin touches six others (leading to staggered rows).

To compare them we use the *packing density*: the proportion of the area of the table covered by coins. The packing density of the square lattice is $\frac{\pi}{4}$ (about 79%); that of the hexagonal one is $\frac{\pi}{\sqrt{12}}$ (about 91 %). So the hexagonal packing seems to be better. But can we be certain we have not missed another, better arrangement?

In 1831, Carl Friedrich Gauss proved that the hexagonal packing is indeed the tightest of the regular packings. These are symmetrical packings built on *lattices*: patterns with double *translational symmetry*. But could some strange irregular arrangement do better? In 1890, Axel Thue finally proved that there was no such arrangement. Thue's circle packing theorem is the 2-dimensional analogue of the **Kepler conjecture**.

The Kepler conjecture (Hales' theorem 1)

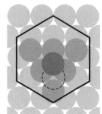

In 1661, Johannes Kepler contemplated a 3-dimensional version of Thue's circle packing theorem: what is the best way of piling up spheres to occupy the minimum space? Fruit-sellers the world over could supply the answer: first you put down one layer of spheres arranged hexagonally according to Thue's circle packing. Then you put a second layer on top, staggered so that the spheres sit at the lowest points possible. Then keep repeating.

In what came to be known as the *Kepler conjecture*, Kepler asserted that this is 'the tightest possible, so that in no other arrangement could more pellets be stuffed into the same container'. On closer inspection, the solution is not unique: there are two choices, depending on whether you place the third directly over the first (the *hexagonal packing*) or staggered (the *face-centred cubic packing*). However, both have the same packing density: $\frac{\pi}{\sqrt{18}}$ (about 74%).

Intuitively, it seemed obvious enough that these should be the optimal solutions. But by 1900 no proof had been found, and David Hilbert highlighted the challenge in his 18th problem. It was not until 1998 that Thomas Hales, with his graduate student Samuel Ferguson, finally produced a proof. However, at 250 pages long, and with large sections relying on computer code and data amounting to over three gigabytes, it posed a massive challenge to check. After four years of work, the team who had been assigned the task gave up, stating that while they were 99% sure the proof was correct, they were unable to fully certify it. Hales is now working on a second generation proof, intended to be verifiable by **proof-checking software**.

Hypersphere packing

The Kepler conjecture was solved (at least with 99% certainty) in 1998. But the same question in higher dimensions remains open. It is not even certain that the closest packing of hyperspheres must be a regular lattice, rather than some less symmetrical arrangement. The regular packings, at least, are well understood up to dimension eight.

Although we do not know what the best packings of hyperspheres are, thanks to the *Minkowski–Hlawka theorem* we do have some idea of what packing density they will produce. This says the optimal packing in *n* dimensions will always achieve a density of at least $\frac{\zeta(n)}{2^{n-1}}$, where ζ is the **Riemann zeta function**.

In 24 dimensions, something remarkable happens. An entirely new way to pack spheres emerges, known as the *Leech lattice* after its 1965 discoverer. It is not quite certain that this is the optimal packing, but in 2004 Henry Cohn and Abhinav Kumar showed that if any other arrangement does improve it, it can do so by only a tiny amount, an increased density of at most 2×10^{-30}.

Hexagonal honeycomb conjecture (Hales' theorem 2)

Suppose you want to divide up a large sheet of paper into cells of area 1 cm^2, using the least possible amount of ink to draw the lines. Probably the first attempt would consist of 1×1 squares. But there are other possibilities: rectangles, tessellating triangles, irregular pentagons, the **Heesch tile**, or any other tesselation which uses just one tile.

For many years the one which involved the lowest total line-length was believed to be the hexagonal lattice. Indeed, this was often tacitly assumed as a fact, even though no-one had proved it. In 1999, Thomas Hales did provide a proof: the hexagonal pattern is indeed the most efficient method.

This explains why bees pack their honey in hexagonal tubes rather than those with square or Heesch tile cross-sections, for example. These tubes require the least amount of wax in proportion to their volume of any system of identical tessellating tubes. An interesting appendix to the story concerns what the bees don't know.

What bees don't know

In 1953, László Fejes Tóth published a paper 'What the bees know and what they do not know', in which he showed that the bees' honeycomb design is not quite optimal, despite their use of the hexagonal honeycomb conjecture.

Bees' hive   Fejes Tóth's hive

Honey cells are open at one end. At the closed end, two layers of hexagonal tubes are separated by a wax partition, consisting of four rhombuses closing off each tube. Fejes Tóth showed that less wax would be needed for a partition comprising two hexagons and two small squares. However, Tóth's design used only 0.35% less wax than the bees'. The compensating benefits of ease of construction and the stability of the hive may explain the bees' choice.

Kelvin's conjecture

Which 3-dimensional shape has the lowest ratio of surface area to internal volume? The answer is the sphere, which explains why soap bubbles are round.

The question becomes more complicated when we want more than one cell. This is a question Lord Kelvin considered in 1887: how to divide up

3-dimensional space into cells of the same size, using a partition of minimal surface area? This is a 3-dimensional analogue of the hexagonal honeycomb conjecture, and Kelvin believed that he had found the optimal arrangement, in what is now called a *Kelvin cell*, essentially a truncated octahedron (a space-filling Archimedean solid), with slightly curved faces.

Weaire–Phelan foam

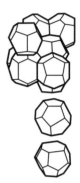

In 1993 Irish physicists Denis Weaire and Robert Phelan refuted Kelvin's conjecture by finding a new structure which improved upon the Kelvin cell, by 0.3%. Their repeating unit is constructed from eight irregular, slightly curved polyhedrons: two dodecahedra (with 12 pentagonal faces), and six 14-hedra, each with two hexagonal and 12 pentagonal faces. This discovery was celebrated in the architecture of the aquatic centre at the Beijing Olympics, in 2008. The architect Kurt Wagner explained that they created a huge Weaire–Phelan foam, and then 'cut the overall shape of the building out of the foam structure'. It is not known whether Weaire-Phelan foam provides the ultimate solution to Kelvin's problem.

The map colouring problem

How many colours do you need to colour a map so that no country borders another of the same colour? This was the question that the British lawyer and mathematician Sir Alfred Kempe addressed in 1879. The question did not concern the geography of the real world: any arrangement of shapes whose 'countries' came in one connected piece (unlike the USA, in which Alaska and Hawaii form disconnected parts) was classed as a 'map'. Countries meeting at a single point are allowed to have the same colour, only ones separated by boundary lines may not.

Kempe's solution was that every conceivable map drawn on a sphere 'can in every case be painted with four colours'. Unfortunately for him, in 1890 Percy Heawood found a problem: it did not mean that four colours could not work, but it did identify a fatal gap in Kempe's argument. Extending his ideas, however, Heawood was able to prove that five colours would always be enough.

The four colour theorem

For over 80 years the map colouring problem stood open: no-one was able to prove that four colours are always enough, nor could anyone construct a map which required five. That was until 1976 when Kenneth Appel and Wolfgang Haken of the University of Illinois published their proof that four colours are indeed always sufficient. Their proof did not rely on mathematical ingenuity alone. It was a mammoth effort, requiring 1000 hours of computer time, and containing 10,000 different diagrams. Appel explained: 'There is no simple elegant answer, and we had to make an absolutely horrendous case analysis of every possibility'.

DIFFERENTIAL GEOMETRY

Gaussian curvature

If we have a surface, we want some way to measure how curved it is. Of course, it may have some flat parts and some highly curved parts. So curvature is a *local* phenomenon. *Gaussian curvature* is a device which uses differential calculus to measure the curvature at a particular point, x. It produces a number $K(x)$. It works by placing an arrow coming out of the surface at x, perpendicular to the surface. This is a *normal vector*. If $K(x) < 0$, then the surface bends *towards* the normal vector in one direction, and *away* from it in another. In other words, the surface is *saddle-shaped* at x. An example is a one-sheeted hyperboloid, which has negative curvature everywhere.

A sphere is an example of a surface with positive curvature everywhere (so $K(x) > 0$ at every point). Starting at our chosen point x, all directions of the surface bend in the same direction, relative to the normal vector.

As expected, a surface may have positive curvature in one place, negative in another and zero at a third. There is a limit to this variety, however, given by the **Gauss–Bonnet theorem**.

Developable surfaces

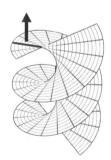

If the Gaussian curvature at a point x is zero, that is, $K(x) = 0$, it does not follow that the surface is completely flat, like a plane. It means there is at least one direction in which the surface is flat. A cylinder has zero curvature at every point, as does any surface which can be unrolled to plane without distortion. These are the *developable surfaces*. Other examples are a cone and a *developable helicoid*.

Theorema Egregium

If I draw a straight line on a piece of paper, and ask if my line is horizonal, your answer would not just depend on the line. It would have to take into account the position of the line in relation to the page or the floor. Qualities such as being horizonal are not intrinsic to a geometric object, but depend on its relationship to the ambient space. Gauss' *Theorema Egregium* says that Gaussian curvature is not like this. It is an intrinsic property of the surface and does not depend on the ambient space. Gauss thought this result 'remarkable' because his original definition does make heavy use of the ambient conditions.

Local and global geometry

There are several ways to view a mathematical surface. One is to pay close attention to its detailed geometry in small regions. Curvature belongs to this *local* world. The **Euler characteristic**, on the other hand, belongs to another realm, namely **topology**. Here, the object is considered *globally*, and changes in small regions are irrelevant. In a wonderful development, these two phenomena are nevertheless intimately related by a fundamental result of 19th-century geometry, attributed to Carl Friedrich Gauss and Pierre Bonnet: the Gauss–Bonnet theorem.

Gauss–Bonnet theorem

If we are working on a surface (*S*) with finite area, and no edges, then **integration** provides a way to take local data at each point of *S* and average it all out, to give one global piece of information about the whole surface. The Gauss–Bonnet theorem says that when you do this with the Gaussian curvature *K*, what emerges is none other than the Euler characteristic $\chi(S)$ (multiplied by a constant 2π).

$$\int_S K = 2\pi \times \chi(S)$$

Because the Euler characteristic is a topological quantity, unaffected by any amount of stretching and twisting, this means that the overall curvature of the surface is similarly fixed. If you bend and pull the surface, you can dramatically alter the curvature at every single point, but these changes all cancel each other out.

Every smooth surface which is a topological sphere (such as the surface of a banana or a frying pan) has Euler characteristic 2. So when you integrate the Gaussian curvature over the whole surface, you will always get the answer 4π.

Geodesics

Everyone knows that the shortest path between two points is a straight line. But what happens if we're working on a surface, such as a sphere, which does not contain any straight lines? In the specific case of a sphere, there is a nice answer: the roles of straight lines are played by *great circles*. These are the largest circles the sphere can hold (obtained by slicing a plane through the centre of the sphere).

In general, the curves which make up the shortest paths on a surface are called its *geodesics*. Shortest paths do not always exist. For example, if we take the ordinary plane, and remove the point $(0,0)$, then there is no shortest route in the punctured plane between the points $(1,0)$, and $(-1,0)$; every route can be shortened by bypassing the hole more closely.

However, geodesics always exist *locally*. In other words, at any point on a surface there are geodesics leaving it, in every direction. On a cyclinder, there are three types

Geodesics on a sphere

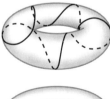
On the punctured plane there are no shortest paths (any path can be made shorter)

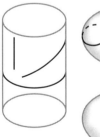

of geodesics: segments of circles, segments of straight lines, and segments of helices. More complicated surfaces can have geodesics which are less easy to describe. On an ellipsoid, for example, they come from intersecting the surface with a suitable hyperboloid.

Cartographic projections

A geographical map is a good example of a mathematical *function*, taking a point on the planet to a point on the piece of paper. But we don't want the function to completely mangle the earth's geography. There are various extra requirements we could add.

1 *Isometric* maps should preserve the distance between any two points. Unfortunately a flat map can never do this.

2 *Equiareal* functions keep areas proportionately the same. An example is the *Albers projection*, which is formed by cutting a cone through the globe, and projecting the points onto it.

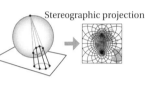
Albers projection

3 *Conformal* functions preserve angles. So two lines which meet on the earth, will meet at the same angle on the map. It is possible to draw a conformal map of the earth, for example by *stereographic projection*.

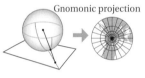
Stereographic projection

4 The *gnomonic* projection preserves the shortest route between any two points (but not the length of that route). This means that it represents great circles on the sphere as straight lines on the page. (Only one hemisphere can be represented.)

Gnomonic projection

5 A *rhumb line* is a path spiralling around the sphere determined only by an initial bearing. These were historically important in navigation. *Mercator projections* are maps which represent rhumb lines as straight lines. They are formed by wrapping the globe in a cylinder, and projecting the points outward.

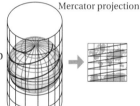
Mercator projection

No map can have more than one of these features; some distortion is inevitable; the question for the cartographer is where the priorities lie.

Isometric maps of the earth

A map of the earth would be *isometric* if the distance between any two points on the earth was the same as that given by the map (suitably scaled down, of course). However, isometric maps, even of portions of the earth, are impossible on flat paper. This is a consequence of the **Theorema Egregium**. Because curvature is intrinsic, it must be preserved by any isometric function (such as one taking a point on the earth to a point on the map). Consequently, any isometric map of the earth has to be curved like a globe, not flat. For a small region, such as a city, this does not create a problem as the earth is approximately flat. There are ways to approximate isometric maps for the whole globe, such as the orange peel map, but fundamentally, the obstacle is immovable.

Orange-peel map

Stereographic projection

Stereographic projection is a way to draw a map of a sphere on a flat plane. The idea is as follows: place the sphere on the plane. Then, given a point x on the sphere, send a beam from the north pole through the sphere, passing through x. The place where that beam crosses the plane is where x is marked on the map. The only point which gets missed off the map is the north pole. The resulting map is *conformal*, that is, angles on the sphere match the corresponding angles on the map.

As well as cartography, stereographic projection is used within mathematics. Starting with the plane of **complex numbers**, performing the procedure backwards results in the *Riemann sphere*, where the north pole represents a new point, known as the 'point at infinity'. (This gives a particularly pleasing way to view conformal maps of the complex numbers, known as *Möbius transformations*.)

TOPOLOGY

Topology

Whereas traditional geometry focuses on rigid objects such as straight lines, angles or curves given by precise equations, topology studies structures at a higher level of abstraction. A shape's *topological* properties are those which can survive any amount of stretching and twisting (but not cutting or gluing). For example a cube can be squashed until it is spherical, but a *torus* (or donut) cannot. So, topologically, a sphere and a cube are equivalent, but a torus is different. Similarly the letter C is topologically equivalent to L, but distinct from B.

Dubbed 'rubber sheet geometry', topology grew into a subject in its own right in the early 20th century, although its roots go back to Leonard Euler's solution of the **Seven Bridges of Königsberg** problem, in 1736. Topology now comprises several large bodies of research. The London underground map is an example of a topological representation, since it ignores the precise geometry of distances and directions, but accurately represents factors such as the ordering of stations, and the intersections of tube (subway) lines.

Möbius strip

Take a rectangular strip of paper, give it a half-twist before gluing the ends together, and you will obtain a *Möbius strip*. Discovered in 1858 by August Ferdinand Möbius, the interest of this object is that it only has one side (or in mathematical terms is *non-orientable*). Möbius strips featured heavily in the work of the artist M.C. Escher. They are mathematically important because many other topological constructions depend on them: notably the **real projective plane and Klein bottle**. Both can be built by gluing the edge of a Möbius strip to itself, in one of two ways.

Orientable surfaces

If you look at just a small portion of a sphere, it looks very much like a 2-dimensional plane. (Indeed, this is the perspective of most human beings for most of their lives.) It is only on zooming out that its *global* spherical nature becomes clear. This is essentially the definition of a *surface*: around every point is a patch of plane (perhaps slightly curved). The question is what are the possible global shapes that can be patched together this way? The whole plane itself is one example; but much focus is on *closed* surfaces, which have a finite area (and come in one piece).

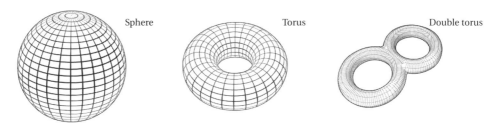

Sphere   Torus   Double torus

In topology we only care about topologically distinct surfaces: a banana-shaped surface, for example, will be classed as a sphere. But a torus (or donut) is a new, non-spherical surface, and the double torus is different again. So here is a method for producing genuinely topologically different surfaces: start with a sphere, and keep pinching holes through it (or equivalently adding on handles). The number of holes is called the surface's *genus*. However, this procedure does not exhaust every possibility: there are the non-orientable surfaces too.

Non-orientable surfaces

The definition of a *surface* is a *local* one: something built from small patches of 2-dimensional plane. Sometimes, mathematics throws up unexpected possibilities. We can patch together pieces of plane in a way which is internally coherent, but which cannot accurately be represented in 3-dimensional space. These are the *non-orientable surfaces*. Technically a definition of a *non-orientable* surface is this: if you put a watermark in a surface, it can be slid around until it is in its starting position, but as a mirror image of the original. This means that the surface must contain a Möbius strip somewhere.

So a way to produce non-orientable surfaces is to start with a sphere, cut slits in it and 'sew in' Möbius strips along their edges. The first two non-orientable surfaces obtained this way are the real projective plane and the Klein bottle.

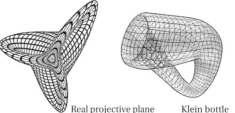

Real projective plane   Klein bottle

Real projective plane and Klein bottle

Start with a square of paper, and draw arrows along the left and right edges, both pointing up. If you glue these edges together, with the arrows matching, you will create a cylinder.

Start again, this time drawing the arrows in opposite directions. Gluing the edges together with the arrow head touching now creates a Möbius strip.

To arrive at a surface without edges, we'll also need to join the two remaining sides: take the configuration for the cylinder, and add a new pair of arrows on the top and bottom, both running from left to right, and you have the pattern for a torus. The sphere can also be created this way (of course two cones joined at their bases form a sphere, topologically speaking).

The two remaining possible configurations represent the *real projective plane* and the *Klein bottle*. If you try to build these shapes out of paper, you'll find that it can't be done in 3-dimensional space without the paper cutting through itself. In a true Klein bottle this does not happen. Nevertheless, making this allowance, some very beautiful models of the Klein bottle and real projective plane can be built. The real projective plane is an important example of a projective space.

The classification of closed surfaces

1 *Orientable surfaces* can be constructed by adding handles to a sphere, to get a torus, then a double torus, and so on.
2 *Non-orientable surfaces* can be formed by starting with a sphere and sewing in Möbius strips: first this produces a real projective plane, then a Klein bottle, and so on.

What if we started with a torus, and sewed a Möbius strip into that? In a piece of mathematical magic, it turns out that this gives us nothing new: this non-orientable surface is the same as a sphere with three Möbius strips sewed in.

One of the first significant results in topology, the classification of closed surfaces (formalized by Brahana in 1921 based on the earlier work of others), says that there are no other surfaces. Every closed surface belongs to one of these two families.

This result can be made more precise using the **Euler characteristic**.

Euler's polyhedral formula

Start with a sphere, and mark some spots (or *vertices*) on it, say V of them. Now join these together with some edges, E of those. (The fine-print is that every vertex must have at least two routes away from it, and edges cannot end, or meet, except at vertices.) This process will have divided the sphere into different regions (or faces). Suppose there are F of those. This simplest example shown here contains one vertex, and one edge, which passes around the sphere once, dividing it into two faces: $V = 1$, $E = 1$ and $F = 2$.

In 1750, Leonhard Euler noticed that in every example the numbers of vertices, edges and faces satisfy the formula: $V - E + F = 2$. (This is sometimes known just as *Euler's formula*: Euler was a man of many formulas.) He did not

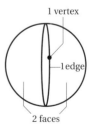
1 vertex
1 edge
2 faces

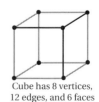
Cube has 8 vertices, 12 edges, and 6 faces

provide a complete proof that this would always be the case, but in 1794 Adrien-Marie Legendre did manage this.

This must be true for every surface which is topologically equivalent to a sphere too: so the vertices, edges and faces of every polyhedron must obey this formula.

Polyhedral formulas on surfaces

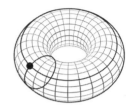

Euler's polyhedral formula does not hold for surfaces, such as a torus, which are not topologically spherical. But we can apply the same technique, taking care that the edges really divide the surface into faces which can be uncurled flat, rather than tubes. For a torus, this can be done with one vertex and two edges, which produces one face. So, in this case: $V - E + F = 0$. This new formula will hold true for any arrangement of vertices, edges and faces on a torus. Trying the same thing for the double torus, we find $V - E + F = -2$.

Similar results hold for non-orientable surfaces too. On the real projective plane, we always get $V - E + F = 1$; on the Klein bottle, $V - E + F = 0$.

Euler characteristic

The polyhedral formulas on surfaces say that if you have a surface, however you divide it up with vertices and edges, the number $V - E + F$ will always come out the same. This fixed number is called the *Euler characteristic* of the surface, symbolized by the Greek letter χ. So if S is a sphere, T is a torus, and D is a double torus, then $χ(S) = 2$, $χ(T) = 0$ and $χ(D) = -2$.

For orientable surfaces, if X is a sphere with g handles added (that is, X has genus g), then $χ(X) = 2 - 2g$. For non-orientable surfaces, if Y is a sphere with n Möbius strips sewn in, then $χ(Y) = 2 - n$.

The Euler characteristic alone cannot identify a closed surface (the torus and Klein bottle both have Euler characteristic zero). But the classification of closed surfaces implies that every surface is completely determined by two pieces of information: whether or not it is orientable, and its Euler characteristic.

Alexander's horned sphere

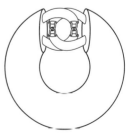

Topologically, many apparently different shapes turn out to be equivalent: a banana is equivalent to a ball, a teacup to a torus, and the letter 'M' to the number '2'. In 1924, the topologist James Alexander pushed this idea to its limit, with the discovery of his *horned sphere*. Topologically, this astonishing object is a sphere, but a truly pathological example of the species. It is constructed from a torus, with a slice cut out, and an arrangement of infinitely many interlocking horns on either side of the slice. The horns keep dividing and interlocking, down to their tips, which form **Cantor dust**.

This came as a shock to topologists of the early 20th century who believed that any sphere should crisply divide 3-dimensional space into two simple parts: an inside and an outside. But the outside of the horned sphere is fiendishly complex. You can draw infinitely many different loops, which cannot slide from one to the other, because they are irretrievably entwined in the horns.

Manifolds

A surface is an object you can divide into patches that each look like a piece of plane (otherwise known as 2-dimensional Euclidean space). The notion of a *manifold* lifts this idea to higher dimensions. An *n-manifold* is an object which can be divided up into patches that look like *n-dimensional* Euclidean space. (How close this resemblance has to be is the difference between plain topology, and differential topology.)

So, a 1-manifold is a curve and a 2-manifold is a surface. There is one obvious example of a 3-manifold: 3-dimensional space itself. But there are others too. The 3-*sphere*, for example is the 3-dimensional analogue of the ordinary 2-sphere, and is the subject of the famous **Poincaré conjecture**, solved in 2003. Closed 3-manifolds are fully classified by the geometrization theorem.

The geometrization theorem for 3-manifolds

Just as the classification of closed surfaces describes every possible closed 2-manifold, so the more recent *geometrization theorem* fully classifies closed 3-manifolds (*closed* meaning that that the manifold has finite volume, and no edges).

In 1982, the Fields medallist William Thurston listed eight special types of 3-manifold that, he thought should be the elementary forms from which all others are built. He described how to chop any closed 3-manifold into pieces, each of which, he believed, should be of his eight types. Thurston's eight manifolds correspond to different notions of *distance*. The commonest are the Euclidean and hyperbolic ones, another is spherical geometry. The remaining five come from certain **Lie groups**.

Thurston showed that many 3-manifolds satisfy his *geometrization conjecture*, but he could not prove that all do. In 2003, Grigori Perelman did prove this, using highly sophisticated methods from **dynamical systems**, and building on the work of Richard Hamilton. The celebrated Poincaré conjecture followed as a consequence.

Simple connectedness

If you draw a loop on an ordinary spherical surface, that loop can be gradually contracted until it is just a single point. This is the definition of being *simply connected*. This property distinguishes a sphere from a torus (donut) where a loop encircling the hole can never be shrunk away. Other surfaces such as cubes are simply connected too, but these are equivalent to spheres. So, topologically, the sphere is the only simply connected 2-dimensional surface. The Poincaré conjecture tackled the question of whether the same thing is true for 3-manifolds instead of surfaces.

Poincaré conjecture (Perelman's theorem)

First posed in 1904 by Jules Henri Poincaré, this conjecture (now elevated to the status of a theorem) occupies a prominent place in modern topology. It had been known for many years that the sphere is the only simply connected surface. Poincaré's question was whether the same was true when we step up one dimension. It was known that the 3-*sphere* (the 3-dimensional analogue of the usual sphere) is simply connected. The missing ingredient was to rule out other undiscovered 3-manifolds which might also be simply connected.

Poincaré's conjecture states that the 3-sphere is indeed the only simply connected 3-manifold. It attracted widespread attention within mathematics, defying numerous attempts to prove it throughout the 20th century, and also in physics where it can be interpreted as limiting the possible shape of the universe. In 2000, the conjecture was listed as one of the Clay Institute's $1,000,000 problems. It was finally proved in 2003 by Grigori Perelman as a consequence of the geometrization theorem. Perelman declined both the prize money and the Fields Medal.

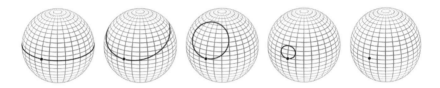

Homotopy

In dimensions other than 3, the Poincaré question needs to be slightly rephrased; simple connectedness is not enough. The appropriate analogue is *homotopy*, which detects whether a manifold contains any holes. Instead of drawing loops, homotopy concerns what happens when you insert a spherical shell into the manifold. The important question is whether this may be contracted to a point, or whether it gets trapped around a hole, just as a loop on a torus is trapped. Spheres of different dimensions can be used to test for holes of different types.

A manifold which contains no holes according to this method is called a *homotopy sphere*. According to the generalized Poincaré conjecture, the ordinary sphere is the only homotopy sphere, in every dimension.

The generalized Poincaré conjecture

We might naïvely expect that the more dimensions we look at the more impenetrable mathematics becomes. But this is not true; in many respects 3- and 4-dimensional space is more awkward to analyse than that of higher dimensions.

The *generalized Poincaré conjecture* says that, in every dimension, the ordinary sphere is the only homotopy sphere. The classification of closed surfaces had already given a positive answer in the 2-dimensional case. In 1961 Steven Smale proved that the generalized Poincaré conjecture is true in all dimensions from 5 upwards, under an additional hypothesis. This triumph won him a Fields medal in 1966. In the same year, Max Newman was able to show that the extra hypothesis was not needed, thereby completing the proof. This just left dimensions 3 and 4. The latter was resolved in 1982 by Michael Freedman (who was also awarded a Fields medal in 1986). So Henri Poincaré's original 3-dimensional conjecture was the last to fall, as it did to Grigori Perelman in 2003.

Differential topology

The topological definition of a manifold allows some fairly wild objects to qualify. In *differential topology*, these requirements are tightened. Only *smooth* manifolds are allowed, so pathologies such as the Koch snowflake and Alexander's horned sphere are eliminated. Similarly, the topological idea of two manifolds being equivalent is quite coarse, allowing one to be pulled and twisted into the shape of the other in pretty violent ways. A finer notion is that of *differential equivalence*: two smooth manifolds are considered the same if one can morph into the other *smoothly* (essentially in a way that can be **differentiated**).

But this raises a subtle possibility: two smooth manifolds could turn out to be topologically identical, but differentially different. Another way to say this is that one underlying topological manifold could support two incompatible smooth structures. This phenomenon is difficult to imagine, not least because it does not actually happen in dimensions 1, 2 or 3. (So the classification of closed surfaces and the geometrization theorem for closed 3-manifolds both remain valid at the level of smooth manifolds.) But in dimension 4 it does occur, in spectacular style.

Aliens from the fourth dimension

Every science-fiction fan knows that the fourth dimension is a crazy place. In the 1980s differential topologists discovered that the truth is even stranger than fiction. In dimensions 1, 2 and 3, the distinction between a *manifold* and a *smooth manifold* is not especially important: every topological manifold can be smoothed, and smooth manifolds which are topologically equivalent are also differentially equivalent.

On entering the fourth dimension, this cosy set-up crashes, badly. In 1983, using ideas from **Yang-Mills** theory, Simon Donaldson discovered a large collection of 4-manifolds which are *unsmoothable*: they do not admit any differential structure at all. Worse, the simplest 4-manifold, 4-dimensional space itself ($\mathbf{R}^4$), came under attack. Michael Freedman found a manifold which is topologically identical to $\mathbf{R}^4$, but differentially different from it. In 1987, Clifford Taubes showed that the situation is even more extreme: there are uncountably infinitely many such manifolds, all differentially inequivalent. These are the *exotic* $\mathbf{R}^4$s.

Exotic spheres

The exotic $\mathbf{R}^4$s are truly an anomaly: in every other dimension (n), there is only one smooth version of Euclidean space ($\mathbf{R}^n$).

In higher dimensions than 4, as in the fourth dimension, it does remain true that smooth manifolds can be indistinguishable topologically but differentially different. (However, unlike the 4-dimensional jungle, there can only ever be finitely many incompatible manifolds in higher dimensions.)

This even happens with spheres: in 1956, by experimenting with the **quaternions** John Milnor discovered a bizarre new 7-dimensional manifold. Later he remembered, 'At first, I thought I'd found a counterexample to the generalized Poincaré conjecture in dimension seven'. On closer inspection, this was not true. His manifold could morph into a sphere, but it could not do so *smoothly*. Topologically it was a sphere but, *differentially*, it was not. This was the first *exotic sphere*.

The smooth Poincaré conjecture in four dimensions

In dimensions 1, 2 and 3 there are no exotic spheres: smooth manifolds which are spherical from a topological perspective, but not from a differential one. In 1963, John Milnor and Michel Kervaire developed *surgery theory*: a powerful way to manipulate high-dimensional manifolds by cutting and gluing. This technological leap allowed them to determine exactly how many exotic spheres there are in every dimension of at least 5. The answer is that there are none in dimensions 5 and 6. But the one that Milnor had discovered in dimension 7 was one of a family of 27. In higher dimensions the answers range from none to arbitrarily large families.

But, as with the case of the exotic $\mathbf{R}^4$s, 4-dimensional space is uniquely awkward. At time of writing, it is not known whether there are any exotic spheres in dimension 4. It is even conceivable that there could be infinitely many. The assertion that there are no 4-dimensional exotic spheres (that is, that every topological sphere is also differentially a sphere), is known as the *smooth Poincaré conjecture in four dimensions*, and is considered a very difficult problem.

KNOT THEORY

Mathematical knots

The *vortex theory of the atom* was an idea in 19th-century physics. It held that atoms were knots in some all-pervading aether. Although the theory was short-lived, it prompted Lord Kelvin and Peter Tait to begin the mathematical investigation of knots, which remains an active area of research today.

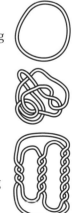

For a mathematician, a *knot* is a knotted piece of string, in which, importantly, the ends have been fused to produce a knotted loop. When more than two pieces of string are involved, it is called a *link*. According to the vortex theory, two knots represented the same chemical element if they are essentially the same. Although this idea is long discredited, the question it poses is a completely natural one. Being 'essentially the same' is a *topological* concept: two knots are equivalent if one can be pulled and stretched into the shape of the other (without cutting or gluing, of course).

The principal aim of knot theory is to find a method to determine whether two knots are equivalent. This is a deceptively deep problem. The **Perko pair** illustrates how difficult the problem is, even for comparatively simple knots. Even the simplest knot of all, the *unknot*, a plain unknotted loop, can be cleverly disguised. Spotting this is called the *unknotting problem*. The **Haken algorithm** provides a theoretical solution to the knot problem, but the search for ever more powerful **knot invariants** goes on.

Knot tables

In the 19th century, Peter Tait began listing all possible knots, according to their number of crossings. By 1877, Tait had listed knots with up to seven crossings. He was joined in his project by the Reverend Thomas Kirkman, a vicar from England, and Charles Little in Nebraska, USA. Communicating by mail, they largely managed to classify knots with eight, nine and ten crossings, and made some inroads into those with 11.

Efforts continued throughout the 20th century but, as the lists grew longer and the knots more complex, identifying those which are really the same became a formidable challenge. In 1998, Hoste, Thistlethwaite and Weeks published a paper entitled 'The first 1,701,936 knots', which was a full classification up to 16 crossings. Although no complete list is yet known beyond this, some important subfamilies have been classified up to 24 crossings, taking current tables to over 500,000,000,000 distinct knots and links.

The Perko pair

One of the early knot tables was compiled by Charles Little in 1885. It featured a list of 166 knots with 10 crossings, including the pair 10_{161} and 10_{162}. Subsequent generations of knot theorists built on this foundation, and for almost 100 years 10_{161} and 10_{162} sat side by side in every knot catalogue and textbook. It was not until 1974 that the New York lawyer and amateur mathematician Kenneth Perko spotted the error: the two are actually different formations of the same knot.

10_{161}

10_{162}

Chiral knots

After the unknot, the simplest knot is the *trefoil*, with its three crossings. Actually, this comes in two variants: left-handed and right-handed. The *figure-of-eight knot* also comes in two mirror-image forms. It's by no means obvious, but these two turn out to be equivalent. With a little pulling around, the left-handed figure of eight can be turned into the right-handed one. Might the same true for the trefoil knot? The answer is no. The trefoil knot is *chiral*, meaning that the two forms are distinct, while the figure-of-eight knot is *achiral*. For more complicated knots, chirality becomes very difficult to detect.

Knot theory has many applications in wider science, and chirality often has physical significance. In chemistry some compounds are chiral, meaning that their molecules come in both left-handed and right-handed varieties, which can have different chemical properties.

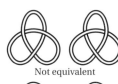

Not equivalent

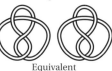

Equivalent

Knot invariants

In 1923, James Alexander found a way of assigning an algebraic expression to each knot. The crucial property is that if two knots are equivalent (no matter how different they appear) they will always produce the same polynomial. The *Alexander polynomial* was the first knot invariant. The *trefoil knot* has polynomial $t^{-1} - 1 + t$, and the *figure-of-eight knot* has $-t^{-1} + 3 - t$. This shows conclusively that the two really are different; no amount of manoeuvring can ever turn one into the other. However, this technique is not perfect. The $(-3, 5, 7)$-pretzel knot has polynomial 1, the same as the unknot, although the two are not equivalent.

The Jones polynomial

For over 50 years, the Alexander polynomial was the best algebraic tool for telling knots apart. However, in 1984, Vaughan Jones noticed an unexpected connection between his own work in analysis and knot theory. His insight blossomed into a brand new knot invariant. Although still not perfect, the *Jones polynomial* holds several advantages over Alexander's. Notably it can almost always identify chirality. The right-handed trefoil knot, for example, has Jones polynomial $s + s^3 - s^4$, while the left-handed one has $s^{-1} + s^{-3} - s^{-4}$.

Jones' discovery quickly found applications in broader science, notably among the knotted DNA molecules of biochemistry. Since his work, the Jones polynomial has been built upon, in the search for yet more powerful invariants. In 1993, Maxim Kontesevich formulated a new mathematical entity, known as the *Kontesevich integral*. It is a seriously complicated object (even the Kontesevich integral for the unknot is difficult to write down). A major open problem in knot theory is *Vassiliev's conjecture*, which implies that the Kontesevich integral can indeed distinguish between any two knots.

The Haken algorithm

In 1970, Wolfgang Haken tackled the question of telling when two knots are equivalent. His tactic was to turn the whole problem inside out. Instead of comparing two knots floating in space, he looked at the knots' *complements*: the 3-dimensional shapes that are left when you remove the knots from the surrounding matter, leaving knot-shaped holes (as if he'd set loosely knotted strings in blocks of glass, and then removed the strings). By telling whether these two objects are topologically the same, the same would go for the knots. Haken developed a method to dissect the two complements in stages, before deciding whether or not they are the same. It was a brilliant idea, but Haken's algorithm still had holes in it when he moved onto other concerns (most notably the **four colour theorem**). However, Sergei Matveev picked it up, and was able to fill in the final gap, in 2003.

Although the Haken algorithm is a spectacular achievement, it is probably too cumbersome ever to be fully implemented on a computer in the real world. So other, more practical, algorithms are used in current knot tabulations, and in approaches to the unknotting problem. Another disadvantage is that the algorithm doesn't leave any fingerprints. Theoretically, it can provide a yes/no answer to the question of whether two specific knots are the same, but it can't identify or describe individual knots. For this, knot invariants are needed.

The unknotting problem

A general aim of knot theory is to tell whether two knots are equivalent. A simpler question is to tell whether a given knot is equivalent to the unknot (the plain unknotted loop). More manageable algorithms than Haken's have been found, specifically for recognizing configurations of the unkot. Even with these, it is unknown whether they can be made to run fast enough ever to be of practical use in the real world, that is, whether the question can be answered in polynomial time (see **Cobham's thesis**). It is known that the unknotting problem is in the complexity class NP, so a positive answer to the **P =NP question** would settle this.

NON-EUCLIDEAN GEOMETRY

Hyperbolic geometry

In the 19th century Carl Friedrich Gauss, Nikolai Lobachevsky and János Bolyai independently discovered a new and unfamiliar possible system for geometry, called *hyperbolic geometry*. The basic Euclidean ingredients survive: distances, angles and areas. But they combine in new and unexpected ways. Crucially, Euclid's **parallel postulate** fails, which was the historical impetus for this discovery. The angles in a triangle now add up to less than π (180°). More weirdly, just knowing the angles in a triangle (say A, B and C) is enough to tell you its area: $\pi -(A + B + C)$. (This is inconceivable in Euclidean space, where there are *similar* triangles of any area you like.)

How can we imagine such an alien space? Various models of the hyperbolic plane have been constructed, notably the **Poincaré disc**. The *Minkowski model* is hyperbolic geometry on one half of a two-sheeted hyperboloid. This plays a central role in special relativity in physics. Most topics of interest in Euclidean geometry have hyperbolic counterparts. There are hyperbolic surfaces and manifolds, hyperbolic trigonometry (involving cosh and sinh), and a well-developed theory of tilings of hyperbolic spaces, for example.

Poincaré disc

Discovered not by Henri Poincaré, but by Eugenio Beltrami in 1868, the *Poincaré disc* is a model of hyperbolic geometry, a construction of a hyperbolic plane designed to be accessible to us Euclidean folk. It takes place within a circle, which represents the infinite boundary of the plane. To the observer inside, the edge of the circle seems infinitely far away. From the outside, distances seem to become ever more compressed the closer you get to the edge. The 'straight lines' in the model are arcs of circles which meet the boundary at right angles, as well as diameters straight through the circle. The disc model of hyperbolic geometry was famously explored by the artist M.C. Escher in several of his illustrations.

Elliptic geometry

Once hyperbolic geometry was discovered the old Euclidean hegemony was blown apart, and a question arose: are there any other possible geometries? In fact, we have been living on one for the last 4 billion years. Hyperbolic geometry breaks the parallel postulate by having more than one line through any point that is parallel to a given line. *Elliptic geometry* says that there are no parallel lines at all; every pair of lines meets. (Euclid's other postulates have to be slightly amended to allow this, but variants of them continue to hold.)

There are subtly different forms of elliptic geometry, but the commonest is the *spherical* variety. Here, the space is the surface of a sphere. The role of straight lines is played by *great circles*: the largest circles the sphere can hold (formed by a plane cutting through the centre of the sphere, these are the sphere's *geodesics*). In elliptic geometry, the angles in a triangle add up to more than 180° (but less than 540°). One such is the triple right-angled triangle.

Biangles

In elliptic geometry, there are no parallel lines: every pair of lines must meet. In this context, the triangle's crown as most elementary polygon is appropriated by the two-sided *biangle*. In spherical geometry, a biangle's two angles are necessarily equal, and its area is determined simply by adding them together. The biangle illustrates that, in elliptic geometry, two points no longer determine a unique line, but there are infinitely many possible lines joining them together. Similarly, on planet Earth, there is no shortest route from the north pole to the south pole.

ALGEBRAIC TOPOLOGY

Hairy ball theorem

You cannot comb a ball.

In the late 19th century, Henri Poincaré discovered what became known affectionately as the *hairy ball theorem*. Suppose you have a sphere with a hair growing out of every point. We want to comb the hairs flat, smoothly around the ball (in other words we have a **vector field** on the ball). Poincaré's theorem says that, however you try to manage it, you will always leave at least one crown. This is a topological theorem, so any surface which is topologically equivalent to a sphere (such as a dog), can also never be combed perfectly.

143

Hairy tori and Klein bottles

The hairy ball theorem raises a question: which shapes can be combed? Dropping down a dimension, it is straightforward to comb a hairy circle without leaving a crown. Athough a sphere cannot be combed, there are two 2-dimensional surfaces which can. The only orientable example is a hairy torus. Double and triple tori are uncombable. A hairy Klein bottle is the only non-orientable combable surface, so the real projective plane, for example, cannot be combed.

The story changes when we move up a dimension. We can comb a hairy 3-sphere (that is the 3-dimensional analogue of the familiar ball). Similarly, we can comb a 5-sphere and 7-sphere and generally an n-sphere where n is odd. But an n-sphere is never combable when n is even.

Geometric fixed points

Suppose you have a piece of paper lying flat on the bottom of a box. If you scrunch, fold, or roll it up, and then throw it back in the box, *Brouwer's fixed point theorem* guarantees that there must be at least one point on the paper which is directly above its original position.

Another example, which inspired Brouwer to make his observation, is if you stir a cup of coffee, at any moment there is a molecule of drink which is exactly in its original place. These are *geometric fixed points*, and Brouwer's theorem guarantees that in many circumstances there will be one.

Brouwer's fixed point theorem

The idea of *Brouwer's fixed point theorem* is that if you take a geometric object and deform it somehow, there must be at least one point whose position is unchanged. There are some important caveats however.

Firstly, in the example of the paper in the box, you cannot tear it. The theorem is easily violated by ripping the paper in two, and swapping the two halves. In mathematical terms, the function must be *continuous*.

The point is directly above where it originally was

Secondly, you must put the paper back entirely in the box (and not spill the coffee). Another example, if you are carrying around a map of a city, there will always be one point on the map which occupies exactly the position that it represents. But if you take the map out of town, this is no longer true. Technically this means that we must be talking about a function from a space X *into itself*.

The final point is even more subtle. If we consider the whole infinite plane, sliding it one unit to the right will move everything, leaving no fixed points. The theorem holds only if X is (topologically) a disc or a ball. That would be a line segment in one dimension, a disc in two dimensions, and a ball in three or more dimensions. In each case, the boundary of X must be included. Shave off the edge, and the theorem no longer holds.

Putting these together, if we have a continuous function $f: X \to X$, from a ball X into itself, then there must be a fixed point, that is, some point x such that $f(x) = x$.

Algebraic topology

Topologists of the early 20th century arrived at a new streamlined language for geometry, based on *simplices*.

- A 0-simplex is a single point.
- A 1-simplex is a line segment bounded by two points.
- A 2-simplex is a triangle bounded by three line segments, and three points.
- A 3-simplex is a tetrahedron bounded by four triangles, six line segments, and four points.
- A 4-simplex is a pentachoron (see **polychora**), bounded by five tetrahedra, ten triangles, ten line segments and five points, and so on.

(These sequences of numbers can be read off **Pascal's triangle**.)

A *complex* is a shape obtained by gluing any number of simplices together, along their edges. What Solomon Lefschetz and others realized is that the data needed here is so minimal, that certain underlying algebraic rules emerge. By adding and subtracting simplices, they produced **groups**. These *homology groups* encode a great deal of data about a shape.

0-simplex
1-simplex
2-simplex
complex
3-simplex

Triangulation

A sphere is not built from basic chunks like simplices. By allowing simplices to be bent and stretched once they are constructed, a much broader range of shapes can be constructed. When a shape can be broken down into these stretched simplices, we say that it has been *triangulated*. A sphere can be triangulated as four triangles. In fact, every manifold in two and three dimensions can be triangulated. For this reason homology groups are powerful tools in practical geometric problems, such as medical imaging.

However, 4-dimensional space is a strange place. Some 4-manifolds cannot be triangulated at all. In higher dimensions, the complete answer is not known.

Euler characteristic is one example of a valuable piece of data which emerges from triangulation.

ALGEBRAIC GEOMETRY

Varieties

The standard way to describe a curve or surface is via an equation, typically involving a **polynomial**. This immediately gives us two perspectives: the geometric curve and the algebraic polynomial. The equation of a circle, for example, is $x^2 + y^2 = 1$. So if we write $P(x, y)$ to stand for the polynomial $x^2 + y^2 - 1$, then the circle is the set of points where P vanishes. That is, it is the collection of all pairs (x, y) such that $P(x, y) = 0$.

This is the basic idea of a *variety*, the set of points where a polynomial (or collection of polynomials) vanish. Geometric operations such as gluing varieties together, or looking at where they overlap, correspond to particular algebraic operations on the polynomials. This is the starting point of the subject of *algebraic geometry*, whose principal aim is to understand varieties as general phenomena, instead of focusing on individual examples (such as conic sections).

Algebraic geometry

Every geometric variety gives rise to a corresponding algebraic structure, a **polynomial ring**. To understand the geometry fully requires knowledge of the algebra, and vice versa. This double-edged attack bore rich fruits over the 20th century.

The primary setting for contemporary geometry is the complex numbers. Here the **fundamental theorem of algebra** guarantees that every variety has its full quota of points. However, modern geometry also applies these techniques in a wide range of other settings.

The ancient subject of **Diophantine equations** searches for integer solutions of polynomials. Algebraic geometry injected new methods to this search, spawning the subject of **Diophantine geometry**. Wiles' proof of Fermat's last theorem is a notable product of this wonderful convergence of geometry and number theory.

Polynomials make sense anywhere where we can add, subtract, and multiply. This opens up geometric questions in unexpected places, such as **finite fields**. The **Weil conjectures** unlocked the secrets of these finite geometries. How algebraic geometry interfaces with algebraic topology is the subject of the **Hodge conjecture**, one of the deepest questions in the subject.

Algebraic geometry is a hugely powerful theory but functions at a fearsomely high level of abstraction, particularly as developed in **Grothendieck's *Éléments de Géométry Algébrique.***

Perspective

How can you represent 3-dimensional space on a 2-dimensional plane? This problem dates back to the cave-painters of the paleolithic era. Of course, there are many artistic obstacles to a fully naturalistic representation. Of particular mathematical interest is the way that things which are further away seem smaller than those which are nearby. Naïve attempts to capture this phenomenon generally look strange and wrong, as in the twisted perspective of ancient Egyptian art. Artists of the Italian renaissance such as Giotto and Brunelleschi made huge progress, discovering how to use vanishing points to draw foreshortened objects.

Vanishing points and vanishing lines

The early use of perspective by artists such as Masaccio involved a single *vanishing point*, typically in the centre of the canvas. If the ground in the painting is imagined as an infinite chessboard, the parallel lines running away from the viewer all converge at this single point.

Later artists such as Vittore Carpaccio experimented by placing this vanishing point in different positions, sometimes even outside the canvas. This modification introduces an extra problem, however. With the central vanishing point, the chessboard lines running away from the viewer converge,

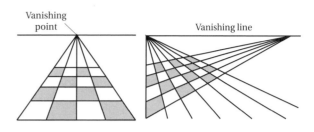

Vanishing point

Vanishing line

and the lines at right angles to them appear horizontal. But relocating the vanishing point means that these perpendicular lines no longer look horizontal. In fact they all converge to a second vanishing point. Straight lines drawn across the chessboard with intermediate angles converge at vanishing points between the two. The line joining the two is the *vanishing line*.

Desargues' theorem

Take two triangles abc and ABC. A theorem named in honour of the father of projective geometry, Gérard Desargues, relates two different notions.

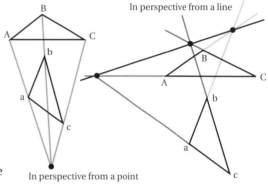

In perspective from a line

In perspective from a point

1 The triangles are *in perspective from a point*, if the lines Aa, Bb, and Cc all converge to a single point.

2 The triangles are *in perspective from a line*, if the three points where AB and ab meet, BC and bc meet, and AC and ac meet, all lie on a straight line.

Desargues' theorem says that these two are equivalent: two triangles are in perspective from a point if and only if they are in perspective from a line.

How to draw a triangle

Desargues' theorem is an important result for the visual artist, as it removes the need to worry about vanishing points. The key point is that the two triangles do not have to be in the same plane.

A relevant case is when ABC is on the floor, and abc is its image on the artist's canvas. We imagine the canvas standing on the floor at 90°. The artist's aim is that the two triangles should be in perspective from a point. But where should she draw abc? Desargues' theorem says that the triangles should also be in perspective from a line. This line must be where the floor and canvas meet.

Floor

Line where floor meets canvas

Canvas

The artist can draw the first point 'a', wherever she likes. Then she has to make sure that each edge of abc extends to hit the floor in the same place that the corresponding side of ABC hits the line of the canvas.

Projective geometry

Any two points on the Euclidean plane define a straight line. Going the other way, it is *almost* true that any two lines meet at a single point. This *duality* between points and lines is an elegant and powerful principle. Unfortunately, it does not quite hold true: pairs of parallel lines never meet.

In art, however, parallel lines can meet, or at least appear to. Parallel lines running away from the viewer eventually converge. What artists call a *vanishing point* corresponds to the mathematician's idea of a 'point at infinity'. The idea is to enlarge the plane by adjoining extra points where parallel lines are deemed to meet.

Once this is done, geometry starts to look different, and in many ways simpler. For example, the three conic sections are reunited, as different views of an ellipse. Of course projective geometry is not Euclidean, since it is engineered for the parallel postulate to fail. But neither is it *non-Euclidean* in the same sense that hyperbolic geometry is. It is best thought of as a closing up (technically a *compactification*) of Euclidean geometry. Every small region is still perfectly Euclidean; it is only when the space as a whole is considered that its overall non-Euclidean nature is revealed.

Homogeneous coordinates

Mathematicians do not like playing fast and loose with the concept of 'infinity'; too much nonsense has been written over the centuries by people playing around freely with this concept. So if projective geometry requires extra 'points at infinity', then we need a system of coordinates for which this is meaningful. August Möbius' homogeneous coordinates provide an elegant solution.

While points on the real plane are given by a pair of Cartesian coordinates such as $(2, 3)$ or (x, y), points on the *projective plane* have three coordinates: $[2, 3, 1]$ or $[x, y, z]$. The difference is that now some coordinates represent the same point, just as equivalent fractions represent the same number. For example $[2, 3, 1]$, $[4, 6, 2]$ and $[-10, -15, -5]$ all represent the same projective point. In general, multiplying each coordinate by a fixed number does not change the point.

All the points of the ordinary plane can be included, by associating (x, y) with $[x, y, 1]$. But there are extra points 'at infinity' too, of the form $[x, y, 0]$. So we can even give the equation of the artists' vanishing line: $z = 0$.

Finite geometry

The approach of algebraic geometry is to study shapes via their defining polynomials. This approach began in the real numbers, with the study of conic sections, for example.

Once the system of complex numbers came of age, this assumed centre stage. But the philosophy of algebraic geometry makes sense in any context where we can add, subtract, multiply and divide. In particular, we can also consider the geometry of **finite fields**.

The basic examples of finite fields are **modular arithmetic**, to a prime base. We can begin with a polynomial such as $x^2 + 2 = 0$ and, instead of solving it in the real numbers, try to solve it in the integers, mod 3.

Here:

$$1^2 + 2 = 3 \equiv 0 \quad (\text{Mod } 3)$$

Also:

$$2^2 + 2 = 6 \equiv 0 \quad (\text{Mod } 3)$$

So, modulo 3, the polynomial $x^2 + 2$ has two solutions, namely 1 and 2. The same polynomial interpreted modulo 5 has no solutions at all.

This is a basic example of starting with a polynomial and counting the points satisfying it in different finite fields. But what is the pattern? As we work over larger and larger finite fields, can the number of solutions flicker around at random or is there some underlying rule? This question was answered in one of the landmarks of 20th-century geometry, Pierre Deligne's proof of the **Weil conjectures**.

Projective geometry also makes sense in this context, yielding objects such as the Fano plane.

The Fano plane

A *projective plane* is a collection of points and lines organized so that:

- any two points lie on a unique line
- any two lines meet at a unique point
- there are four points, no three of which lie on one line.

A standard way of manufacturing projective planes is through homogeneous coordinates; this produces the real and complex projective planes. We can play the same trick starting with a finite field. The smallest field of all is the two element field F_2 (there not being any true field with one element). In this case the result is the *Fano plane*, consisting of just seven points, connected by seven lines, named after the Italian geometer Gino Fano.

Weil's zeta function

Finite fields come in towers, with a prime number at their base. The first tower is based on 2 and consists of the fields F_2, F_4, F_8, F_{16}, … (of sizes 2, 4, 8, 16, and so on). The tower with 3 at its base consists of F_3, F_9, F_{27}, F_{81}, … For any prime number p, there is a tower F_p, F_{p^2}, F_{p^3} … where the first m storeys make up F_{p^m}.

A *variety* is a geometric object defined by polynomials. For instance, a circle is defined by $x^2 + y^2 = 1$. Given such a polynomial or variety V, we can ask how many points there are satisfying it at each level, F_{p^m}. Call this number N_m. So V has N_1 points in F_p, N_2 points in F_{p^2}, N_3 in F_{p^3}, and so on. Certainly the variety cannot lose points as it ascends the tower, so $N_1 \leq N_2 \leq N_3 \leq \dots$ This sequence $N_1, N_2, N_3, \dots$ is the key to the geometry of finite fields, and is what André Weil set out to understand. His idea was to encode it into a single *L-function*, ζ, know as Weil's *zeta function*.

The Weil conjectures [Deligne's theorem]

Weil's zeta function is an abstruse object. The *Weil conjectures*, however, assert that it is far simpler to understand than it first appears.

His first conjecture said that ζ is determined by finitely many pieces of data. This is a crucial fact, as it means the sequence $N_1, N_2, N_3, \ldots$ does not jump around at random, but is governed by a fixed, predictable pattern.

The two remaining conjectures pin down this pattern precisely. Significantly, the second identifies those places where ζ is 0. In the simplest case, it says that all the zeroes of ζ lie on the critical line $\mathrm{Re}(z) = \frac{1}{2}$. This is highly reminiscent of the **Riemann hypothesis**.

The Weil conjectures were a driving force behind the huge expansion of algebraic geometry in the 20th century. The first was proved by Alexander Grothendieck in 1964, and the others by Pierre Deligne in 1974.

Hodge theory

By the mid 20th century, geometry had come a long way from anything Euclid would have recognized. At the same time, geometers' concerns were still traditional ones: what sorts of shapes may exist? Answers to this question had been hugely enriched by algebra, in two different directions: firstly the polynomial equations of algebraic geometry, and secondly the groups of **algebraic topology**. The first gives us varieties as the fundamental notion of shape. Extending these slightly are *algebraic cycles*, built by formally adding varieties together, and multiplying them by rational numbers.

On the other side is the topological set-up of **simplices**. A critical difference here is that these objects are constructed from the real, rather than the complex numbers. Another distinction is that these are flexible topological objects, not tied down by polynomial equations. Again, simplices are formally added together into *topological cycles.*

The setting for the meeting of these two powerful theories is projective geometry over the complex numbers. The question William Hodge addressed in his 1950 speech to the International Congress of Mathematicians is: when do these two different ideas produce the same result? When is a topological cycle equivalent to an algebraic one?

The Hodge conjecture

A partial answer for the central problem of Hodge theory is easy to give. When viewed over the real numbers, the complex numbers have dimension 2 (which is why the Argand diagram looks like a 2-dimensional plane). Similarly any complex variety must have even numbered dimension, from a real perspective. So the first criterion is that to be algebraic, the cycle must have even dimension.

However, this is not enough. Hodge was also a master of calculus. His work on **Laplace's equation** provided him with the language to describe some particularly stable topological cycles, now called *Hodge cycles*. He conjectured that in these he had found the right topological description of algebraic cycles. Certainly every algebraic cycle is a Hodge cycle. The $1,000,000 question (since the Clay Institute announced the prize in 2000) is whether the converse is also true.

No-one doubts the importance of Hodge's theory. Whether Hodge's conjecture provides the right answer, though, is open to question. In 1962 Atiyah and Hirzebruch showed that the conjecture is false when limited to cycles over the integers, instead of cycles over rational numbers. Andre Weil did not believe in Hodge's conjecture, and thought geometers would be better off searching for a counterexample. Until a proof or a counterexample is found, it remains, in the words of Alexander Grothendieck, 'the deepest conjecture in the analytic theory of algebraic varieties'.

Grothendieck's *Éléments de Géométrie Algébrique*

Cartesian coordinates first opened up geometry to algebraic methods, and with great success, as illustrated by the classification of conic sections and Newton's cubics. Modern algebra is naturally a very abstract subject. The introduction of varieties allowed this abstraction to cross over to geometry, allowing a much more general line of attack, rather than one focused on individual examples. During the 1960s, the subject underwent a further gigantic upheaval. The revolution was instigated by Alexander Grothendieck. Perhaps the greatest master of abstraction, his four-volume treatise *Éléments de Géométrie Algébrique* reformulated the entire subject of algebraic geometry at a much deeper level. Central to his work was the replacement of varieties by schemes.

Schemes

The motivation for Grothendieck's approach was a difference between the two languages of algebra and geometry. The fundamental geometric objects were varieties, defined by polynomials. These polynomials simultaneously give rise to an algebraic object: a **polynomial ring**. Grothendieck realized that only a very limited type of ring can arise in this fashion, however. The commonest ring of all is **Z**, the ring of integers. But there is no variety which has **Z** as its ring. His bold idea was that the geometric techniques which had been developed should work with any ring, even where there was no underlying variety. He called these new structures *schemes*.

Schemes are highly abstract; many do not have obvious geometric interpretations. Working with them poses a formidable technical challenge. The pay-off is that the **category** of schemes is far better behaved than that of varieties. This leads to a more coherent and powerful overall theory, as witnessed by the proof of the Weil conjectures.

DIOPHANTINE GEOMETRY

Diophantine geometry

Although a topic in number theory, the study of Diophantine equations had its roots in geometry, in the study of Pythagorean triples. In the 1940s, the geometer André Weil realized that the sophisticated techniques of algebraic geometry could have deep implications here. This was the start of the reunification of number theory and geometry that took place over the 20th century.

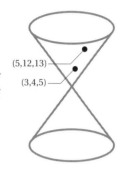

(5,12,13)

(3,4,5)

Considered geometrically, the equation $x^2 + y^2 = z^2$ defines a surface (specifically a double cone). In number theory, this is the condition for a Pythagorean triple. What Euclid was able to show, then, is that this cone contains infinitely many points whose coordinates are all whole numbers.

In general, polynomial equations (such as Fermat's $x^n + y^n = z^n$) define a variety. The number theorist's question is whether this includes any points (x, y, z) which have integer coordinates. Usually it is enough to search for points whose coordinates are rational numbers. Hence looking for rational points on varieties is a preoccupation for modern number theorists.

Rational points on varieties

A *variety* is a geometric object defined by polynomial equations. The first examples are curves: 1-dimensional varieties. In fact, curves are also the key to higher dimensions, since any variety can be covered by a family of curves.

Curves are subdivided by their complexity, or genus. Simple curves such as conic sections have genus 0, and must either have infinitely many rational points (as is the case with the circle $x^2 + y^2 = 2$) or none at all (as happens for $x^2 + y^2 = 3$). These days, we can tell which is the case without too much difficulty. On the other hand, more complicated curves with genus 2 or more can only ever have a finite number of rational points. This deep fact was conjectured by Louis Mordell in 1922, and finally proved by Gerd Faltings in 1983.

Faltings' theorem was a huge step forward. But it did not quite close the book on the matter of rational points on curves. Between the simple and complex, sit the enigmatic elliptic curves with genus 1. These may have either a finite or an infinite number of rational points....

How to tell the difference is the question addressed by one of the most important open problems in the subject, the **Birch and Swinnerton-Dyer conjecture**.

Elliptic curves

There is something special about the equation $y^2 = x^3 - x$. It describes a curve, but an unusual one: you can *add* its points according to the following rule. If a and b are points on the curve, draw a straight line joining the two. This line must intersect the curve at a third point c. We say that $a + b + c = 0$. So $a + b = -c$. (The zero of the group is represented by an extra 'point at infinity', which corresponds to vertical lines; see **projective geometry**.)

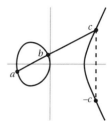

Only for *elliptic curves* (not to be confused with ellipses) does this rule work so smoothly that the curve becomes a **group**. Because the resulting groups are difficult to predict, they are a mainstay of modern **cryptography**.

In general, the equation of an elliptic curve is $y^2 = x^3 + Ax + B$, for some A, B and C. These are comparatively simple curves, which nevertheless continue to defy understanding. Elliptic curves play central roles in modern number theory, notably in Wiles' proof of Fermat's last theorem and **Langland's programme**. Their properties are the subject of the celebrated Birch and Swinnerton-Dyer conjecture.

Rational solutions of elliptic curves

Starting with an equation such as $y^2 = x^3 + 1$, a number theorist's first instinct is to ask whether there are any whole numbers which satisfy it (such as $3^2 = 2^3 + 1$). Recently it has been convenient to extend the search to cover rational numbers which obey the equation. A key question for such an equation is whether it has infinitely many rational solutions, or just a limited, finite number.

The most elementary equations where this question is not yet fully understood are the elliptic curves, such as $y^2 = x^3 + 1$. Solving this problem for elliptic curves is an important goal of contemporary number theory. During the 1960s Peter Birch and Henry Swinnerton-Dyer made a conjecture which would quantify the rational solutions of these important equations.

Birch and Swinnerton-Dyer conjecture

From any elliptic curve (call it E), there is a procedure to define a corresponding **L-function**, L. Birch and Swinnerton-Dyer claim that this function encodes the details of E's rational solutions. In particular, they believe the function can detect whether E has infinitely many rational points, or only finitely many. Their conjecture implies that if $L(1) = 0$ then the curve has infinitely many rational points, and if $L(1) \neq 0$ then it does not.

The best progress to date is due to Victor Kolyvagin in 1988. When combined with subsequent results of Wiles and others, Kolyvagin's theorem proves half the conjecture, namely that if $L(1) \neq 0$ then E only has finitely many rational points. The remainder now comes with a $1,000,000 price tag, courtesy of the Clay Foundation.

Modular forms
The slick, modern world of **complex analysis** seems a very long way from the hoary problems of Diophantine equations. However, just as number theory and geometry were drawn together over the 20th century, more recently complex analysis has also been pulled into the mix. A key notion is that of a *modular form*. A modular form is a

function which takes complex numbers from the upper half-plane as inputs and gives complex numbers as outputs. Modular forms are notable for their high level of symmetry.

The sine function is a *periodic* function, meaning that it repeats itself. If you take a step of 2π to the right, the function looks the same. Modular forms satisfy similar, but more intricate rules, where the symmetry is determined not by a single number such as 2π but by 2×2 matrices of complex numbers.

Modular forms have starred in two of the greatest stories in modern mathematics: the proof of Fermat's last theorem (via the modularity theorem), and the investigation of **the monster** (as thrown up by the **classification of finite simple groups**).

Modularity theorem

In the 1950s, Yukata Taniyama and Goro Shimura made a conjecture that proposed to bridge two very different areas of mathematics. It concerned two entirely different types of object. The first are the elliptic curves, a preoccupation of number theorists. The second come from the world of complex analysis, the modular forms. Taniyama and Shimura claimed that elliptic curves and modular forms are essentially the same things, though described in utterly different languages. They claimed that **L-functions** should provide a dictionary for translating between the languages of analysis and number theory.

The Taniyama–Shimura conjecture appeared insurmountably difficult. All the same, it became the focus of intense interest in the 1980s when Gerhard Frey and Kenneth Ribet showed that a proof of the conjecture would imply Fermat's last theorem. In 1995, Andrew Wiles proved a major chunk of the theorem with Richard Taylor. This was enough for Wiles to deduce Fermat's last theorem. In its entirety, the conjecture was proved in Breuil, Conrad, Diamond and Taylor in 2001. The *modularity theorem*, as it is now known, is one plank in an overarching design called Langlands' program.

Langlands' program

Mathematics is full of *conjectures*: assertions that some mathematicians believe should be true, but no-one has been unable to prove. Some, such as Poincaré's, are later elevated to theorems. Others are disproved and then jettisoned (although even false conjectures, such as Hilbert's program, can be powerful engines of progress). Not many mathematicians, however, have put their name to such a grand, unifying vision as that laid out by Robert Langlands. Writing to André Weil in 1967, he addressed a massive question: what happens when two mathematical worlds, algebra and analysis, meet?

The modularity theorem shows that certain elements from the two worlds are closely related. This was predicted by Langlands, but his broader program goes much further. To express it, he needed to go beyond modular forms to the *automorphic forms*, complex functions whose symmetries are described by larger matrices. Central to Langlands' project are L-functions, which convert algebraic data coming from **Galois theory** into analytic functions in the complex numbers. Langlands believes that, as this divide is crossed, key concepts on one side marry up with key concepts on the other side.

Langlands' number field conjectures

In addition to the proof of the modularity theorem, there has been major progress towards realizing Langlands' vision. The algebraic side comes in three parts: *local* fields, *function* fields and *number* fields, which are of particular importance in analysis, geometry and number theory, respectively. By 2000, Langlands' conjectures had been verified in two out of the three, with Laurent Lafforgue winning a Fields medal for his 300-page solution of the case of function fields.

Only one part remains unproved, but it is monumental. Langland's conjectures for number fields continue to stand defiantly open, and represent a massive challenge to today's number theorists.

ALGEBRA

AS COMMONLY UNDERSTOOD, the starting point of algebra is
the rather mystifying replacement of numbers with letters. One
purpose of this move is to allow the expression of general
rules, such as the fact that $a \times b = b \times a$, whatever the numbers
a and b. A more sophisticated example is the binomial theorem.

Using letters to stand for unknown numbers also opened
up the science of equation solving. A simple example is to
find a number x, such that $4 + x = 6$. The quest to solve more
complex *cubic* and *quartic equations* was the driving concern
for mathematicians of the Italian renaissance. Then came Abel

and Galois' paradigm-shattering work on the *quintic equation,* which lifted algebra to new heights. At this stage, the process of replacing familiar objects with more abstract analogues reoccurs. Here the familiar number systems are replaced with more general algebraic structures. Especially important are *groups.* At this level magnificent classification theorems are possible, notably those of finite simple groups and simple Lie groups.

Modern abstract algebra supplies much of the machinery for other areas of mathematics, from geometry and number theory to quantum field theory.

LETTERS FOR NUMBERS

Letters for numbers

For many people, the moment mathematics becomes uncomfortable is when letters start appearing where previously there were only numbers. What is the point of this? The first purpose is to use a letter to stand for an unknown number. So '$3 \times y = 12$' is an **equation**, which says that y is a number which when multiplied by 3 equals 12. In this example, the role of the y could equally well be played by a question mark: '$3 \times ? = 12$'. But this is less practical when we have more than one unknown quantity. For example: which two numbers add together to equal 5, and multiply together to equal 6? Using letters to represent these numbers produces a pair of **simultaneous equations:** $x + y = 5$ and $x \times y = 6$.

Usually traditional times-symbols $\times$ are left out when writing algebra, or replaced with dots. So our equation could be written '$3y = 12$' or '$3 \cdot y = 12$'. Notice that the number is conventionally placed before the letter, so '$3y = 12$' rather than '$y3 = 12$'.

Variables and substitution

As well as standing for specific but unknown numbers, letters can be used as *variables*, in place of *any* number. Here's an example: if I am jogging down the road, the formula $d = 4t + 5$ might express the relationship between my distance from my house (d, measured in metres) and the time since I started (t, measured in seconds). Here d and t no longer represent individual unknown numbers to be discovered. Instead they can vary; if we want to know my distance after 2 seconds, we *substitute* the value $t = 2$ into the formula, and discover that $d = 4 \times 2 + 5 = 13$ metres.

Asking how long it will take until I reach the end of the road 21 metres away means substituting in $d = 21$, to get $21 = 4t + 5$. Now we can solve this equation, to arrive at the answer $t = 4$ seconds.

Brackets

Brackets are used to separate strings of arithmetic into chunks. So $(3 + 2) \times 8$ means first add 3 and 2, and then multiply the result by 8. To work through more complicated calculations, you always start with the innermost brackets. So:

$$((5 \times (2 + 1)) + 1) \div 4 = ((5 \times 3) + 1) \div 4 = (15 + 1) \div 4 = 16 \div 4 = 4$$

To avoid having to write out endless brackets, there is a convention for the order of unbracketed operations: BIDMAS (BEDMAS).

BIDMAS

What is $2 + 3 \times 7$? It may seem a trivial question, but only one of these can be correct:

(i) $2 + 3 \times 7 = 5 \times 7 = 35$

(ii) $2 + 3 \times 7 = 2 + 21 = 23$

BIDMAS (or sometimes BEDMAS) is a rule to avoid this sort of confusion, by reminding us of the correct order of arithmetical operations:

Brackets **I**ndices (Exponents) **D**ivision **M**ultiplication **A**ddition **S**ubtraction

Since **M** comes before **A**, (i) above is wrong. Brackets trump all else, so if (i) is what was intended, it can be expressed as $(2 + 3) \times 7 = 5 \times 7 = 35$.

Similarly, $6 \div 3 - 1 = 2 - 1 = 1$, but $6 \div (3 - 1) = 6 \div 2 = 3$.

If everyone used a lot of brackets $((2 \times 5) - ((6^4) \div 3))$, there would be no need for BIDMAS. But to avoid wasting ink and making a mess, it is a useful convention.

Common factors and expanding brackets

(i) $3 \times (5 + 10) = 3 \times 15 = 45$

(ii) $3 \times 5 + 3 \times 10 = 15 + 30 = 45$

It is no coincidence that (i) and (ii) produce the same answer. If I have three $5 bills, and three $10 bills, the total is the same whether I consider this as three lots of $15, or calculate how much I have in $5 bills and add it to how much I have in $10 bills.

Technically, the law of *distributivity* assures us that equality always holds in these situations. This is more often encountered under the guise of *expanding brackets* and *taking out common factors*. Both involve a pair of brackets with addition or subtraction happening inside, being multiplied by something outside: $3 \times (5 + 10)$

These are opposite procedures. *Expanding a bracket* involves multiplying everything inside it by whatever is outside, and then adding up:

$$3 \times (5 + 10) = 3 \times 5 + 3 \times 10$$

Going back the other way, if a string of terms are added up, and they are all divisible by a particular number, we can take out this *common factor*:

$$20 + 28 = 4 \times 5 + 4 \times 7 = 4 \times (5 + 7)$$

This may all seem straightforward when only numbers are being used. But, once variables are involved, these techniques become essential for simplifying formulas. For example, $ax + 4x = x(a + 4)$ shows a common factor of x being taken out.

Here, a pair of brackets is expanded: $2a(x - y^2) = 2ax - 2ay^2$.

Squaring brackets

If we have addition inside a bracket, being multiplied by something outside, the technique of expanding brackets tells us how to cope. For example: $(1 + 2) \times 2 = 1 \times 2 + 2 \times 2$. Does a similar thing happen when the bracket is raised to a power? It is a common mistake to believe so.

Unfortunately, however, $(1 + 2)^2 \neq 1^2 + 2^2$, the first being equal to $3^2 = 9$, and the second to $1 + 4 = 5$. So squaring brackets is more complicated. But, if we interpret the squaring as multiplication, the rules of multiplying brackets do still apply:

$$(1 + 2)^2 = (1 + 2) \times (1 + 2) = 1(1 + 2) + 2(1 + 2) = 1 + 2 + 2 + 4.$$

Using a variable is more illuminating:

$$(1 + x)^2 = (1 + x)(1 + x) = 1(1 + x) + x(1 + x) = 1 + x + x + x^2 = 1 + 2x + x^2$$

If we have two variables, the answer comes out as:

$$(y + x)^2 = y^2 + 2yx + x^2$$

The binomial theorem

What of brackets raised to powers higher than squares? Opening up more brackets by hand quickly becomes laborious. If we persevere, the results are as follows:

$$(1 + x)^3 = 1 + 3x + 3x^2 + x^3$$
$$(1 + x)^4 = 1 + 4x + 6x^2 + 4x^3 + x^4$$
$$(1 + x)^5 = 1 + 5x + 10x^2 + 10x^3 + 5x^4 + x^5$$

There are two-variable versions too, for example:

$$y^4 + 4y^3x + 6y^2x^2 + 4yx^3 + x$$

So to calculate $(7 + 2a)^4$ we just make the substitution in this formula. But what is the meaning of these sequences of numbers, 1, 4, 6, 4, 1, and so on?

These are the **binomial coefficients**. The coefficients of $(1 + x)^4$ are given by $^4C_0 = 1$, $^4C_1 = 4$, $^4C_2 = 6$, $^4C_3 = 4$ and $^4C_4 = 1$.

The discovery of the binomial theorem is commonly attributed to Blaise Pascal, though it was known by earlier thinkers, including Omar Khayyam. A convenient way to remember the binomial coefficients is Pascal's triangle.

Pascal's triangle

In the 17th century, the French mathematician Blaise Pascal discovered the triangle that bears his name:

```
              1
            1   1
          1   2   1
        1   3   3   1
      1   4   6   4   1
    1   5  10  10   5   1
            etc.
```

160

Each row begins and ends with 1, and each number in between is the sum of the two numbers above it. Pascal's triangle is principally used as a way of calculating binomial coefficients. If you want to expand $(1 + y)^4$, the fifth row of the triangle tells you the answer:

$$1 + 4y + 6y^2 + 4y^3 + y^4$$

There are many other patterns concealed in this triangle too. For instance, if you look at the diagonals, the first is just 1s, and the second simply counts: 1, 2, 3, 4, 5, … But the third diagonal (1, 3, 6, 10, …) lists the **triangular numbers**. After this are higher-dimensional analogues of **polygonal numbers**, the next being the *tetrahedral numbers*. Deleting the even numbers from the triangle produces a pattern which gets ever closer to the **fractal** *Sierpińksi triangle*, as more and more rows are added.

EQUATIONS

Equations

Although a widely feared term, an *equation* is nothing more than an assertion that one thing is equal to another. So $1 + 1 = 2$ is an equation, as is $E = Mc^2$, and anything where an equals sign sits between two mathematical expressions. Commonly, equations involve **polynomials**.

When manipulating an equation, the rule is to treat it like a pair of scales: keep it balanced. This means that, to remain true, you must always perform the same action to both sides: if you add 6 to the left-hand side, or multiply it by 24, you must do the same thing to the right, or else end up with something false. This is obvious when only numbers are involved:

$$13 - 2 = 11 \text{ so } (13 - 2) + 6 = 11 + 6 \quad \text{and similarly } (13 - 2) \times 24 = 11 \times 24.$$

The same applies when unknowns are involved. So for instance, if $3x + 1 = 16$, then subtracting 1 from both sides gives $3x = 15$, and dividing both sides by 3, we get $x = 5$, which is the *solution* to the original equation.

Polynomials

A polynomial is an expression where unknown numbers (represented by letters such as x and y) are added and multiplied. So, for example, $3x^2 + 15y + 7$ is a polynomial in two variables: x and y. The numbers 3 and 15 are called the *coefficients*: 3 is the coefficient of x^2 and 15 is the coefficient of y. In this polynomial, 7 is the *constant term*, which involves no x or y.

Polynomials are important in geometry: setting the polynomial equal to zero defines a condition on the coordinates of a point on the plane, (x, y). The set of all points satisfying this condition produces a geometric object, such as a straight line, or in this case a **parabola**.

If a polynomial has just one variable, this produces an equation $x^2 + 3x + 2 = 0$. Every number x either satisfies it or not. *Solving* this equation involves finding all x for which the equation is true: the *roots* of the polynomial.

The *degree* of a polynomial is the highest power it contains: so $x^2 + 3x + 2$ has degree 2, or in other words is a **quadratic equation**. The simplest polynomials are those of degree 1: the *linear* polynomials.

The **fundamental theory of algebra** says that (in most cases) a polynomial of degree n will have n roots in the **complex numbers**.

Linear equations
As soon as humans could perform basic addition and multiplication, they were equipped to address questions such as 'what number when doubled and added to three gives nine?', a useful skill for dividing up resources. It is probable, therefore, that our species has been solving equations such as $2x + 3 = 9$ since Paleolithic times. This equation is of the simplest kind: it has one unknown (x), and that is never squared, or square rooted, or otherwise complicated. We call such equations *linear*, because they come from the **equation of a straight line**: $y = 2x + 3$. The key to solving these is to perform the same action to both sides, until x is left on its own: first subtract 3 from both sides: $2x = 6$. Then divide both sides by 2, to get $x = 3$.

What about equations such as $4x + 20 = 4$? Writing in around AD 250, Diophantus of Alexandria considered this type of equation 'absurd'. So the full story had to wait for the introduction of **negative numbers** by Brahmagupta in the seventh century AD. The same method could then extend to solve any linear equation in one unknown: starting with $4x + 20 = 4$, first subtract 20 from both sides: $4x = -16$. Then dividing both sides by 4 gives the answer: $x = -4$.

Factorizing polynomials
If we are trying to solve an equation such as $x^3 - 7x + 6 = 0$, we are searching for values of x for which this equation is true. For example, 2 is such a solution since $2^3 - 7 \times 2 + 6 = 0$ but 3 is not a solution, as $3^3 - 7 \times 3 + 6 = 12 \neq 0$. It turns out that:

$$x^3 - 7x + 6 = (x - 1)(x^2 + x - 6)$$

(This can be checked by expanding the brackets.) If the original equation is to hold, it must be that $(x - 1)(x^2 + x - 6) = 0$. So if we substitute the value of 1 for x, then the first bracket is zero, and thus the equation holds. It follows that $x = 1$ is a solution to the original equation.

Now two numbers multiplied together can only produce zero if one of them is zero. So if $(x - 1)(x^2 + x - 6) = 0$, then it must follow that either $x = 1$ or $x^2 + x - 6 = 0$. This second polynomial can be split up further, into $(x - 2)(x + 3)$. So the original equation can be rephrased as $(x - 1)(x - 2)(x + 3) = 0$. This is its *factorized* form. At this point the solutions can just be read off: they are 1 and 2, as already expected, and -3. Can there be any other solutions? The answer is no. For any other value of x, the expression $(x - 1)(x - 2)(x + 3)$ must be non-zero.

The factor theorem

The *factor theorem* makes precise the argument around factorizing polynomials. A number a is a root of a polynomial P, if and only if $(x - a)$ divides P, that is, there is another polynomial Q such that $P = (x - a) \times Q$. To completely solve a polynomial we can split it into the form $(x - a)(x - b) \ldots (x - c)$. Then its solutions are exactly the numbers a, b, …, c. In the previous example these are 1, 2, and -3.

Quadratic equations

Quadratic equations differ from linear equations by involving x^2 as well as plain x-terms. These were studied by the ancient Babylonians for calculating the dimensions of fields of certain areas. For instance, a rectangular field might have one side 5 metres longer than the other, and an area of $36\,m^2$. So calculating its dimensions amounts to solving the equation $x \times (x + 5) = 36$, that is, $x^2 + 5x - 36 = 0$.

The method for solving quadratic equations originates with the seventh-century Indian mathematician Brahmagupta, and ninth-century Persian scholar Muhammad ibn Mūsā al-Khwārizmī, from whose book on the subject, *Hisāb al-jabr w'al-muqābala* ('Compendium on calculation by completion and balancing'), we get the word *algebra*.

The quadratic formula

In modern notation, any quadratic equation can be rearranged into the form $ax^2 + bx + c = 0$, for some values of a, b, c. This will typically have two solutions, and these can be found by a formula imprinted in the minds of generations of mathematics students: $\frac{-b \pm \sqrt{b^2 - 4ac}}{2a}$.

The two different solutions come from the '$\pm$'. So the equation $x^2 + 5x - 36$ has solutions $\frac{-5 \pm \sqrt{5^2 - 4 \times 1 \times (-36)}}{2 \times 1}$ which is $\frac{-5 \pm 13}{2}$, that is 4 and -9. This formula comes from completing the square.

Completing the square

Some quadratic equations are easier to solve than others: just taking square roots is enough to see that $x^2 = 9$ has solutions $x = 3$ and $x = -3$. The equation $x^2 + 2x + 1 = 9$ may seem more complicated. But if we spot that it is equivalent to $(x + 1)^2 = 9$, then taking square roots gives $x + 1 = \pm 3$, giving solutions of $x = -4$ and $x = 2$.

Completing the square is a method to turn any quadratic equation into one of this type. The first step is to arrange the equation so that the x^2-term has coefficient 1. For instance, starting with $2x^2 + 12x - 32 = 0$, divide both sides by 2 to get $x^2 + 6x - 16 = 0$. The second stage involves the coefficient of the x-term. Dividing this by 2 and then squaring it gives the number which 'completes the square', as we shall see. So, from $x^2 + 6x - 16 = 0$, we divide 6 by 2 to get 3, and square this to get 9. We want this number to be the only constant on the left-hand side: $x^2 + 6x + 9$. So we have to change the right-hand side to keep the equation balanced: $x^2 + 6x + 9 = 25$.

The pay-off for all this manipulation is that the left-hand side is now a square: $(x + 3)^2$. So the equation can be solved by taking square roots: $x + 3 = \pm 5$, and the solutions are $x = -8$ and $x = 2$. Applying this method to the general equation $ax^2 + bx + c = 0$ produces the formula $\frac{-b \pm \sqrt{b^2 - 4ac}}{2a}$.

Cubic equations

Following the quadratics, next in the hierarchy of equations in one unknown are the *cubic equations*: those which also involve x^3, such as $x^3 - 6x^2 + 11x - 6 = 0$. The first serious analysis of cubics was made in the 11th century by the Persian poet and polymath Omar Khayyam, using **conic sections**.

In 16th-century Italy, cubic and quartic equations became the great problems of the age. Mathematicians such as Girolamo Cardano, Niccolò Fontana, Scipione del Ferro and Lodovico Ferrari gambled their reputations on public bouts of equation solving. Cardano published the general solution to the cubic in his 1545 book *Ars Magna*, crediting del Ferro with its discovery. Their work was a major force in the acceptance of negative numbers and the development of complex numbers. In general, a cubic equation has three complex solutions, and it always has at least one among the real numbers.

The cubic formula

The formula for the general cubic $x^3 + ax^2 + bx + c = 0$ is significantly more complicated than that for the quadratic. To give it, we first define $q = \frac{-a^3}{27} + \frac{ab}{6} - \frac{c}{2}$ and $p = q^2 + \left(\frac{b}{3} - \frac{a^2}{9}\right)^3$. Then the first solution is given by:

$$x = \sqrt[3]{q + \sqrt{p}} + \sqrt[3]{q - \sqrt{p}} - \frac{a}{3}$$

The other two solutions are:

$$x = \left(\frac{-1 \pm \sqrt{3}\,i}{2}\right)\sqrt[3]{q + \sqrt{p}} + \left(\frac{-1 \pm \sqrt{3}\,i}{2}\right)\sqrt[3]{q - \sqrt{p}} - \frac{a}{3}$$

Here, i is the imaginary unit (see **complex numbers**), and $\frac{-1 \pm \sqrt{3}\,i}{2}$ gives the two cube roots of 1, besides 1.

Quartic equations

Beyond the cubics lie the *quartic equations*, which additionally involve x^4. An example is $x^4 + 5x^3 + 5x^2 - 5x - 6 = 0$. Without complex numbers, which were yet to gain acceptance, these equations posed a serious conundrum for the renaissance algebraists such as Lodovico Ferrari. Some quartics such as $x^4 + 1 = 0$ have no solutions in the real numbers, and others have four. Nevertheless, they persevered, prepared simply to assume negative and complex numbers as their working required, hoping they would later cancel out.

In 1545 Girolamo Cardano published *Ars Magna*, whose 40 chapters included an account of Ferrari's method for solving quartic equations. It is even more cumbersome than that for the cubic.

The quartic formula

To solve $x^4 + ax^3 + bx^2 + cx + d = 0$, first let $e = ac - 4d$ and $f = 4bd - c^2 - a^2d$. Then use the formula for cubic equations to solve $y^3 - by^2 + ey + f = 0$.

This cubic must have a solution in the real numbers. Pick one such and call it y.

Next, let $g = \sqrt{a^2 - 4b + 4y}$ and $h = \sqrt{y^2 - 4d}$. Then the four solutions to the original quartic can be found by solving two quadratic equations:

$$x^2 + \tfrac{1}{2}(a + g)x + \tfrac{1}{2}(y + h) = 0 \quad \text{and} \quad x^2 + \tfrac{1}{2}(a - g)x + \tfrac{1}{2}(y - h) = 0$$

Quintic equations

The formulas for finding the solutions of cubic and quartic equations are certainly fiendish. Between 1600 and 1800, the assumption among the mathematical community was that the formulas for quintic, sextic, septic equations would become ever more complicated, and mathematicians toiled away trying to discover them. Leonhard Euler admitted 'All the pains that have been taken in order to resolve equations of the fifth degree, and those of higher dimensions … have been unsuccessful'.

The story took a surprising twist in the 19th century, in the hands of one of mathematics' most brilliant and tragic figures: Niels Abel. The formulas for quadratics, cubics and quartics involve addition, subtraction, multiplication, division and taking *radicals* (roots): $\sqrt{\ }, \sqrt[3]{\ }, \sqrt[4]{\ }$, and so on. Working in the mathematical backwater of Norway, Abel produced a six-page manuscript, in which he proved that the search was in vain: there is no corresponding formula for quintic equations or for those of any higher degree. These equations will always have solutions; that is the fundamental theorem of algebra. But there is no single method using radicals which will allow you to find them.

This is sometimes known as the *Abel-Ruffini theorem*, as Paolo Ruffini had arrived at the same conclusion (although his 500-page tome contained an incomplete proof). Tragically, Abel did not survive to witness the algebraic revolution he had began, dying penniless at the age of 26.

Insoluble equations

Throughout history, the adoption of new number systems has always produced solutions to previously insoluble polynomial equations. The introduction of negative numbers allowed equations such as $x + 6 = 4$, which Diophantus considered 'absurd', to be solved in exactly the same manner as other linear equations. The **real numbers** included irrationals such as $\sqrt{2}$, which provided solutions to problems like $x^2 = 2$. The complex numbers are built around a new number i, a solution to the hitherto intractable equation $x^2 = -1$.

With this accomplished, the question was: are there any insoluble polynomials left? The *fundamental theorem of algebra* gives a triumphant answer: no.

The fundamental theorem of algebra

In a piece of mathematical magic, proved by Carl Friedrich Gauss in his doctoral thesis in 1799, it turns out the complex numbers do not merely provide the solution to the equation $x^2 = -1$. Every polynomial built from complex numbers must always have a solution, also in the complex numbers. For example, $x^5 + 2ix = -4$ must have a solution in the complex numbers.

Indeed, a polynomial of degree n (that is, with highest term x^n) will usually have n different solutions. Occasionally, however, these solutions can double up, as in the case of $(x - 1)^2 = 0$ which has just one solution: $x = 1$. The fundamental theorem of algebra has several proofs, four of which were discovered by Gauss. All invoke the power of **complex analysis**.

Simultaneous equations

An equation with one variable x, such as $3x + 4 = 10$, can be solved. But it doesn't make sense to try to 'solve' an equation in two variables, such as $x + y = 4$. There are infinitely many solutions: $x = 2$ and $y = 2$ is one, $x = 1.5$ and $y = 2.5$ is another, $x = 1001$ and $y = -997$ a third, and so on. In fact, $x + y = 4$ defines an infinitely long straight line, and the coordinates of any point on it will solve the original equation.

However, if we introduce a second equation $x - y = 2$, then it does make sense to search for a solution: can we find numbers x and y which satisfy both $x + y = 4$ and $x - y = 2$? Graphically, what we are looking for is the place where the two lines cross.

There are two main methods for tackling this sort of problem. The first is by *elimination*: we add or subtract the equations from each other, until one variable vanishes. In this case adding the two equations eliminates y (the $+y$ and $-y$ cancel out), leaving $2x = 6$. We easily solve this, to get $x = 3$. Now we substitute that value back in to one of the original equations, say $x + y = 4$, which becomes $3 + y = 4$, and we can solve this to get $y = 1$. So the solution is $x = 3$, $y = 1$.

Starting again with $x + y = 4$ and $x - y = 2$, an alternative method to use is *substitution*. First we alter one equation to give one variable in terms of the other: we can alter $x + y = 4$ to $x = 4 - y$. Then we substitute this into the other equation: where previously we saw x, we replace it with $4 - y$. So $x - y = 2$ becomes $(4 - y) - y = 2$, which is $4 - 2y = 2$, so $2y = 2$, giving $y = 1$. Putting this back into one of the original equations, say $x + y = 4$ gives $x + 1 = 4$, which gives $x = 3$.

Thankfully, we get the same answer using either method.

Larger systems of equations

As a general rule, to solve a system in three unknown quantities, you need three equations, and similarly for higher numbers. So the system $x + y + z = 0$, $x - y - z = 2$, $x - 2y + z = 3$ can be solved by adapting the usual methods for pairs of simultaneous equations. Larger systems than this are best tackled using matrices (see **matrices and equations**).

There are two things that can go wrong, however. If we start with the simultaneous equations $x + y = 1$ and $2x + 2y = 2$, and try to solve them, we will not get very far. The reason for this is that they are not really two different equations, more like the same one twice. This is easy to spot with just two equations, but the same phenomenon can happen more subtly with larger systems. For instance, $x + y + z = 6$, $2x - y + z = 3$ and $x + 4y + 2z = 15$ cannot be solved uniquely. On closer inspection, this is because the third equation is not really new, but comes from the first two (triple the first, and subtract the second). Systems like this are called *dependent*. They do have solutions: for example, $x = 1$, $y = 2$, $z = 3$ is a solution. In fact, they have infinitely many solutions, lying along the line.

The opposite problem is when the system has no solutions at all. For example, we cannot hope to find any solutions to the system $x + y = 1$ and $x + y = 2$. These are *inconsistent*. In geometric terms, they represent parallel lines (so there is no use looking for the place where they cross). In three dimensions, we may get parallel planes such as $x + y + z = 1$ and $x + y + z = 2$.

In more complicated systems, we can end up with *skew lines*, such as those given by $z = x + y = 1$ and $z = x - y = -1$. These are not parallel, but nor do they cross; they pass each other in 3-dimensional space, without meeting.

Polynomial rings

Originally, numbers were just tools for counting objects. In time, humans came to see their order and beauty, and were inspired to investigate them further for their own sake. Over the 20th century, a similar shift of perspective happened regarding polynomials. Once, a polynomial was just a convenient way to formalize a problem which involved some unknown number. A more modern, abstract approach sees polynomials as objects in their own right, which can be added, subtracted and multiplied together.

So the collection of all polynomials with integer coefficients and just one variable (X) forms a **ring**, called $\mathbf{Z}[X]$. Alternatively we could look at the ring of polynomials in two variables with complex coefficients: $\mathbf{C}[X, Y]$. There are many other possibilities. Like different number systems, these new structures have hidden depths, and their study is of huge importance in contemporary algebra, number theory, and geometry.

VECTORS AND MATRICES

Vectors

In the geometry of the plane, objects such as $\binom{3}{4}$ are known as *vectors*. Essentially, they are a system for giving directions from one point to another.

The top row gives the distance to travel right, and the bottom row the distance to travel up. So $\binom{3}{4}$ translates as 'three right and four up'. A negative number on the top row means left, and on the bottom down. So $\binom{-2}{-1}$ translates as 'two left and one down'.

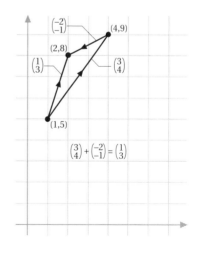

Starting at the point $(1, 5)$ and following the vector $\binom{3}{4}$ takes us to $(4, 9)$. If we then take the vector $\binom{-2}{-1}$, we arrive at $(2, 8)$. On the other hand, the trip from $(1, 5)$ to $(2, 8)$ can be given in one step, by $\binom{1}{3}$. So it makes sense to say that $\binom{3}{4} + \binom{-2}{-1} = \binom{1}{3}$. This illustrates how to *add* vectors: just add the numbers in the corresponding positions.

The same rule applies to vectors in three dimensions such as $\begin{pmatrix} 1 \\ 2 \\ 3 \end{pmatrix}$, and similarly for four, five and higher dimensions.

Parallelogram law

A vector has direction and magnitude. It can conveniently be written as a straight arrow, of a particular length. It does not have any particular start or end point, however. (This makes vectors perfect for modelling quantities such as velocity.) You can start the same vector $\begin{pmatrix} 1 \\ 4 \end{pmatrix}$ at the origin $(0,0)$, or at the point $(-100, 101)$.

If we start at the origin, and apply $\begin{pmatrix} 1 \\ 4 \end{pmatrix}$, we get to the point $(1, 4)$. If we then apply $\begin{pmatrix} 3 \\ 2 \end{pmatrix}$, we get to the point $(4, 6)$. Starting again, and applying these vectors in the opposite order, we also arrive at $(4, 6)$, via $(3, 2)$. This illustrates that, for any two vectors **u** and **v**, **u** + **v** = **v** + **u**, that is, vector addition is *commutative*. This may not be a profound truth, but it is an important one, and is known as the *parallelogram law* (see **quadrilaterals**).

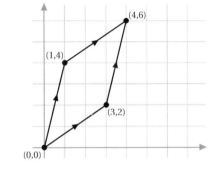

Although vectors were not studied explicitly until the 19th century, the parallelogram law had already been recognized by Heron of Alexandria in the first century AD.

Length of a vector

A vector has *magnitude* or *length*, as well as direction. The way to find the length of a vector is through **Pythagoras' theorem**. So the length of $\begin{pmatrix} 3 \\ 4 \end{pmatrix}$ is $\sqrt{3^2 + 4^2} = 5$. The length of a vector **v** is denoted $\|\mathbf{v}\|$. The lengths of vectors are governed by the triangle inequality, and are generalized by the *dot product*.

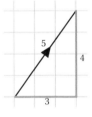

The triangle inequality

Everyone knows that is further to travel along two sides of a triangle than to take a shortcut along the third. This piece of common sense makes its appearance in mathematics as the *triangle inequality*, occupying an axiomatic status in several areas. Any sensible notion of *distance* should obey this rule, whether that be in an exotic **hyperbolic** space, or in ordinary Euclidean space.

The triangle inequality is also important in the study of vectors. If **v** and **u** are vectors, it is not true that the length of **v** + **u** is equal to the lengths of **v** and **u** added together. For example, if $\mathbf{v} = \begin{pmatrix} 3 \\ 4 \end{pmatrix}$ and $\mathbf{u} = \begin{pmatrix} 0 \\ -4 \end{pmatrix}$, then $\|\mathbf{v}\| = 5$ and $\|\mathbf{u}\| = 4$, but $\|\mathbf{v} + \mathbf{u}\| = 3$. So we cannot say, in general, that $\|\mathbf{v} + \mathbf{u}\| = \|\mathbf{v}\| + \|\mathbf{u}\|$. However, the length of **v** + **u** cannot be more than the total lengths of **v** and **u**. So $\|\mathbf{v} + \mathbf{u}\| \leq \|\mathbf{v}\| + \|\mathbf{u}\|$, which says exactly that the third side of the triangle is shorter than the sum of the other two sides.

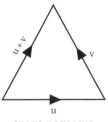

$\|v + u\| \leq \|v\| + \|u\|$

The dot product

There is a way to combine two vectors, not to get a third vector, but a number which describes their relationship. This is called the *dot product* (or *scalar product*). It works by multiplying together the corresponding entries in the vector, and adding them up. So $\begin{pmatrix} 1 \\ 2 \end{pmatrix} \cdot \begin{pmatrix} 3 \\ 4 \end{pmatrix} = 1 \times 3 + 2 \times 4 = 3 + 8 = 11$.

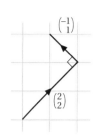

The dot product has several useful properties. Firstly it incorporates the length of a vector: $\|\mathbf{v}\| = \sqrt{\mathbf{v} \cdot \mathbf{v}}$. Secondly, it can detect whether two vectors are perpendicular, or *orthogonal*: $\mathbf{u} \cdot \mathbf{v} = 0$ whenever $\mathbf{u}$ and $\mathbf{v}$ are at right angles. For example, $\begin{pmatrix} -1 \\ 1 \end{pmatrix} \cdot \begin{pmatrix} 2 \\ 2 \end{pmatrix} = -1 \times 2 + 1 \times 2 = 0$. Expanding on this, the dot product can also be used to find the angle between two vectors.

The angle between two vectors

The dot product provides a convenient method for finding the angle between any two vectors. If the angle θ is between the vectors $\mathbf{u}$ and $\mathbf{v}$, the formula connecting them is:

$$\cos \theta = \frac{\mathbf{u} \cdot \mathbf{v}}{\|\mathbf{u}\| \|\mathbf{v}\|}$$

So, taking the example of $\mathbf{u} = \begin{pmatrix} 1 \\ 1 \end{pmatrix}$ and $\mathbf{v} = \begin{pmatrix} 2 \\ 0 \end{pmatrix}$, we get:

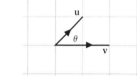

$\mathbf{u} \cdot \mathbf{v} = 1 \times 2 + 1 \times 0 = 2$. Also, $\|\mathbf{u}\| = \sqrt{1^2 + 1^2} = \sqrt{2}$, and $\|\mathbf{u}\| = \sqrt{2^2} = 2$.

Putting these values in the formula

$$\cos \theta = \frac{2}{\sqrt{2} \times 2} = \frac{1}{\sqrt{2}}$$

gives a result of $\theta = \frac{\pi}{4}$, or $45°$.

Cauchy–Schwarz inequality

Mathematics is full of *inequalities*, which state that one quantity must be less than another. One of the most widely used was discovered by Augustin Cauchy in 1821, and later extended by Herman Schwarz. It says that, for any four numbers a, b, x, y:

$$(ax + by)^2 \le (a^2 + b^2)(x^2 + y^2)$$

Or alternatively

$$|ax + by| \le \sqrt{(a^2 + b^2)(x^2 + y^2)}$$

This can be expressed very neatly in terms of the dot product of two vectors. Putting $\mathbf{u} = \begin{pmatrix} a \\ b \end{pmatrix}$, and $\mathbf{v} = \begin{pmatrix} x \\ y \end{pmatrix}$, the inequality becomes:

$$|\mathbf{u} \cdot \mathbf{v}| \le \|\mathbf{u}\| \|\mathbf{v}\|$$

This version of the Cauchy–Schwarz inequality follows from the formula for the angle between two vectors, as $\cos \theta$ must be between 0 and 1.

It extends to larger collections. Suppose you have two equally sized collections of real numbers a, b, …, c and x, y, …, z. Then

$$(ax + by + \cdots + cz)^2 \le (a^2 + b^2 + \cdots + c^2)(x^2 + y^2 + \cdots + z^2)$$

The Cauchy–Schwarz inequality appears in many placed in mathematics, including in the theory of **probability distributions**.

The cross product

The dot product of two vectors is not a vector, but a scalar: an ordinary number. There is also a way to combine two vectors **u** and **v** to give a third **u** × **v**, known as the *cross product*. Algebraically, it is rather delicate, and only works in 3-dimensional space. It also fails to satisfy some criteria we expect a product to obey. For example **u** × **v** ≠ **v** × **u** (in fact **u** × **v** = −**v** × **u**). It is defined as follows:

$$\begin{pmatrix} a \\ b \\ c \end{pmatrix} \times \begin{pmatrix} x \\ y \\ z \end{pmatrix} = \begin{pmatrix} bz - cy \\ cx - az \\ ay - bx \end{pmatrix} \quad \text{So for example} \quad \begin{pmatrix} 2 \\ 0 \\ 0 \end{pmatrix} \times \begin{pmatrix} 0 \\ 3 \\ 0 \end{pmatrix} = \begin{pmatrix} 0 \\ 0 \\ 6 \end{pmatrix}.$$

Geometrically, **u** × **v** is always perpendicular to both **u** and **v**. The direction is given by the *right-hand rule*: if you point your thumb and forefinger along the directions of **u** and **v** respectively, then your middle finger extends in the direction of **u** × **v**.

The length of **u** × **v** is $\|\mathbf{u} \times \mathbf{v}\| = \|\mathbf{u}\| \, \|\mathbf{v}\| \sin \theta$, where θ is the angle between **u** and **v**. (If the two vectors are parallel, then the angle is zero, and the result is always zero.)

Despite being algebraically awkward, the cross product is important in physical problems such as understanding electromagnetic fields. (See **Maxwell's equations**.)

Matrices

A *matrix* is an array of numbers such as $\begin{pmatrix} 1 & 2 \\ 3 & 4 \end{pmatrix}$ or $\begin{pmatrix} 1 & 1 & 0 \\ 0 & 1 & 1 \end{pmatrix}$.

Matrices can come in any rectangular form, but square matrices are particularly important. You can only add matrices which are the same size. To do this, just add the corresponding entries. So:

$$\begin{pmatrix} 1 & 2 \\ 3 & 4 \end{pmatrix} + \begin{pmatrix} 5 & 7 \\ 6 & 8 \end{pmatrix} = \begin{pmatrix} 6 & 9 \\ 9 & 12 \end{pmatrix}$$

Matrices are efficient ways to encapsulate geometric transformations, such as rotations and reflections, which involve taking the coordinates of a point and recombining them in some way.

Multiplying a vector by a matrix

To start with a simple example, we can multiply the single row matrix $(2 \quad 3)$ by the vector $\begin{pmatrix} 4 \\ 5 \end{pmatrix}$ to get a 1×1 matrix. It works by taking each number in the matrix in turn, and multiplying it by the corresponding number in the vector, and adding all these together. So:

$$(2 \quad 3)\binom{4}{5} = (2 \times 4 + 3 \times 5) = (8 + 15) = (23)$$

This is the fundamental technique. With a larger matrix we do the same thing, taking each row in turn. So:

$$\begin{pmatrix} 2 & 3 \\ 6 & 7 \end{pmatrix}\binom{4}{5} = \begin{pmatrix} 2 \times 4 + 3 \times 5 \\ 6 \times 4 + 7 \times 5 \end{pmatrix} = \binom{23}{59}$$

The same process will always work, as long as the width of the matrix is the same as the height of the vector. So:

$$\begin{pmatrix} 1 & 2 & 3 \\ 3 & 2 & 1 \end{pmatrix}\begin{pmatrix} 4 \\ 5 \\ 6 \end{pmatrix} = \begin{pmatrix} 1 \times 4 + 2 \times 5 + 3 \times 6 \\ 3 \times 4 + 2 \times 5 + 1 \times 6 \end{pmatrix} = \binom{32}{28}$$

An important matrix is $\begin{pmatrix} 1 & 0 \\ 0 & 1 \end{pmatrix}$, which is the 2×2 *identity matrix*. This has the property that it leaves every vector as it is: $\begin{pmatrix} 1 & 0 \\ 0 & 1 \end{pmatrix}\binom{a}{b} = \binom{a}{b}$. The 3×3 identity matrix is $\begin{pmatrix} 1 & 0 & 0 \\ 0 & 1 & 0 \\ 0 & 0 & 1 \end{pmatrix}$

Usually denoted *I*, the identity matrix is the *multiplicative identity*, meaning that it plays the same role in multiplying matrices as the number 1 does in multiplying numbers.

Multiplying matrices
Multiplying two matrices is easy, once you

know how to multiply a vector by a matrix. To calculate $\begin{pmatrix} 1 & 2 \\ 3 & 4 \end{pmatrix}\begin{pmatrix} 5 & 6 \\ 7 & 8 \end{pmatrix}$, just split $\begin{pmatrix} 5 & 6 \\ 7 & 8 \end{pmatrix}$ into two vectors $\binom{5}{7}$ and $\binom{6}{8}$ and evaluate each separately:

$$\begin{pmatrix} 1 & 2 \\ 3 & 4 \end{pmatrix}\binom{5}{7} = \binom{19}{43} \quad \text{and} \quad \begin{pmatrix} 1 & 2 \\ 3 & 4 \end{pmatrix}\binom{6}{8} = \binom{22}{50}$$

Then put the matrix back together, so:

$$\begin{pmatrix} 1 & 2 \\ 3 & 4 \end{pmatrix}\begin{pmatrix} 5 & 6 \\ 7 & 8 \end{pmatrix} = \begin{pmatrix} 19 & 22 \\ 43 & 50 \end{pmatrix}$$

After a little practice, this can be done without splitting up the matrix.

The thing to remember is that, when you evaluate an entry, you use the corresponding row from the first matrix and the column from the second. So, focusing on the bottom central number in this evaluation, we get:

$$\begin{pmatrix} 1 & 2 & 3 \\ 2 & 4 & 6 \\ 3 & 6 & 9 \end{pmatrix}\begin{pmatrix} 4 & 5 & 6 \\ 2 & 3 & 4 \\ 1 & 2 & 3 \end{pmatrix} = \begin{pmatrix} \dots & & \dots \\ & \vdots & \\ \dots & 3 \times 5 + 6 \times 3 + 9 \times 2 & \dots \end{pmatrix} = \begin{pmatrix} \dots & & \dots \\ & \vdots & \\ \dots & 51 & \dots \end{pmatrix}$$

Unlike multiplying numbers, it is generally untrue that *AB = BA*, so matrix multiplication is *non-commutative*.

Determinants

The *determinant* of a square matrix is a number associated to it, which encodes, in a rather subtle way, useful information about it. For a 2×2 matrix, $\begin{pmatrix} a & b \\ c & d \end{pmatrix}$, the determinant is defined to be $ad - bc$.

We can write 'det A' or $|A|$ as short-hand for the determinant of A; an example is:

$$\det \begin{pmatrix} 1 & 2 \\ 3 & 4 \end{pmatrix} = 1 \times 4 - 2 \times 3 = 4 - 6 = -2$$

Larger square matrices also have determinants, but calculating them is a little more involved:

$$\det \begin{pmatrix} a & b & c \\ d & e & f \\ g & h & i \end{pmatrix} = a \cdot \det \begin{pmatrix} e & f \\ h & i \end{pmatrix} - b \cdot \det \begin{pmatrix} d & f \\ g & i \end{pmatrix} + c \cdot \det \begin{pmatrix} d & e \\ g & h \end{pmatrix}$$

Extending this process allows the calculation of 4×4 and larger matrices, although the process becomes increasingly time-consuming.

One useful feature is that the determinant respects matrix multiplication. For any matrices A and B:

$$\det (AB) = \det A \det B$$

So if you know the determinant of A and of B, you immediately know the determinant of AB, without having to work through the whole process.

The most important information the determinant carries is whether or not it is 0. If det $A = 0$, then A has no *inverse*, that is, there is no other matrix B, such that $AB = I$ (where I is the identity matrix). If det $A \neq 0$, then A does have an inverse.

Inverting matrices

If we have a matrix A, a fundamental question to ask is whether or not it has an *inverse*, that is, whether there is another matrix B, where $BA = I$. If such a B exists, it means that any process encoded by A can be undone, by B. If no such B exists it means that A entails a fundamental loss of information, and so cannot be undone.

To invert a 2×2 matrix $A = \begin{pmatrix} a & b \\ c & d \end{pmatrix}$, we proceed as follows.

1 Find its determinant, det $A = ad - bc$. If this is zero, then stop, because A does not have an inverse.
2 Form a new matrix, called the *adjugate* of A, adj $A = \begin{pmatrix} d & -b \\ -c & a \end{pmatrix}$.
3 Divide each entry in this new matrix by det A, to get A^{-1}, the inverse of A:

$$A^{-1} = \begin{pmatrix} \dfrac{d}{ad - bc} & \dfrac{-b}{ad - bc} \\ \dfrac{-c}{ad - bc} & \dfrac{a}{ad - bc} \end{pmatrix}$$

Or more briefly, $A^{-1} = \frac{1}{\det A} \operatorname{adj} A$. This also holds for 3×3 and larger matrices, but both the adjugate and determinant become more complicated to calculate.

The adjugate of a matrix

The procedure for inverting 3×3 and larger matrices is essentially the same as for 2×2 matrices, but finding the *adjugate* is a little trickier. Given a matrix $A = \begin{pmatrix} 1 & 2 & 3 \\ 3 & 2 & 1 \\ 2 & 1 & 3 \end{pmatrix}$, it works as follows:

1 First, form a new matrix by swapping the rows and columns of the original. So the first row becomes the first column, the second row becomes the second column, and so on. This is called the *transpose* of A.

$$A^{\mathrm{T}} = \begin{pmatrix} 1 & 3 & 2 \\ 2 & 2 & 1 \\ 3 & 1 & 3 \end{pmatrix}$$

2 Next, we focus on one entry of A^{T}. Striking out the whole row and column of that entry leaves a 2×2 matrix. For example, taking the 3 on the top row of A^{T}, removing its row and column:

$\begin{pmatrix} 1 & 3 & 2 \\ 2 & 2 & 1 \\ 3 & 1 & 3 \end{pmatrix}$, leaves the matrix $\begin{pmatrix} 2 & 1 \\ 3 & 3 \end{pmatrix}$. The determinant of this smaller matrix is called the *minor* of the entry. In this case $\det \begin{pmatrix} 2 & 1 \\ 3 & 3 \end{pmatrix} = 2 \times 3 - 3 \times 1 = 3$.

3 The next matrix is formed by replacing every entry in A^{T} with the corresponding minor. This produces:

$$\begin{pmatrix} 5 & 3 & -4 \\ 7 & -3 & -8 \\ -1 & -3 & -4 \end{pmatrix}$$

4 The final step is to change the signs according to the rule $\begin{pmatrix} + & - & + \\ - & + & - \\ + & - & + \end{pmatrix}$, where $+$ means do not change the sign, and $-$ means do. So we finally arrive at:

$$\mathrm{adj}\,A = \begin{pmatrix} 5 & -3 & -4 \\ -7 & -3 & 8 \\ -1 & 3 & -4 \end{pmatrix}$$

If we want to use this to find the inverse of A, we apply the formula $A^{-1} = \dfrac{1}{\det A}\,\mathrm{adj}\,A$, and divide every entry of the adjugate by the determinant of A, in this case, -12. So:

$$A^{-1} = \begin{pmatrix} \dfrac{-5}{12} & \dfrac{1}{4} & \dfrac{1}{3} \\ \dfrac{7}{12} & \dfrac{1}{4} & \dfrac{-2}{3} \\ \dfrac{1}{12} & \dfrac{-1}{4} & \dfrac{1}{3} \end{pmatrix}$$

Transformation matrices

Transformations such as rotations and reflections involve taking the coordinates of a point, and recombining them in some way. For example, reflecting a point in the line $y = x$ involves swapping its coordinates. So $(1, 2)$ becomes $(2, 1)$. Reflecting in the x-axis involves changing the sign of the y-coordinate: $(1, 2)$ becomes $(1, -2)$. Rotating by 90° about the origin (anticlockwise as always) takes $(1, 2)$ to $(-2, 1)$.

This sort of manipulation can succinctly be expressed by matrix multiplication. To do this, it is easiest to think in terms of vectors. So, instead of a point, such as $(1, 2)$, we will work with the vector which gives directions to it from the origin, namely $\binom{1}{2}$. Then reflection in the line $y = x$ is given by multiplying by the matrix $\begin{pmatrix} 0 & 1 \\ 1 & 0 \end{pmatrix}$. So:

$$\begin{pmatrix} 0 & 1 \\ 1 & 0 \end{pmatrix} \begin{pmatrix} 1 \\ 2 \end{pmatrix} = \begin{pmatrix} 2 \\ 1 \end{pmatrix}$$

Similarly, reflection in the x-axis is described by multiplying by the matrix $\begin{pmatrix} 1 & 0 \\ 0 & -1 \end{pmatrix}$. So:

$$\begin{pmatrix} 1 & 0 \\ 0 & -1 \end{pmatrix} \begin{pmatrix} 1 \\ 2 \end{pmatrix} = \begin{pmatrix} 1 \\ -2 \end{pmatrix}$$

Rotating by 90° about the origin is described by $\begin{pmatrix} 0 & -1 \\ 1 & 0 \end{pmatrix}$. So:

$$\begin{pmatrix} 0 & -1 \\ 1 & 0 \end{pmatrix} \begin{pmatrix} 1 \\ 2 \end{pmatrix} = \begin{pmatrix} -2 \\ 1 \end{pmatrix}$$

Rotation matrices

The commonest rotations are by 90°, 180° and 270° about the origin. These are described by the matrices $\begin{pmatrix} 0 & -1 \\ 1 & 0 \end{pmatrix}$, $\begin{pmatrix} -1 & 0 \\ 0 & -1 \end{pmatrix}$ and $\begin{pmatrix} 0 & 1 \\ -1 & 0 \end{pmatrix}$ respectively.

What of more general rotations? Matrices can only really cope with rotations about the origin. (Rotating about any other point can be broken down into a rotation about the origin, and a translation.)

The general form for the matrix of rotation by an angle of θ, is :

Rotate by θ

$\begin{pmatrix} \cos\theta & -\sin\theta \\ \sin\theta & \cos\theta \end{pmatrix}$. So rotating by 30° is given by $\begin{pmatrix} \frac{1}{2} & \frac{\sqrt{3}}{2} \\ \frac{\sqrt{3}}{2} & -\frac{1}{2} \end{pmatrix}$.

The determinant of a rotation matrix is always 1, since $\cos^2\theta + \sin^2\theta = 1$.

Reflection matrices

Matrices can express reflection in any line which passes through the origin. The commonest examples are the lines $y = x$, $y = -x$, $y = 0$ and $x = 0$. These are described by the matrices $\begin{pmatrix} 0 & 1 \\ 1 & 0 \end{pmatrix}$, $\begin{pmatrix} 0 & -1 \\ -1 & 0 \end{pmatrix}$, $\begin{pmatrix} 1 & 0 \\ 0 & -1 \end{pmatrix}$ and $\begin{pmatrix} -1 & 0 \\ 0 & 1 \end{pmatrix}$ respectively.

For other rotations, every line through the origin is uniquely defined by its *gradient* or, equivalently, its angle to the x-axis. The x-axis itself (that is, the line $y = 0$) has angle 0°, the line

$y = x$ has angle 45°, the y-axis has angle 90° and $y = -x$ has angle 135°. At 180°, we are back at the x-axis. If the angle to the x-axis is θ, then the gradient is $\tan\theta$, and so the equation of the line is given by $y = (\tan\theta)x$. Reflection in this line is described by the matrix:

$$\begin{pmatrix} \cos 2\theta & \sin 2\theta \\ \sin 2\theta & -\cos 2\theta \end{pmatrix}$$

The line $y = \dfrac{1}{\sqrt{3}}x$ has an angle of 30°, so reflecting in this line corresponds to multiplying by the matrix:

$$\begin{pmatrix} \dfrac{1}{2} & \dfrac{\sqrt{3}}{2} \\ \dfrac{\sqrt{3}}{2} & -\dfrac{1}{2} \end{pmatrix}$$

The determinant of any reflection matrix is always -1.

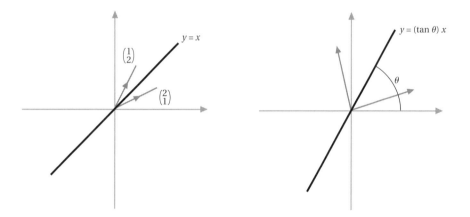

Enlargement and shearing matrices

Enlargements, centred on the origin, can be encoded as matrices in a very straightforward way. To enlarge the vector $\begin{pmatrix} 2 \\ 1 \end{pmatrix}$ by a scale factor of 3, say, each coordinate needs to be multiplied by 3 to give $\begin{pmatrix} 6 \\ 3 \end{pmatrix}$. This is equivalent to multiplying by the matrix $\begin{pmatrix} 3 & 0 \\ 0 & 3 \end{pmatrix}$, since $\begin{pmatrix} 3 & 0 \\ 0 & 3 \end{pmatrix}\begin{pmatrix} 2 \\ 1 \end{pmatrix} = \begin{pmatrix} 6 \\ 3 \end{pmatrix}$.

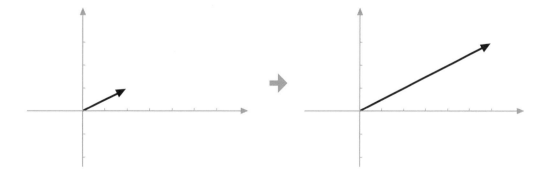

In general, an enlargement by a scale factor a is given by the matrix $\begin{pmatrix} a & 0 \\ 0 & a \end{pmatrix}$.

A different sort of transformation is given by the matrix $\begin{pmatrix} 1 & 1 \\ 0 & 1 \end{pmatrix}$. This is an example of a *shear*. Shearing always keeps one line fixed (in this case the x-axis) and slides other points parallel to this line, in proportion to their distance away from it. This proportion is called the *shear factor* (in this case 1). It is notable that shearing always preserves the area of a shape.

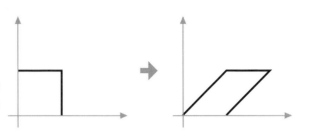

Groups of matrices

Matrix multiplication works best for matrices which are square. In this case, any two matrices of the same size can be multiplied, and there is a special matrix of each size, the *identity*, which leaves every other untouched. For 3×3 matrices, this is written as $\begin{pmatrix} 1 & 0 & 0 \\ 0 & 1 & 0 \\ 0 & 0 & 1 \end{pmatrix}$.

However, the set of all 3×3 matrices does not form a **group**. The problem is that not every matrix has an inverse. In particular, if the determinant of a matrix A is zero, then A has no inverse. Luckily, this is the only obstacle. If we rule out the matrices with zero determinant, then the remaining matrices do form a group, called the *general linear group* of degree 3. There are other, smaller groups lurking inside here too. Restricting to matrices which have determinant 1 produces another group, the *special linear group*.

Focusing on 2×2 matrices, the collection of just rotation and reflection matrices forms a group, called the *orthogonal group* of degree 2. In higher dimensions, the range of possible transformations grows, but orthogonal groups remain important.

These different matrix groups are the prime examples of **Lie groups**, and a painstaking analysis of them resulted in the **classification of simple Lie groups**. Beyond this, by replacing the numbers in these matrices with elements from a **finite field**, we can find the families which feature in the **classification of finite simple groups**.

Matrices and equations

As well as being central to geometry, matrices are useful for condensing a whole system of **simultaneous equations** into a single one. Take the pair of equations $2x + y = 7$ and $3x - y = 3$. These can be expressed together as:

$$\begin{pmatrix} 2 & 1 \\ 3 & -1 \end{pmatrix} \begin{pmatrix} x \\ y \end{pmatrix} = \begin{pmatrix} 7 \\ 3 \end{pmatrix}$$

Now, the **inverse** of the matrix $\begin{pmatrix} 2 & 1 \\ 3 & -1 \end{pmatrix}$ is $\begin{pmatrix} \frac{1}{5} & \frac{1}{5} \\ \frac{3}{5} & -\frac{2}{5} \end{pmatrix}$. If we multiply both sides by this, the two

matrices on the left cancel out, and we are left with $\begin{pmatrix} x \\ y \end{pmatrix} = \begin{pmatrix} \frac{1}{5} & \frac{1}{5} \\ \frac{3}{5} & -\frac{2}{5} \end{pmatrix} \begin{pmatrix} 7 \\ 3 \end{pmatrix}$. Working this out gives

$\begin{pmatrix} x \\ y \end{pmatrix} = \begin{pmatrix} 2 \\ 3 \end{pmatrix}$, so the solution to the original system is $x = 2$ and $y = 3$. This approach can be extended to solving bigger sets of equations with more unknowns.

GROUP THEORY

The group axioms

If we are adding **integers** together, two things are obvious:

- There is a special number 0 which does nothing when added to any other.
- If you add any number n to its negative $-n$, you get 0.

A third observation is slightly subtler, and is suggested by an example: $(12 + 5) + 6 = 17 + 6 = 23$, which gives the same answer as $12 + (5 + 6) = 12 + 11 = 23$. So the way we choose to bracket the numbers doesn't matter. It is because of this property of *associativity* that we can write $12 + 5 + 6$ without ambiguity.

These simple facts, and the philosophy of **abstraction** give the idea of an abstract algebraic structure called a *group* as a collection of objects which can be combined in pairs, satisfying three *axioms*:

1 There is one special object, the *identity*, which does nothing when combined with any other.
2 Every object has an *inverse* which combines with it to produce the identity.
3 The process of combination is *associative*.

Groups

Groups are ubiquitous in abstract algebra. A *group* is a collection of objects, along with some way to combine them, satisfying the group axioms. The integers under addition is one example, as described above.

Multiplying numbers can also produce a group: this time the identity will be 1, and the inverse of a number q will be its **reciprocal** $\frac{1}{q}$. But the reciprocal of 2 is $\frac{1}{2}$, which is not an integer. So the integers do not form a group under multiplication. If we extend to the rational numbers, then we almost have a group. But there's still one problem. The number 0 has no inverse, as there is no number you can multiply by 0 to get 1. So we'll exclude that, to arrive at the *group of non-zero rational numbers*. This forms a group under multiplication.

These examples are both infinite, but finite groups also exist, including many **symmetry groups** and permutation groups.

Another observation about the integers is that $17 + 89 = 89 + 17$, so the order of combination doesn't matter. Groups such as this are called *Abelian* after the Norwegian algebraist Niels Abel. Many non-Abelian groups also exist, including groups of matrices under multiplication.

Permutation groups
How many different ways can the set $\{1, 2, 3\}$ be ordered? **Factorials** supply the answer: $3! = 3 \times 2 \times 1 = 6$. The six orderings are: $1, 2, 3$; $1, 3, 2$; $2, 1, 3$; $2, 3, 1$; $3, 1, 2$; $3, 2, 1$.

For each reordering (such as $2, 3, 1$ or $1, 3, 2$) there is a corresponding function, called a *permutation*, which shows how to rearrange the digits to get from $1, 2, 3$ to the new ordering:

1	→	2
2	→	3
3	→	1

or

1	→	1
2	→	3
3	→	2

These permutations can be performed one after the other. Putting these two together gives:

1	→	2	→	3
2	→	3	→	2
3	→	1	→	1

which simplifies to

1	→	3
2	→	2
3	→	1

The *identity* permutation does not move anything:

1	→	1
2	→	2
3	→	3

Each permutation has an inverse. The inverse of

1	→	2
2	→	3
3	→	1

is

1	→	3
2	→	1
3	→	2

because when you put these together, you get the identity:

1	→	2	→	1
2	→	3	→	2
3	→	1	→	3

So these six permutations form a **group**. This is called the *symmetric group* on three elements, or S_3. Similarly we can construct the symmetric group on any number of elements.

Cycle notation

Cycle notation is a convenient short-hand for describing permutations. Instead of writing out the table

1	→	2
2	→	3
3	→	1

we can just use (1 2 3) to indicate that this permutation takes 1 to 2, 2 to 3, and 3 completes the cycle back to 1.

Similarly the function

1	→	1
2	→	3
3	→	2

can be written as (1)(2 3), since 1 is on a cycle on its own, and 2 and 3 form a cycle of length 2. Usually the trivial cycle (1) is left out, and the permutation is just written as (2 3).

Librarian's nightmare theorem

If customers borrow books one at a time, and return them one place to the left or right of the original place, what arrangements of books may emerge? The answer is that, after some time, every conceivable ordering is possible. The simplest permutations are the *transpositions*, which leave everything alone except for swapping two neighbouring points. The question is; which more complex permutations can be built from successive transpositions? The answer is that every permutation can be so constructed.

In cycle notation, (1 3 2) is not a transposition, as it moves three items around: 1 to 3, 3 to 2, and 2 to 1. But this has the same effect as swapping 1 and 2, and then swapping 2 and 3. That is to say, (1 3 2) = (1 2)(2 3). The librarian's nightmare theorem guarantees that every permutation can similarly be expressed as a product of transpositions.

Alternating groups

The librarian's nightmare theorem says that every permutation can be broken down into transpositions. This representation is not unique, however. There are many ways to build a particular permutation from transpositions. For example, in the symmetric group S_5, (1 3 2) = (1 2)(2 3) and also (1 3 2) = (2 3)(1 2)(2 3)(1 2). Through all possible representations, something does remain fixed. If a permutation is a product of an even number of transpositions, then every representation must comprise an even number. Similarly, a permutation which is a product of an odd number of transpositions can only be written as a combination of an odd number. This fact divides permutations into the *odd* ones and the *even* ones.

The collection of even permutations is particularly important, as it forms a subgroup, called the *alternating group* on *n* elements, A_n. When n ≥ 5, this group is a **simple group**. The fact that A_5 is the first non-Abelian simple group is central to **Galois' theorem**.

Cayley tables

Children are usually expected to memorize multiplication tables:

×	1	2	3	4	5	6	7	8	9	10
1	1	2	3	4	5	6	7	8	9	10
2	2	4	6	8	10	12	14	16	18	20
3	3	6	9	12	15	18	21	24	27	30
4	4	8	12	16	20	24	28	32	36	40
5	5	10	15	20	25	30	35	40	45	50
6	6	12	18	24	30	36	42	48	54	60
7	7	14	21	28	35	42	49	56	63	70
8	8	16	24	32	40	48	56	64	72	80
9	9	18	27	36	45	54	63	72	81	90
10	10	20	30	40	50	60	70	80	90	100

Of course the table is incomplete, as there are infinitely many integers. In multiplication modulo 5 (see **modular arithmetic**), for example, this obstacle disappears. Here, we can fill in all the values:

×	1	2	3	4
1	1	2	3	4
2	2	4	1	3
3	3	1	4	2
4	4	3	2	1

Named after the 19th-century British mathematician Arthur Cayley, this table completely defines the group of non-zero integers under multiplication modulo 5. (Notice that each element of the group appears exactly once in each column and in each row, making Cayley tables examples of **Latin squares**.)

For another example, take the symmetry group of an equilateral triangle. As well as the identity, written 'ι', this contains a rotation by 120°, call it R. The other rotation is by 240°, the result of doing R twice, so R^2. There is also a reflection in the vertical line, call this T. The remaining two reflections are the results of doing T followed by either R or R^2, so we call them TR and TR^2. Then the full multiplication table shows how these six elements interact:

	ι	R	R²	T	TR	TR²
ι	ι	R	R²	T	TR	TR²
R	R	R²	ι	TR²	T	TR
R²	R²	ι	R	TR	TR²	T
T	T	TR	TR²	ι	R	R²
TR	TR	TR²	T	R²	ι	R
TR²	TR²	T	TR	R	R²	ι

Every finite group can be captured in a Cayley table in this way.

Isomorphisms

Arthur Cayley understood that the pattern within one of his tables contained the abstract essence of a group. The names of the elements, and the geometric scenarios that gave rise to it, are of secondary importance. To say that two groups are *isomorphic* means that they are essentially the same (even if they arose in completely different situations). All that is required to turn one into the other is a systematic changing of the labels.

In cycle notation, the symmetric group on {0,1,2} has Cayley table:

	ι	(0 1 2)	(0 2 1)	(0 1)	(0 2)	(1 2)
ι	ι	(0 1 2)	(0 2 1)	(0 1)	(0 2)	(1 2)
(0 1 2)	(0 1 2)	(0 2 1)	ι	(1 2)	(0 1)	(0 2)
(0 2 1)	(0 2 1)	ι	(0 1 2)	(0 2)	(1 2)	(0 1)
(0 1)	(0 1)	(0 2)	(1 2)	ι	(0 1 2)	(0 2 1)
(0 2)	(0 2)	(1 2)	(0 1)	(0 2 1)	ι	(0 1 2)
(1 2)	(1 2)	(0 1)	(0 2)	(0 1 2)	(0 2 1)	ι

On closer inspection, this is a disguised version of the group of symmetries of the equilateral triangle. The following relabelling changes one group into the other:

$$
\begin{aligned}
\iota &\longrightarrow \iota \\
(0\ 1\ 2) &\longrightarrow R \\
(0\ 2\ 1) &\longrightarrow R^2 \\
(0\ 1) &\longrightarrow T \\
(0\ 2) &\longrightarrow TR \\
(1\ 2) &\longrightarrow TR^2
\end{aligned}
$$

This relabelling is called an *isomorphism*. Of course, for two groups to be isomorphic they must have the same number of elements. But this is not sufficient. The addition modulo 6 also produces a group of six elements, but fundamentally different from the one above:

	0	1	2	3	4	5
0	0	1	2	3	4	5
1	1	2	3	4	5	0
2	2	3	4	5	0	1
3	3	4	5	0	1	2
4	4	5	0	1	2	3
5	5	0	1	2	3	4

The term *isomorphic* can apply to algebraic structures other than groups (such as **rings** or **fields**), but always carries the same meaning: that two structures are essentially identical, with only a renaming of the elements needed to turn one into the other.

Simple groups

Just as a prime number is one which cannot be divided into smaller numbers, so a *simple group* cannot be broken into smaller groups. In the case of finite groups, there is an analogue of the fundamental theorem of arithmetic too: the *Jordan–Hölder theorem* of 1889, which says that every finite group is built from simple groups, in a unique way. The classification of finite simple groups gives a complete account of these fundamental building blocks.

For infinite groups, the situation is not so straightforward, since not every group can be broken down into indivisible pieces in this precise way. However, certain special cases can be tackled, a particularly important case being the **classification of simple Lie groups**.

The classification of finite simple groups

One of the most spectacular mathematical feats of the 20th century, the *classification of finite simple groups* was the culmination of a mammoth team project, spread across 500 papers by more than 100 different mathematicians worldwide. Efforts to create a 'second generation proof' in one place are continuing, and will stretch to 12 volumes (at time of writing six have been published). The final theorem gives precise descriptions of 18 infinite families of finite groups.

The first of these is the family of cyclic groups: addition modulo p, where p is prime. Next come the alternating groups, and the remaining families are **symmetry groups** of certain finite geometric structures. The theorem also describes 26 individual groups, known as the *sporadic groups*, the largest of which is **the monster**.

Ultimately, this remarkable theorem states that the 18 families and 26 individual groups together comprise the entire collection of finite simple groups: there are no others. These, then, are the atoms from which every finite group is built.

The monster

Weighing in at 808,017,424,794,512,875,886,459,904,961,710,757,005,754,368,000,000,000,000 elements, *the monster* is the largest of the 26 sporadic finite simple groups. There are, of course, finite groups of unlimited size. The interesting thing about the monster is that it is a one-off, not part of any larger family or pattern. Predicted in 1973, and constructed by Bernd Fischer and Robert Griess in 1980, the monster was initially seen as a curiosity: a freakish combinatorial possibility.

In 1979, however, John Conway and Simon Norton were surprised to find its fingerprints in the unconnected area of **modular forms**. Dubbing this unexpected phenomenon *monstrous moonshine*, they made a bold conjecture that these two worlds are actually intimately related. In a tour de force in 1992, Richard Borcherds proved the moonshine conjecture using deep ideas from quantum field theory. Having completed the proof, he described his feelings as 'over the moon'.

Lie groups

The **symmetry group** of a square contains four rotations: by 0°, 90°, 180° and 270°. In contrast, a circle can be rotated by 1° or 197.8°, or any amount, to leave it looking the same. This means that the symmetry group of the circle is infinite. Beyond this however, it makes sense to talk about two rotations being 'near' to each other: rotating by 1° is nearly the same as leaving the circle fixed. Rotating by 0.1° or 0.01° are closer still, so these rotations are getting closer to the group's identity: the trivial symmetry which moves nothing. This group allows for gradual, continuous changes rather than the discrete symmetry groups of polygons or tessellations.

In fact the symmetry group of the circle starts to look rather like the circle itself: you can rotate by any angle, and 360° takes you back to the start. But there are also infinitely many reflections, corresponding to choosing different diameters of the circle as the mirror lines. Two reflections can be close to each other (if their lines are nearly the same), but a reflection can never be close to the identity. So this symmetry group comes in two distinct components: rotations which can slide towards the identity, and reflections which cannot.

This is called the second *orthogonal group*. (The third corresponds to the symmetries of the sphere, and so on.) The orthogonal groups are the first examples of Lie groups: groups which are also **manifolds**, named after their Norwegian discoverer Sophus Lie (pronounced 'Lee'). The symmetries of any smooth manifold produce a Lie group, making these of great importance in physics. Often, as with the orthogonal groups, a Lie group can be made concrete as a group of matrices.

The classification of simple Lie groups

In 1989, the quantum theorist A. John Coleman wrote an article entitled 'The greatest mathematical paper of all time'. His nomination for this majestic title went to an 1888 work by Wilhelm Killing. In it, among other technological breakthroughs, Killing laid the groundwork for a full classification of simple Lie groups.

Lie groups today play central roles in physics, and are mathematically profound for the way they tie together the algebra of group theory, with deep ideas from differential geometry. Simple Lie groups are especially important: they are those which cannot be broken down into smaller Lie groups.

Subsequently Élie Cartan completed the momentous project begun by Killing. As Jean Dieudonné wrote, it was 'made possible only by his uncanny algebraic and geometric insight and that has baffled two generations of mathematicians'. The result of Killing and Cartan's efforts is a list of four infinite families of simple Lie groups deriving from matrix groups. Additionally there are five individual groups, the so-called *exceptional* Lie groups, which come from the **quaternions** and **octonions**. Every simple Lie group must then be either a member of one the four families, or equal to one of the five exceptional groups.

The atlas of Lie groups

Since Killing and Cartan's classification of simple Lie groups, we have known that every simple Lie group must be one from the list. A tremendous achievement though this was, it is not the end of the story, since it does not tell us the inner workings of these groups. The best way to understand such gigantic, abstract objects is to approximate them by things we know far better, namely groups of matrices. This is the subject of **representation theory**. The *atlas of Lie groups* is an ongoing project to accumulate the representation theory of all Lie groups.

E_8

The most complicated of the exceptional simple Lie groups is known as E_8. It describes the symmetries of an object in 57-dimensional space. E_8 itself is 248-dimensional. The analysis of the matrix representations of E_8 was completed in 2007, by a team of mathematicians and programmers, led by Jeffrey Adams at the University of Maryland, USA. The information they found describing E_8 occupies a massive 60 gigabytes. By comparison, the human genome is less than one gigabyte in size.

The extension problem

In chemistry, elements such as carbon and hydrogen are combined into *compounds*. A formula such as C_5H_{12} tells us that each molecule of the compound contains five carbon and twelve hydrogen atoms. This information is not enough to pin down the new chemical exactly. Elements can combine in different ways called *isomers*. Two isomers can both have the formula C_5H_{12} but, because these atoms are connected in different ways, the result can be two compounds with very dissimilar chemical properties.

The same thing is true in group theory. To understand a group it is not enough to know its make-up in terms of simple groups; two groups can combine in many different ways. For example, start with the group of addition modulo 2:

	0	1
0	0	1
1	1	0

Taking two copies of this and joining them together gives two possibilities. The first is to add pairs from the group. This produces the so-called *Klein 4 group*:

	(0,0)	(0,1)	(1,0)	(1,1)
(0,0)	(0,0)	(0,1)	(1,0)	(1,1)
(0,1)	(0,1)	(0,0)	(1,1)	(1,0)
(1,0)	(1,0)	(1,1)	(0,0)	(0,1)
(1,1)	(1,1)	(1,0)	(0,1)	(0,0)

The alternative produces addition modulo 4:

	0	1	2	3
0	0	1	2	3
1	1	2	3	0
2	2	3	0	1
3	3	0	1	2

Understanding the different possibilities for combining two groups is called the *extension problem*. In general it is intractably difficult. But particular special cases, such as finite groups of prime size, are subjects of intensive study. One technique deployed by Marcus du Sautoy and others, is the use of **L-functions** to encode information about the groups.

Solvable groups
The easiest groups to understand are *Abelian* groups: those where it is always true that $xy = yx$. All the familiar number systems satisfy this, but not every group is Abelian. Matrix groups, permutation groups, and symmetry groups are typically not. Second best are groups which, though not Abelian themselves, are built from Abelian simple pieces. These are the *solvable* groups. Groups which are built from non-Abelian simple pieces, such as A_n, are not solvable. Évariste Galois' great insight was that groups are useful for studying equations. Whether or not the resulting group is solvable is a significant question, addressed in Galois' theorem.

ABSTRACT ALGEBRA

Abstract algebra
Higher algebra differs from the plain letters-for-numbers variety, mainly in that calculations are performed in settings which are far more abstract than the familiar systems of the integers or rational numbers. The role of the number system (such as the integers) is played by a *structure*: a collection of objects (called its *elements*) which take the place of individual numbers. The whole set-up is governed by some precisely defined axioms, such as the **group axioms**. These structures can then be studied in depth, purely by investigating the logical consequences of the axioms.

Algebraic structures

The range of abstract algebraic structures studied today is so vast as to be bewildering, even to professional mathematicians. The most important examples, however, are **groups**, **rings** and **fields**. At first glance, this dry, formal approach may seem totally disconnected from any sort of reality. But many ordinary mathematical objects, such as the integers, complex numbers and matrices fit into one or more of these boxes, so to study abstract structures is also to study these more common systems.

Beyond that, these structures have a habit of appearing in unexpected places; an abstract approach allows a swathe of problems to be solved at a stroke, instead of the same work having to be repeated in many slightly different contexts. Best of all is when abstract structures can be *classified*. This means a list of all structures satisfying the axioms can be given explicitly. Examples are the classification of finite simple groups and the classification of simple Lie groups.

Rings

Groups provide a wonderful way to study either addition or multiplication in the abstract. But, for ordinary numbers, both of these processes go on at the same time. A new way to abstract this set-up was also needed. The solution that mathematicians of the early 20th century arrived at is the notion of a *ring*. This is a collection of objects which can be added, subtracted and multiplied together, following precise axioms governing how these processes interact. The most important example is, of course, **Z** the ring of integers. But there are many others too: rings of matrices and polynomials appear throughout algebra and geometry, and rings of functions play an important role in analysis.

All fields are also rings. Similarly all rings are groups (when you forget about their multiplication and focus on addition and subtraction).

Fields

In a ring we have addition, subtraction and multiplication. But something is missing: division. This is evident in the ring of integers. Dividing one integer (say 5) by another (7) does not usually give an integer answer. So we cannot hope to be able to divide without leaving the structure.

A *field* is a place which also permits division. (There is an exception though: even in a field, you can never have **division by 0**.) The most common fields are the rational numbers, the real numbers and the complex numbers. Another important class is that of finite fields. Non-examples are the 'field' with one element, the **quaternions** and the **octonions**.

Prime fields

In arithmetic modulo 4 (see **modular arithmetic)**, the number 1 cannot be divided by 2. The reason is that the multiples of 2 are 0, 2, 4, 6, 8, 10, … Each of these, when divided by 4, leaves remainder 0 or 2. So, no whole number, when multiplied by 2, gives an answer congruent to 1 mod 4. This means that when we look at the numbers {0, 1, 2, 3} under arithmetic modulo 4, we can add, subtract and multiply them, but not divide them. So this structure is a ring, but not a field.

In modulo 5, however, any number can be divided by any other (except 0 as always). For instance, $1 \div 2 \equiv 3 \bmod 5$, because $2 \times 3 \equiv 1 \bmod 5$. This means that the collection of numbers $\{0, 1, 2, 3, 4\}$ does form a field (called $\mathbf{F}_5$). You can do addition, subtraction, multiplication and division in this structure. The crucial difference is that 5 is prime, but 4 is not. In the same way, arithmetic modulo any prime p produces a finite field, called $\mathbf{F}_p$. But this never works for non-primes.

Finite fields

The prime fields $\mathbf{F}_p$ are not quite the only finite fields. Just as we can extend the real numbers to the complex numbers by adding in a square root of -1, so we may extend the field $\mathbf{F}_p$ by incorporating some new elements.

For any n, there is a new field $\mathbf{F}_{p^n}$, with exactly p^n elements. So, in particular, there is a field $\mathbf{F}_4$ with four elements, but it is not the same as $\{0, 1, 2, 3\}$ under arithmetic modulo 4. These finite fields are remarkably useful objects, not least in **cryptography**.

The 'field' with one element

According to the axioms, every field must contain an element which does nothing when multiplied by any other (usually called '1' for obvious reasons) and an element which does nothing when added to anything else ('0'), and these must be different: $0 \neq 1$. One of the most basic facts about fields then, is that all of them contain at least two elements. (In fact the smallest field, $\mathbf{F}_2$, has exactly two.)

So there can be no field with only one element. However, this logical impossibility didn't stop Jacques Tits from opening discussions on the subject in 1957. Nor has it prevented numerous mathematicians since then from developing a body of knowledge around the non-existent entity known as $\mathbf{F}_1$ (or perhaps more appropriately as $\mathbf{F}_{un}$, 'un' being the French for 'one').

This is not (only) a joke. These mathematicians view $\mathbf{F}_1$ as a notional limit object for fields, as infinity is for the natural numbers. The guiding principle is that $\mathbf{F}_1$ turns combinatorial objects into geometric ones. According to this philosophy, the set of integers can be imagined as a curve over $\mathbf{F}_1$, and plain unstructured sets of objects come to resemble **varieties**.

Galois theory

The rational numbers, real numbers and complex numbers are all examples of fields. Although self-contained, these number systems are closely related: rational numbers are all real numbers, and real numbers are all complex numbers. An important thing which can happen when we move from a smaller field to a bigger one is that equations which previously had no solutions can gain some. $x^2 - 2 = 0$ has no rational solutions, but when we move to the real numbers we find one: $\sqrt{2}$. Similarly, $x^2 + 1 = 0$ has no solutions among the real numbers, but by stepping into the complex numbers we find one: i.

Building on Abel's work on **quintic equations**, in the early 19th century, Évariste Galois drew together several central themes of algebra. From his works, the subject of *Galois theory*, has grown and grown. In modern terms, Galois theory is the study of how one field can live inside another. A crucial aspect is to decide what new equations can be solved in the bigger field that could not be solved in the smaller.

Symmetries and equations

When passing from a smaller field (such as the real numbers) to a larger one (such as the complex numbers), the new field brings with it new symmetries. For example, *complex conjugation* is a symmetry of the complex numbers, not present in the real numbers. This symmetry flips over i and −i, the two solutions of the equation $x^2 + 1 = 0$. Galois theory can be thought of as the study of symmetries of solutions of equations. These symmetries form a group. Évariste Galois realized that a critical question is whether or not the resulting group is **solvable**.

Like Abel, Galois' life was both spectacular and tragically short. A committed revolutionary, he was once arrested for threatening the king's life (though later acquitted). Aged just 21, he was killed in a duel, in circumstances which remain mysterious.

Galois' theorem

Galois addressed the question of when an equation can be solved by radicals (**roots**). The **quadratic formula** is a formula involving just $+, -, \times, \div$, $\sqrt{}, \sqrt[3]{}, \sqrt[4]{}, \ldots$ which will give the solutions to any quadratic equation. The cubic and quartic formulas are similar.

Galois' theorem says that such a formula exists if and only if the corresponding group is solvable.

The symmetric groups S_1, S_2, S_3 and S_4 are solvable, which explains why linear, quadratic, cubic and quartic equations all have formulas for their solution. From then on the groups $S_5, S_6, S_7, \ldots$ are not solvable. When you split S_n into its simple pieces, you encounter the non-Abelian simple group A_n (see **alternating groups**). Hence quintic (and higher degree) equations are not solvable by radicals.

Representation theory

The trouble with algebra, it is often said, is that it is so abstract. It is not only the baffled student who thinks so; this is a problem of which professional algebraists are all too aware. A group, for example, is a very abstract object. For finite groups, we can get a handle on them by writing out the **Cayley table** (as long as it is not too big). With this done, the inner workings of the group are laid bare. For infinite groups there is no such easy way in. More concrete groups are much easier to work with. Matrix groups are the prime examples. All we need to do is to learn the laws of matrix multiplication, and the group becomes far more accommodating.

Representation theory is about bringing abstract algebra down to earth. The best case is when we can identify that our group actually *is* a matrix group (that is, it is **isomorphic** to one). Even when this is not true, we can often find matrix groups which are reasonable approximations to our group. These are called the group's *representations*. Representation theorists aim to piece together the original group from the information provided by its representations. Representation theory began with groups, but the study of rings and other structures by these methods is a major theme of modern algebra.

Category theory

'No man is an island', said the poet John Donne, and the same is true of algebraic structures. If we wish to study a particular group, for example, it is often productive to investigate its relationships with other groups. The classical example is to break the group down into smaller groups. But other, subtler relationships are also possible. These relationships are best expressed as **functions** between groups. So, a certain amount of information resides not with the individual group, but at a higher level, in the collection of all groups, and the functions between them. This is an example of a **category**.

Something rather remarkable can happen; you can prove things using very general techniques, which do not seem to rely on the properties of groups at all, but just the category of objects with its functions. The topologists Saunders Mac Lane and Samuel Eilenberg turned what had previously been dismissed as 'abstract nonsense' into a subject in its own right. They were led to this through their development of **algebraic topology**, which associates groups to topological objects. The best way of imagining this, they realized, is as a *functor* between the category of topological spaces and that of groups.

With mathematicians contemplating ever more abstract and arcane objects (the **schemes** of modern algebraic geometry being a prime example), the importance of category-theoretic methods is rapidly growing.

When are two things the same?

This is a surprisingly deep question, with many answers, depending on precisely what is meant by 'the same'. Different approaches to geometry provide different answers there, but the question is important in algebra too.

The tightest concept of sameness is that of *equality*: two things are the same if they are not really two at all, but one. (In non-mathematical language, *equality* has a different meaning, and things, such as men and women, can be equal without being the same.)

Equality is the right notion of sameness for set theory; in algebra this is too restrictive. In group theory, **isomorphisms** better capture the idea of being identical. Two groups may not literally be the same thing, but if they are isomorphic, then they have the same Cayley tables and are identical in every important respect.

In some contexts, notably algebraic geometry, this is still too tight. Kiiti Morita considered two rings to be essentially the same if they had exactly the same representations. This does not imply that they are isomorphic, but being *Morita equivalent* is enough to guarantee that many important properties carry over from one to the other.

The derived category

The best language for addressing the question of when two things are the same is that of category theory. It was here that an even more profound notion of sameness was discovered, by Jean-Louis Verdier and Alexander Grothendieck. It is called the *derived category*, and generalizes Morita equivalence. If two objects produce the same derived category this means that, ignoring the superficial details, at their core, they are in many ways the same. The derived category is an important component of **Langlands' program**, as well as modern algebra and algebraic geometry.

PIGEONHOLE PRINCIPLE • THE SIZE OI
LE • FACTORIALS • PERMUTATION •
ONS OF A SET • BELL NUMBERS • PA
• GRAPHS • SEVEN BRIDGES OF I
T PATH PROBLEM • THE CHINESE PO
LLINGSALESMANPROBLEM • THE
GRAPH THEORY • ERD S NU
RAMSEY NUMBERS • THE

DISCRETE MATHEMA

PERHAPS THE MOST IMPORTANT DIVIDING LINE in mathematics is between continuous and discrete systems. The continuous world is smooth and well suited to geometry and analysis. In contrast, a discrete situation comes in separate pieces. The first thing we can do with such an arrangement is *count*. Of course, this is where mathematics began, many thousands of years ago. In a more sophisticated guise, the science of counting is known as *combinatorics*. This body of knowledge includes many techniques for calculating the number of ways of arranging a collection of objects.

Graph theory has countless applications throughout mathematics. In this context, a graph is simply a discrete

TICS

collection of points, some of which are joined together with edges. This is the right way to express many important optimization questions, such as the travelling salesman and Chinese postman problems. Graphs also relate to questions of pure geometry, through the subject of topological graph theory.

It is always exciting when a simple question reveals unexpected levels of complexity. The subject of *Ramsey theory* leads the way on this, containing some of the most computationally intensive problems in all of mathematics. Yet its underlying ideas are very easy and natural.

COMBINATORICS

The pigeonhole principle

Suppose 1001 pigeons are housed in 1000 coops. It must be that at least one coop contains two or more pigeons. If this seems obvious, it is. But it is also extraordinarily useful. In general, the *pigeonhole principle* says that if n objects are distributed among m boxes, and $n > m$, then at least one box must contain two or more objects. As an example, people have four different blood types: O, A, B and AB. Imagining these as boxes, the pigeonhole principle guarantees that from any group of five or more people, at least two will share the same blood type.

A more technical rendering is as follows: if $f : A \rightarrow B$ is a **function** where B is a finite set and A is strictly larger than B, then there must be some elements x and y of A where $f(x) = f(y)$.

For example, we can define the *parity* of a whole number to be 1 if it is odd, and 0 if it is even. The pigeonhole principle guarantees that in any collection of three numbers, at least two will have the same parity. The reason that this simple idea is so useful is that it is *non-constructive*. It asserts the existence of x and y without having to go to the trouble of taking blood tests.

The size of the union

Suppose we have two collections of objects A and B. How many objects are there all together? That is, how many objects does the **union**, $A \cup B$, contain? It is tempting to count the objects in A and those in B, and add the two numbers together. If we use $|A|$ to denote the size of A, what this suggests is that $|A \cup B| = |A| + |B|$.

This will sometimes be true, but not always. My table has two musical instruments on it and two wooden things on it. However, there are not four objects in total, but three: a metal flute, a wooden ukulele and a wooden spoon. The difficulty arises when A and B overlap, that is, have non-empty *intersection*, $A \cap B$. When we count each of A and B and add the results together, we are actually double-counting those objects that are in both A and B. So we have not worked out $|A \cup B|$ as we hoped, but $|A \cup B| + |A \cap B|$. This can be corrected, by subtracting $|A \cap B|$ again:

$$|A \cup B| = |A| + |B| - |A \cap B|$$

The inclusion–exclusion principle

The formula for the size of the union tells us the size of the union of two sets $A \cup B$. What happens if we have three sets, and want to know the size of $A \cup B \cup C$? We can apply the formula twice, and after a little manipulation we arrive at:

$$|A \cup B \cup C| = |A| + |B| + |C| - |A \cap B| - |B \cap C| - |A \cap C| + |A \cap B \cap C|$$

This is a rather long formula, but there are only three components to it: first add up the sizes of the individual sets, then subtract the intersections of pairs, and then add the common intersection of all three.

To make this logic clearer, we rewrite A, B, C as A_1, A_2, A_3. Then, by using **sum notation**, we can get a more concise formula:

$$|A_1 \cup A_2 \cup A_3| = \sum_{i \leq 3} |A_i| - \sum_{i < j \leq 3} |A_i \cap A_j| + |A_1 \cap A_2 \cap A_3|$$

This suggests how to extend this technique to n sets $A_1, A_2, A_3, \ldots, A_n$. First add the sizes of the individual sets A, then subtract the intersections of pairs, and then add the intersections of threes, subtract that of fours, and so on. In formula form this is:

$$|A_1 \cup A_2 \cup \ldots \cup A_n| = \sum_{i \leq n} |A_i| - \sum_{i < j \leq n} |A_i \cap A_j| + \sum_{i < j < k \leq n} |A_i \cap A_j \cap A_k| - \cdots + (-1)^{n-1} |A_1 \cap A_2 \cap \ldots \cap A_n|$$

Factorials

Suppose we have five mathematicians: Abel, Bernoulli, Cantor, Descartes and Euler. If we want to rate them in order of greatness, how many orderings are possible? There are five choices for first place: A, B, C, D or E. Once this position is filled, there are four remaining possibilities for second place, and then three for third, and two for fourth. Once positions 1 to 4 have been filled there is just one person left, who must go fifth. As this argument suggests, the number of possible orderings of five people is given by $5 \times 4 \times 3 \times 2 \times 1 = 120$. We write this as 5!, read as 'five factorial'. The factorial function is fundamental in combinatorics. From it are built many other, subtler functions, starting with those for permutations and combinations. In general:

$$n! = n \times (n - 1) \times (n - 2) \times \ldots \times 2 \times 1$$

The convention is that $0! = 1$. This may seem strange, but stops us having to insert a lot of caveats and special cases in our statements. For $n \geq 0$, the factorial function satisfies the recurrence relation: $(n + 1)! = (n + 1) \times n!$

Factorials grow very quickly. Starting at 0, the sequence of factorials begins: 1, 1, 2, 6, 24, 120, 720, 5040, 40320, 362880, 3628800. By 60! we have reached the number of atoms in the universe.

Permutations

Suppose we have gold, silver and bronze medals for mathematics, and a shortlist of five candidates: Abel, Bernoulli, Cantor, Descartes and Euler (A, B, C, D and E). How many different possible outcomes are there?

The answer depends on the rules of the award. The question is: may someone receive more than one medal? Suppose they may. Then the number of choices for each medal is five, so the number of total possible outcomes is $5 \times 5 \times 5 = 5^3 = 125$.

However, if we insist that the medals go to different people, then there are again five choices for the gold-medal winner, but then four for the silver, and three for the bronze, giving a total of $5 \times 4 \times 3 = 60$. We call this the number of *permutations* of three objects from five, written as $^5P_3$. In general ^nP_r is the number of possible sequences of r objects from a collection of n. There is a formula for this in terms of factorials:

$$^nP_r = \frac{n!}{(n - r)!}$$

In the above example, happily this matches the argument above: $\frac{5!}{(5-3)!} = \frac{120}{2} = 60$.

Combinations

When we are calculating permutations, the *order* of the selected objects matters: E, B, D is not the same as D, E, B. However, sometimes we may not care about the order. In these circumstances, we want a *combination* rather than a permutation.

This time, we have three identical gold medals, and intend to award them from a shortlist of Abel, Bernoulli, Cantor, Descartes and Euler (A, B, C, D and E). Therefore the ordering of the three recipients does not matter: E, B, D is the same as D, E, B. How many possible outcomes are there now? We know there are $^5P_3 = 60$ different possible orderings of the recipients. These fall into clusters of six which are identical from our new perspective, such as {B, E, D}, {B, D, E}, {D, B, E}, {D, E, B}, {E, B, D}, {E, D, B}. What we want to know is the number of clusters. This is given by $\frac{60}{6} = 10$. A general formula for combinations is provided by binomial coefficients.

Binomial coefficients

In general the number of sets of size r which can be selected from a set of size n is written as $\binom{n}{r}$, or sometimes ^nC_r. The formula is:

$$\binom{n}{r} = \frac{^nP_r}{r!} = \frac{n!}{r!(n-r)!}$$

These combinations also appear as the coefficients in the **binomial theorem**, and can be read off **Pascal's triangle**.

Partitions of a set

If we have four fruits: an apple, a banana, a cherry and a date (a, b, c, d). How many ways can we divide these onto identical plates? The simplest case is to put them onto four plates individually {a}, {b}, {c}, {d}. There is only one way to do this. If instead we divide them as one pair and two singles such as {a, b}, {c}, {d}, then there are $\binom{4}{2} = 6$ choices for the pair. Alternatively, we could divide them as two pairs, such as {a, b}, {c, d}. There are three ways to do this. There are four ways to divide them into a triple and a single {a, b, c}, {d}. Finally we could put them all on one plate: {a, b, c, d}. Adding these all together, we get $1 + 6 + 3 + 4 + 1 = 15$ possibilities. This is the fourth *Bell number B(4)*.

Bell numbers

Named after Eric Temple Bell, the nth *Bell number B(n)* is the number of ways of dividing up a set of n objects into subsets. The sequence of Bell numbers begins: 1, 2, 5, 15, 52, 203, 877, 4140, … Unfortunately, there is no simple formula for Bell numbers. But there are several complicated ones, including Dobiński's formula:

$$B(n) = \frac{1}{e} \sum_{k=0}^{\infty} \frac{k^n}{k!}$$

Partitions of integers

Suppose we have five identical coins. How many ways could we divide these up into smaller collections? (The question differs from the partitions of a set described by the Bell numbers, because the coins are identical. So if we were to divide them into one set of size 2 and three of size 1, then we would not care *which* two coins go together, because they are all interchangeable.)

Ultimately, the question reduces to this: in how many ways can 5 be expressed as a sum of positive whole numbers? The answer in this case is seven:

$$1 + 1 + 1 + 1 + 1 = 2 + 1 + 1 + 1 = 2 + 2 + 1 = 3 + 1 + 1 = 3 + 2 = 4 + 1 = 5$$

Writing $P(n)$ to stand for the number of partitions of n, this tells us that $P(5) = 7$.

There is no straightforward way to write an exact formula for $P(n)$. But these values can be recovered from Euler's partition function.

In 1918, G.H. Hardy and Srinivasa Ramanujan found a formula which estimates the value of $P(n)$:

$$P(n) \sim \frac{1}{4n\sqrt{3}}\, e^{\pi \cdot \sqrt{\frac{2n}{3}}}$$

As n gets larger and larger, this estimate gets better and better. (Technically, the true value divided by the estimate will get closer and closer to 1, so they are *asymptotically* equal.)

Euler's partition function

Rather than an explicit formula for $P(n)$, Leonhard Euler found a *generating function*: a nice algebraic description of the series $\Sigma P(n)x^n$. From his formula, the individual numbers can be worked out. His function is:

$$\prod_{r \geq 1} (1 - x^r)^{-1}$$

To get at the numbers $P(n)$, we have to expand each bracket using the **generalized binomial theorem**:

$$(1 + x + x^2 + x^3 + x^4 + \cdots)(1 + x^2 + x^4 + x^6 + \cdots)(1 + x^3 + x^6 + \cdots)(1 + x^4 + \cdots)$$

The idea of having to expand infinitely many brackets is somewhat alarming. But we can begin the task, at least. The constant term must be 1 (that is $1 \times 1 \times 1 \times \ldots$). The x-term will similarly just be x. For the x^2-term, we need to look at the first two sets of brackets, and it comes out as $2x^2$. Similarly, for the x^3 we look at the first three brackets. Continuing like this, we gradually build up the series:

$$1 + x + 2x^2 + 3x^3 + 5x^4 + 7x^5 + 11x^6 + 15x^7 + \cdots$$

This tells us that the first few partition numbers are 1, 1, 2, 3, 5, 7, 11, 15, …

GRAPH THEORY

Graphs

The word 'graph' has several meanings in mathematics. In *graph theory*, all that is meant is a collection of points or *vertices*, some of which are joined together by lines or *edges*. A *path* is then a string of edges connecting one point to another.

These simple structures can capture the essence of many more complicated scenarios. They are ubiquitous in **topology** and combinatorics, and form the best language for analysing problems like the **four colour theorem**. Graphs feature in many *optimization problems* such as the **shortest path probelm**, **Chinese postman problem**, and the **travelling salesman problem**. In these cases, graphs often come with an extra piece of structure: each edge has a number, its *weight*, attached. These problems involve finding a route of minimal weight.

Seven bridges of Königsberg
Now part of Russia and known as Kaliningrad, the Prussian city of Königsberg has an illustrious mathematical history. It was the home of David Hilbert, Rudolf Lipschitz and Christian Goldbach, among others. The city itself was also at the centre of a problem, whose solution by Leonhard Euler sparked the birth of several mathematical disciplines. The River Pregel divides in two as it passes through the city, and different parts of the city were connected by seven bridges, as shown. The question was: was it possible to walk around the city, crossing each bridge exactly once?

Euler's first insight in 1735 was that the geometrical details were extraneous: everything that mattered could be captured by a simplified diagram consisting only of *vertices* (standing for the various landmasses) joined by *edges* (representing the bridges). This insight marked the birth of *graph theory*. (Technically, this arrangement is a *multi-graph*, since it has some vertices which are joined by more than one edge, not allowed for in an ordinary graph.) Euler then understood that the answer to the problem was no. His solution came from analysing the degrees of the vertices of the graph.

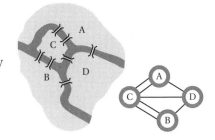

Degree of a vertex
In a graph, the *degree* of a vertex is the number of edges coming out of it. Leonhard Euler reasoned that, if a solution existed to the Bridges of Königsberg, most vertices would need to have an even number of edges coming out of them: half to act as entries, and the same number as exits. This would have to be true for all apart from at most two vertices: the start and end of the path. Since the Königsberg graph has four vertices, and all are of odd degree, Euler concluded that no such route is possible.

The shortest path problem
Anyone who has used a route-finding program online or a satellite-navigation system in their car is familiar with the *shortest path problem*. In the context of a weighted graph, the question is: given two vertices A and B, how do we find the path of least weight between the two?

In applications, the *weight* of each edge could be its length, or it could be the time taken to travel it (compare a 10 mile stretch of single-track road against an 11 mile stretch of motorway). In telecommunications, weight could represent bandwidth. There are many solutions to this problem, including

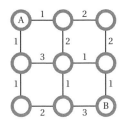

Dijkstra's algorithm of 1959. This starts at A and calculates minimum weight paths first to the vertices adjacent to A, then to those increasingly distant from A, until it reaches B.

The Chinese postman problem

A postman needs to deliver letters to every street in his area. What is the shortest route he can take? This is the Chinese postman problem ('Chinese' because it was first studied by Mei-ko Kwan in 1962, though related to Euler's older bridges of Königsberg problem).

A suitable model of the problem is a weighted graph, where the weight of each edge represents its length. The idea is to find a circuit of minimum overall weight which passes along each edge. If the graph has only vertices of *even degree* (that is, with an even number of edges joined) then it will have an *Eulerian circuit*, a route that goes along each edge exactly once, and this will be the optimal solution. If it does not, some edges will have to be traversed more than once. But which ones, and in which order?

Kwan came up with an **algorithm** to solve the problem. Essentially, it modifies the graph into one in which all vertices have even degree, and then takes an Eulerian circuit of the new graph.

Pen and paper puzzles

Which of the shapes pictured can you draw without lifting your pen from the paper or repeating any edges?

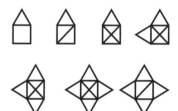

Considering each picture as a graph, the question becomes: which of these graphs have an *Eulerian path*, that is, a route which crosses every edge exactly once? Based on Leonhard Euler's work on the Bridges of Königsberg, the answer is graphs that have at most two vertices of odd degree.

The travelling salesman problem

A salesman needs to travel to ten different towns: what is the shortest route he can take? As for the Chinese postman problem, this can be formalized graph-theoretically, with the towns represented by vertices, and the roads between them as weighted edges. This time, the best case scenario is a *Hamiltonian path*: a route around the graph which passes through each vertex exactly once. But, again, such a path may not exist.

When the number of vertices is small, trial and error may produce an optimal result. With bigger graphs this quickly becomes impractical. Unlike the Chinese postman problem, there is no simple algorithm to solve the general case: as Merrill Flood commented in 1956, 'the problem is fundamentally complex'. In the language of modern **complexity theory** the problem is **NP-complete**. However, special cases of the problem are intensively studied, and are used in the optimization of countless processes, from train scheduling to genome-sequencing. The largest instance of the problem solved to date is an 85,900-city challenge posed by Gerhard Reinelt in 1991, and completed in 2006 by David Applegate and colleagues, requiring a total of 136 years of CPU time.

The three utilities problem

I have received an extraordinary number of letters respecting the ancient puzzle that I have called 'Water, Gas and Electricity'. It is much older than electric lighting, or even gas but the new dress brings it up to date. The puzzle is to lay on water, gas and electricity, from W, G and E, to each of the three houses, A, B and C, without any pipe crossing another.

So wrote Henry Dudeney, Britain's pre-eminent puzzlist, in 1913. It is not too difficult to get to a situation where just one more pipe is required, say from W to B.

Dudeney himself provided a solution, of sorts, in which this missing pipeline passes *underneath* house A. This clever answer shows the importance of making the rules of the game crystal clear (which explains why mathematicians have such a reputation for pedantry). In mathematical terms, the question concerns a graph consisting of two sets of three vertices (W, G, E and A, B, C), where each vertex is connected to all three opposing vertices, but none on its own side.

This graph goes by the name of $K_{3,3}$. The question becomes whether $K_{3,3}$ can be drawn on a flat sheet of paper. Dudeney's answer, with its ingenious detour through the third dimension, is now ruled out. Indeed the puzzle, correctly stated, has no solutions at all: the task is impossible. However, it can be solved on a torus (see **topological graph theory**.)

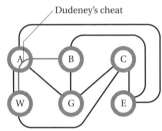

Dudeney's cheat

Planar graphs

When can a graph be drawn on a sheet of paper without any of its edges crashing through each other? Graphs which can be drawn in this way are called *planar*, meaning they can be represented in the 2-dimensional plane. A graph is *complete* if every pair of edges is connected by an edge. The complete graph on four vertices is known as K_4: it is planar. But K_5, the complete graph on five vertices, is not: however you arrange it, two edges will always collide.

The three utilities problem gives us another non-planar graph, called $K_{3,3}$. In unpublished work in 1928, Lev Pontryagin proved that every non-planar graph encodes either K_5 or $K_{3,3}$ (though not necessarily in a straightforward fashion). Surprisingly, these two small graphs are the *only* obstacles to planarity. This celebrated fact is usually known as *Karatowski's theorem*, after its first publication by Kazimierz Kuratowski in 1930.

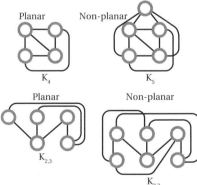

Planar   Non-planar

K_4 K_5

Planar   Non-planar

$K_{2,3}$ $K_{3,3}$

Topological graph theory

Planar graphs are defined by the fact that they do not encode either K_5 or $K_{3,3}$. Both these graphs can be drawn on a **torus**, as indeed can K_7 and $K_{4,4}$. It is still an open question exactly what graphs may be drawn on a torus, or on other surfaces. A complete answer to this question is a goal of *topological graph theory*.

Erdős numbers

The roaming Hungarian mathematician Paul Erdős was one of the most prolific in history: at his death in 1996 he had published around 1500 papers, more even than Leonhard Euler (although page-for-page Euler wrote more). Shunning worldly possessions, Erdős travelled the globe, sleeping on friends' sofas. He is reputed to have believed that 'a mathematician is a machine for turning coffee into theorems'.

In 1969, in tribute to his eccentric friend, Casper Goffman playfully coined the idea of an *Erdős number*, which quickly became folklore. The idea is based on a graph whose vertices are individual mathematicians (past and present), where two are joined by an edge if they have co-authored a paper together. The resulting graph is not *connected*: there are pairs of mathematicians who can never be joined by a path (for instance, if one of them only ever wrote solo papers.)

People who co-authored a paper with Erdős have an *Erdős number* of 1. (There are 511 such known.) Mathematicians who wrote a paper with one of these, but not with Erdős, have Erdős number 2, and so on. (Erdős, uniquely, has an Erdős number of 0.) People who can never be reached by this process have infinite Erdős number. Calculating an individual's Erdős number amounts to solving the shortest path problem between them and Paul Erdős.

RAMSEY THEORY

Ramsey's theorem

Frank Ramsey was a mathematician whose insights not only produced a single theorem, but spawned an entire subject, *Ramsey theory*. He is like Évariste Galois in this regard and, like Galois, he died tragically young, at the age of 26. Ramsey's theorem of 1930 is a classic for finding pattern and structure amid disorder.

Look at sets of five distinct whole numbers: {1, 13, 127, 789, 1001} is one example. Every such set is assigned one of four colours: red, white, green or blue. So {1, 13, 127, 789, 1001} might be red, and {2, 13, 104, 789, 871} green. It does not matter on what basis these colours are assigned: the critical thing is that every collection of five numbers has one, and only one, colour. *Ramsey's theorem* says that there will always be some infinite subcollection of whole numbers (call it A) which is monochromatic: all sets of five elements from A are the same colour.

There is nothing magic about the numbers 4 and 5. The theorem guarantees that for any numbers n and m, if all sets n of distinct numbers are assigned one of m colours, then there will be some infinite monochromatic subset.

The dinner party problem

How many people need to be invited to dinner to guarantee that there will be three guests who are all either mutually acquainted, or mutual strangers?

To translate this into mathematics, we can express the dinner party as a coloured graph. Draw a dot to represent each guest, and join two with a green edge if they are acquainted, or a red edge if they are not. The answer to the problem is not five, since it is possible to concoct a 5-vertex graph which contains neither a red triangle nor a green one.

However, any such graph of six vertices must contain a monochromatic triangle. This translates back to a solution to the original problem. This innocent question is the starting point for one of the most intractable problems in mathematics: Ramsey numbers.

Ramsey numbers

It is useful for mathematical purposes (if not for planning parties) to extend the dinner party problem as follows: how many people need to be invited to guarantee that there will either be m guests who are mutually acquainted, or n who are mutual strangers?

A finite version of Ramsey's theorem asserts that this problem does always have a solution. Let $R(m, n)$ be the minimum number of guests required to solve the problem. The values $R(m, n)$, for different values of m and n, are called *Ramsey numbers*. Though simple to define, the exact values of Ramsey numbers remain deeply mysterious. Even the value of $R(5, 5)$ is as yet unknown, though established to be between 43 and 49. *Ramsey theory* is the study of this problem, and the many variations on this theme. It remains an active topic of research, with relevance to computer science and game theory.

The happy ending problem

A collection of points is said to be in *general position* if no three lie on a straight line. Draw five dots on a piece of paper, making sure that they are in general position. The *happy ending problem*, proved by Esther Klein in 1932, says that, no matter how you do this, you will always be able to pick four dots which form the corners of a **convex quadrilateral**. (Convexity is the critical requirement here, as any four points in general position form a quadrilateral.)

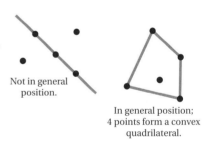

Not in general position.

In general position; 4 points form a convex quadrilateral.

After the happy ending

You might wonder where the happy ending was in the happy ending problem. Actually, the name does not come from the mathematics, but from the problem's human history. Having solved the original problem, Esther Klein set to work on generalizing it. The new, much more difficult question was: how many points in general position are needed to guarantee there will be five which form a convex pentagon? And similarly for a convex hexagon, or in general, a convex n-gon?

Klein and George Szekeres worked on this with Paul Erdős, and were able to prove that these questions do have a solution. For any number n, there is some number $K(n)$, so that any collection of $K(n)$ points in general position must contain a convex n-gon. Klein and Szekeres' collaboration not only culminated in a proof of this theorem, but in their marriage. This was the *happy ending*, and it was Erdős who named the problem so.

Actually calculating the values of $K(n)$ is a seriously tough challenge. It is straightforward that $K(3) = 3$, and the original happy ending problem of 1932 asserts that $K(4) = 5$. Shortly after this, Endre Makai proved that $K(5) = 9$. The only other value known for certain is $K(6) = 17$, proved in 2006 by Szekeres and Lindsay Peters. It is conjectured that, for all n, $K(n) = 2^{n-2} + 1$, but it will surely require a major technical advance to prove this.

Golomb rulers
A ruler is a straight, notched stick, where each notch is marked with a number, representing its distance from the end. Of course, the distance between any two notches is the difference between the two numbers.

While an ordinary ruler has marks at 0 cm, 1 cm, 2 cm, 3 cm, 4 cm, and so on, a *Golomb ruler* has only some of these markings. The principle, conceived by Solomon Golomb, is that no two notches should be the same distance apart as any other pair.

So a ruler notched at $(0, 1, 2)$ is not Golomb, as 0 and 1 are the same distance apart as 1 and 2. The ruler $(0, 1, 3)$ is Golomb, however. Moreover, this is an *optimal* Golomb ruler as it is the shortest possible, with three marks. $(0, 2, 3)$ is also an optimal Golomb ruler of length 3. But, as it is the mirror image of $(0, 1, 3)$, they are not counted as genuinely different.

Optimal Golomb rulers
$(0, 1, 3)$ is the unique optimal Golomb ruler with three notches. If we try to add an extra mark: $(0, 1, 3, 4)$ and $(0, 1, 3, 5)$ both fail, and $(0, 1, 3, 6)$ is not Golomb either since 0 and 3 are the same distance apart as 3 and 6. $(0, 1, 3, 7)$ is Golomb, but it is *not* optimal, there is a shorter ruler with four notches: $(0, 1, 4, 6)$. There are two optimal Golomb rulers with five notches: $(0, 1, 4, 9, 11)$ and $(0, 2, 7, 8, 11)$.

This suggests that locating optimal Golomb rulers is difficult. You cannot simply extend one to find the next. For longer rulers, there is an extra difficulty: how do you know if your n-marked ruler is optimal? This is compounded by there being no obvious rule to tell how many optimal rulers there should be. The only approach, it seems, is to compare all possible rulers with n notches, and pick the shortest.

Finding Golomb rulers
Current methods for searching for optimal Golomb rulers require serious amounts of computing time. A distributed computing project run by distributed.net has found the longest currently known optimal Golomb ruler, with 26 notches:

$$(0, 1, 33, 83, 104, 110, 124, 163, 185, 200, 203, 249, 251, 258,$$
$$314, 318, 343, 356, 386, 430, 440, 456, 464, 475, 487, 492)$$

Golomb rulers have uses in cryptography and communications technology, both of which were investigated by Solomon Golomb in the 1960s.

ANALYSIS

IF YOU DRAW A GRAPH of a moving object's distance from
a point against time, the gradient of this graph represents its
speed, or the rate of change of distance. This relationship had
been known since the time of Archimedes. What was missing
was an understanding of the underlying mathematical laws
of gradients and rates of change. This void was filled in the
17th century, with the development of *differential calculus.*
It was a mathematical triumph, but a human tragedy. Isaac
Newton and Gottfried Leibniz were perhaps the two greatest
scientists of the age, and both claimed the discovery. The
ensuing dispute was one of the most divisive and acrimonious
in the history of science, and culminated with Newton, in his
capacity as President of the Royal Society, publicly denouncing
Leibniz for plagiarism.

• GEOMETRIC PROGRESSIONS • SESSA'S
MOFASEQUENCE•EXPONENTIALGROWTH
CONVERGING AND DIVERGING SEQUENCES
ERIES • THE HARMONIC SERIES DIVERGES
EMENT THEOREM • GENERALIZED BINOMIAL
HEOREM • ACHILLES AND THE TORTOISE •
ADOXES • THE RATIONAL NUMBERS HAVE
EM • DISCRETENESS AND CONTINUITY •
TA • DIFFERENTIABILITY • THE GRADIENT
WITH SECANTS • DIFFERENTIATING FROM
VATIVE SONG • STANDARD DERIVATIVES •
E • THE CHAIN RULE • PROOF OF THE CHAIN
RENTIATION • MAXIMA AND MINIMA • THE
VE TEST • RATES OF CHANGE OF POSITION
RIVATIVES • PARTIAL DIFFERENTIATION
DVANCED INTEGRATION STEP

In any event, it was not until the 19th century that calculus was given a truly firm foundation, through the work of Augustin-Louis Cauchy and Karl Weierstrass. Newton and Leibniz had both relied on fictional objects called 'infinitesimals'. In Weierstrass' hands these were replaced with the notion of a *limit*. This new approach was more robust, but also more technically demanding, opening up very delicate issues of *convergence* and *continuity*. The new subject became known as *analysis* to indicate its level of difficulty.

One reward for this hard work is a much richer picture of the complex numbers. Taking pride of place in this scene is the *exponential function* and in particular one of mathematics' crown jewels: *Euler's formula*.

SEQUENCES

Sequences

A *sequence* is an infinite list of numbers: 1, 1, 2, 5, 15, 52, 203, 877, 4140, 21147, ... This example is an *integer sequence*, as only whole numbers are involved. One can also have real or complex-valued sequences. The sequences that we are most familiar with are those which follow some discernible pattern or rule, allowing us to predict what comes next. Mathematically, there is a formula for their nth term.

However, any infinite list of numbers qualifies as a valid sequence, even if there is no rule apparent. (This is formalized by considering a real-valued sequence as a function $f : \mathbf{N} \to \mathbf{R}$.)

Sequences feature in every branch of mathematics, as well as the wider world. Their cousins are **series**: sequences which we add up as we go along.

Arithmetic progression

$$5, 8, 11, 14, 17, 20, 23, \ldots$$

This sequence is an example of an *arithmetic progression*, one of the commonest types of sequence. Their defining characteristic is that successive terms are produced by repeatedly adding on the same number. This is called the *common difference*; in the above sequence it is 3. More generally, we might denote it d.

If we write a for the first term of the sequence (5 in the case above), then a general arithmetic progression is:

$$a, a + d, a + 2d, a + 3d, a + 4d, a + 5d, \ldots$$

So the formula for the nth term is:

$$a_n = a + (n - 1)d$$

The 100th term of the above sequence is therefore $5 + 99 \times 3 = 302$.

We can produce a new type of sequence called an *arithmetic series* by adding up the terms of an arithmetic progression as we go along. Starting from the above example we get:

$$5, 13, 24, 38, 55, 75, 98, \ldots$$

The nth term of this new sequence is:

$$na + \frac{n(n - 1)d}{2}$$

So, if we add up the first 100 terms of the first sequence, we will get:

$$100 \times 5 + \frac{100 \times 99 \times 3}{2} = 15{,}350$$

Geometric progressions

2, 6, 18, 54, 162, 486, …

A *geometric progression* is a sequence where successive terms are formed by repeatedly multiplying by some fixed number, the *common ratio*. In this example, the common ratio is 3. If we write r for the common ratio, and a for the first term, then a general geometric progression can be written as:

$$a, ar, ar^2, ar^3, ar^4, \ldots$$

So the nth term is ar^{n-1}, and in the above example, the 16th term will be 28,697,814. Because the common ratio is greater than 1, this sequence grows very quickly. Indeed, this is a classic example of **exponential growth**.

If the common ratio is smaller than 1, then the sequence rapidly tends towards its limit of 0 (see **limits of sequences**). An example is:

$$4, \frac{4}{3}, \frac{4}{9}, \frac{4}{27}, \frac{4}{81}, \frac{4}{243}, \ldots$$

If the terms are added up as we go along, the result is a **geometric series**.

Sessa's chessboard

The beginnings of the game of chess are a matter of historical conjecture. The game is believed to date back to ancient India, some time before AD 600. Although no-one can be certain, there is a legend about an original creator, the wise man Sessa.

One day, Sessa visited the King to demonstrate his invention. The King was so delighted with the clever game that he decided to reward Sessa with whatever he should ask for. Sessa stated that he would like 1 grain of wheat on the first square of his chessboard; on the second square, he would like 2 grains, on the third 4 grains, and on the fourth 8 grains. On each square he wanted double the number of grains on the previous square.

The King was angry that his generous offer should be so insulted, and strode out of the room, commanding a courtier to see to it that Sessa got his wish. The courtier set to work. On the 11th square he had to put 1024 grains, on the 21st over a million. By the 51st square, the courtier would have had to place the entire global wheat harvest for 2007; and there are 64 squares on the chessboard.

Sessa's chessboard is an example of the dramatic consequences of exponential growth.

What comes next?

What comes next in the sequence 1, 2, …? The obvious answer is 3, but on reflection it could equally well be 4. If we take a longer sequence, such as 1, 2, 4, 8, 16, is there still ambiguity?

Counting the number of pieces into which the pictured circles overleaf are divided, we get a surprising answer: 31. Of course it would not be incorrect to say that the next number is 32, if we instead interpret the sequence as powers of 2. The conclusion is that there is no predetermined 'next term', until we understand the mathematics behind the sequence. Similarly, 1, 2, 4, 8, 16, 1001, … is another perfectly good start to a sequence, as is 1, 2, 4, 8, 16, $-\sqrt[3]{\pi}$, … In different contexts these might be exactly what is required.

The moral is that just writing the first few terms of a sequence followed by '…' is ambiguous and inadequate. As the philosopher Ludwig Wittgenstein noted, this sober assessment does not come naturally to humans. When presented with the start of a sequence such as 1, 5, 11, 19, 29, …, our tendency is to analyse it and believe that we have 'understood' it when we can see how to continue it.

The circle sequence illustrates the risks of relying too heavily on this sort of insight. To avoid this ambiguity, it is better to give the underlying rule directly. The standard way to do this is by giving the nth term.

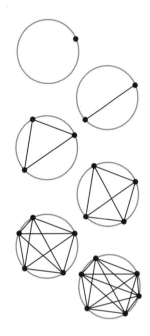

nth term of a sequence

To discuss general sequences we need some notation to save us from having to say 'the first term', 'the second term', and so on. Mathematicians generally use subscripts to keep track:

$$a_1, a_2, a_3, a_4, …$$

The most famous sequence of all begins 1, 2, 3, 4, … Here the first term is 1, the second is 2, the third is 3, and, for each n, the nth term is n. We can write this as $a_n = n$. An even simpler sequence is 1, 1, 1, 1, …. In this case, the first term is 1, the second term is 1. Indeed, for every n, the nth term is 1, or $a_n = 1$. Consider the more complicated sequence 3, 5, 7, 9, 11, … This can be captured by saying that, for every n, $a_n = 2n + 1$. This is an arithmetic progression.

It may seem pedantic to keep repeating the phrase 'for every n'. But many sequences are too complicated to be pinned down by just one formula. For example, consider 2, 0, 8, 0, 32, 0, 128, 0, … Here the sequence is described by $a_n = 2^n$ when n is odd, but $a_n = 0$ when n is even. The circle sequence above (see **what comes next**) has its nth term as the maximum number of pieces a circle can be cut into, by the chords between n points. It begins 1, 2, 4, 8, 16, 31, 57, 99, … and its nth term is given by $a_n = \frac{1}{24}n^4 - \frac{1}{4}n^3 + \frac{23}{24}n^2 - \frac{3}{4}n + 1$.

Exponential growth

Sequences such as $a_n = 2n$ or $a_n = n^4$ model *polynomial growth*, because the nth term of the sequence is given by a **polynomial** in n.

In the story of Sessa's chessboard, the courtier needed to place 2^{n-1} grains of wheat on the nth square. The resulting sequence is defined by $a_n = 2^{n-1}$. Because n features as a power, this is an example of *exponential growth*, much faster than polynomial growth.

If n features as a negative power, such as $b_n = 3^{-n}$, or equivalently as a positive power of a number less than 1, $b_n = \left(\frac{1}{3}\right)^n$, then we have instead *exponential decline*, which falls away to zero very quickly. Faster sequences can be built using **towers of exponentials**, and so on.

Limits of sequences

The *harmonic sequence* is $1, \frac{1}{2}, \frac{1}{3}, \frac{1}{4}, \frac{1}{5}, \frac{1}{6}, \dots$ That is to say its nth term is $\frac{1}{n}$. As we move along this sequence, we get closer and closer to 0, but of course never arrive. Zero is called the *limit* of this sequence.

However, in the domain of analysis, precision is needed. In a literal sense, it is also true that the harmonic sequence gets 'closer and closer' to -2. The difference is that the sequence will eventually get, and remain, as close to 0 as you like. After some point, every remaining term will be within one millionth of 0. At a later point, every remaining term will be within one billionth, or as close as we like. In contrast, the sequence never dips below 0, so can never get within even a distance of 2 of -2. This is formalized through the notion of convergence.

Convergence

Like the **epsilon and delta** definition of continuity, the formal definition of a limit of a sequence is intricate. Consider the harmonic sequence, which has a limit of 0. We can formalize this by saying that, given any positive number, no matter how small (call it ε), there is some point (N) along the sequence, after which every remaining term is within ε of 0. That is, for every $n \geq N$, we have $\left|\frac{1}{n}\right| < \varepsilon$. Mathematicians usually use the Greek letter ε (epsilon) to stand for indeterminate, but very small numbers. (Paul Erdős used to call children *epsilons* for this reason.)

The sequence (a_n) has l as its limit if the following holds:

> For any positive number ε (no matter how small), there is some point (N) along the sequence, after which every remaining term is within a distance ε of l. That is, for every $n \geq N$, we have $|a_n - l| < \varepsilon$.

A consequence is that a sequence can only ever have one limit. Those which have a limit are said to *converge*.

Converging and diverging sequences

The sequence $\frac{1}{2}, \frac{2}{3}, \frac{3}{4}, \frac{4}{5}, \dots$ has a limit, namely 1. Sequences with limits are said to *converge*. We also say that $\frac{n}{n+1}$ *tends to* 1, as n *tends to* infinity. This is written $\frac{n}{n+1} \to 1$ as $n \to \infty$.

On the other hand, Sessa's sequence 1, 2, 4, 8, 16, 32, … does not have a limit. It simply grows and grows. This sequence is said to *diverge to infinity*, written $2^n \to \infty$ as $n \to \infty$.

Similarly, the sequence $-3, -6, -9, \dots$ *diverges to minus infinity*, or $-3n \to -\infty$ as $n \to -\infty$. (In all of these, 'infinity' is solely used as short-hand to describe the eventual behaviour of the sequence.)

Many sequences neither converge nor diverge. For example, the sequence 1, -1, 1, -1, 1, -1, … cycles endlessly between 1 and -1.

SERIES

Series

A sequence is an infinite list of numbers: $1, \frac{1}{2}, \frac{1}{4}, \frac{1}{8}, \ldots$ It *converges* if it gets ever closer to some fixed number (in this case 0). A *series* is obtained when we add the numbers up as we go along: $1 + \frac{1}{2} + \frac{1}{4} + \frac{1}{8} + \cdots$ Again, this converges if it gets ever closer to some fixed number. This series converges to 2.

We use the large sigma notation to describe series, as for finite **sums**. Since this series consists of successive powers of $\frac{1}{2}$ and converges to 2, we might write:

$$\sum_{n=0}^{\infty} \frac{1}{2^n} = 2$$

Series occupy a central place in modern analysis, but they are quirky characters. Telling whether or not a series converges can be tricky, as shown by the deceptive **harmonic series**. Even when we know that a sequence does converge, actually finding the limit can be extremely difficult. It is by no means obvious that the series $1 - \frac{1}{3} + \frac{1}{5} - \frac{1}{7} + \frac{1}{9} - \cdots$ should converge to $\frac{\pi}{4}$. The whole topic is rather delicate, as **Riemann's rearrangement theorem** illustrates.

Geometric series

A geometric progression is a sequence of the form a, ar, ar^2, ar^3, ar^4, $\ldots$ A specific example is $3, \frac{3}{4}, \frac{3}{16}, \frac{3}{64}, \frac{3}{256}, \ldots$ If we add up the terms as we go along, we get a *geometric series*: Σar^n and $\Sigma \frac{3}{4^n}$.

The *partial sums* for these are $S_n = a + ar + ar^2 + ar^3 + \cdots + ar^{n-1}$. So in the given example, $S_1 = 3$, $S_2 = 3 + \frac{3}{4}$, $S_3 = 3 + \frac{3}{4} + \frac{3}{16}$, $S_4 = 3 + \frac{3}{4} + \frac{3}{16} + \frac{3}{64}$, and so on. There is a convenient formula for these, namely:

$$S_n = a \times \frac{1 - r^n}{1 - r}$$

For the given example, $S_n = 3 \times \frac{1 - \frac{1}{4^n}}{1 - \frac{1}{4}} = 4 - \frac{1}{4^{n-1}}$.

This general formula for S_n is valid for any geometric series. But, when r is between 0 and 1, the geometric series converges to a limit of:

$$\sum ar^n = \frac{a}{1 - r}$$

In our example, this comes out as $\Sigma \frac{3}{4^n} = 4$.

The above formula for S_n is not too difficult to derive. Since we know that:

$$S_n = a + ar + ar^2 + ar^3 + \cdots + ar^{n-1}$$

If we multiply this by r we get:

$$rS_n = ar + ar^2 + ar^3 + \cdots + ar^{n-1} + ar^n$$

Subtracting these two equations, most of the terms cancel out. So, $S_n - rS_n = a - ar^n$, which simplifies to the formula we want.

Harmonic series

It is obvious that some series diverge to infinity. For example $1 + 2 + 3 + 4 + 5 + \cdots$ has no hope of settling on a finite limit. For other series it is not so easy to tell. The *harmonic series* is obtained by adding together the terms of the harmonic sequence $1, \frac{1}{2}, \frac{1}{3}, \frac{1}{4}, \frac{1}{5}, \frac{1}{6}, \ldots$ (Its name derives from the harmonics in music.) So the series is $1 + \frac{1}{2} + \frac{1}{3} + \frac{1}{4} + \frac{1}{5} + \frac{1}{6}, \cdots$ or $\Sigma \frac{1}{n}$.

Here, the individual terms certainly get closer and closer to zero. This is a necessary condition for the series to converge, but as it happens it is not sufficient.

The harmonic series diverges

Around 1350, Nicole Oresme proved the unexpected fact that the harmonic series actually diverges to infinity. Start with the original series:

$$1 + \tfrac{1}{2} + \tfrac{1}{3} + \tfrac{1}{4} + \tfrac{1}{5} + \tfrac{1}{6} + \tfrac{1}{7} + \tfrac{1}{8} + \tfrac{1}{9} + \tfrac{1}{10} + \tfrac{1}{11} + \tfrac{1}{12} + \tfrac{1}{13} + \tfrac{1}{14} + \tfrac{1}{15} + \tfrac{1}{16} + \tfrac{1}{17} + \cdots$$

If we decrease each term, and still get something which diverges, then the original must diverge too. So we decrease each term as follows:

$$1 + \tfrac{1}{2} + \tfrac{1}{4} + \tfrac{1}{4} + \tfrac{1}{8} + \tfrac{1}{8} + \tfrac{1}{8} + \tfrac{1}{8} + \tfrac{1}{16} + \tfrac{1}{16} + \tfrac{1}{16} + \tfrac{1}{16} + \tfrac{1}{16} + \tfrac{1}{16} + \tfrac{1}{16} + \tfrac{1}{16} + \tfrac{1}{32} + \cdots$$

Now we can group the terms of this new series into a procession of halves:

$$1 + \tfrac{1}{2} + \underbrace{\tfrac{1}{4} + \tfrac{1}{4}}_{\frac{1}{2}} + \underbrace{\tfrac{1}{8} + \tfrac{1}{8} + \tfrac{1}{8} + \tfrac{1}{8}}_{\frac{1}{2}} + \underbrace{\tfrac{1}{16} + \tfrac{1}{16} + \tfrac{1}{16} + \tfrac{1}{16} + \tfrac{1}{16} + \tfrac{1}{16} + \tfrac{1}{16} + \tfrac{1}{16}}_{\frac{1}{2}} + \tfrac{1}{32} + \cdots$$

This new series will keep adding half and half and half, and so it will eventually grow bigger than any number you care to name. Therefore the harmonic series will do the same.

This is a very surprising result, because the terms of the harmonic sequence get so small that the series hardly seems to be growing at all. In fact, as Leonhard Euler showed, if we add the first n terms of the series, we get approximately $\ln n$. To reach just 10, we have to add the first $12,367$ terms. To reach 100, we have to add around 1.5×10^{43} terms. So although this series diverges to infinity, it does so very slowly indeed.

Brun's constant

Leonhard Euler investigated a series based on the primes: $\frac{1}{2} + \frac{1}{3} + \frac{1}{5} + \frac{1}{7} + \frac{1}{11} + \cdots$ If this series converged, its limit would certainly be of interest. However, Euler was able to prove that it does not converge. Like the harmonic series, this prime series diverges to infinity. (This provides an alternative proof that there are infinitely many prime numbers.)

In 1919, Viggo Brun looked at what would happen to this series if he focused on **twin primes**, instead (that is, prime numbers which are 2 apart). *Brun's series* comes from adding up the reciprocals of all pairs of twin primes:

$$\left(\frac{1}{3} + \frac{1}{5}\right) + \left(\frac{1}{5} + \frac{1}{7}\right) + \left(\frac{1}{11} + \frac{1}{13}\right) + \left(\frac{1}{17} + \frac{1}{19}\right) + \cdots$$

Brun's theorem was that this series does indeed converge to a finite limit. This remarkable number is known as *Brun's constant*, and has been pinned down to approximately 1.90216.

Does this mean that the **twin prime conjecture** is false? Not necessarily, although that is certainly possible. It does imply that twins are very sparse among the primes, even if there are infinitely many of them.

Riemann's rearrangement theorem

Addition is a process that applies to finite collections of numbers. It is tempting to think of series as being 'infinitely many numbers added together', and even to write statements like $1 + \frac{1}{2} + \frac{1}{4} + \frac{1}{8} + \frac{1}{16} + \cdots = 2$. But ultimately this is short-hand which bypasses the crucial concept, that of a *limit*.

There are several ways in which series are unlike ordinary addition. One is that when we add finitely many numbers together, the order of addition does not make a difference. For series this is not true. For example, the series below converges to $\ln 2$.

$$1 - \frac{1}{2} + \frac{1}{3} - \frac{1}{4} + \frac{1}{5} - \frac{1}{6} + \frac{1}{7} - \frac{1}{8} + \frac{1}{9} - \cdots$$

We can re-order this in a clever way:

$$1 - \frac{1}{2} - \frac{1}{4} + \frac{1}{3} - \frac{1}{6} - \frac{1}{8} + \frac{1}{5} - \frac{1}{10} - \frac{1}{12} + \cdots$$

Grouping certain terms together, we get:

$$\left(1 - \frac{1}{2}\right) - \frac{1}{4} + \left(\frac{1}{3} - \frac{1}{6}\right) - \frac{1}{8} + \left(\frac{1}{5} - \frac{1}{10}\right) - \frac{1}{12} + \cdots$$

When we work out the brackets, this becomes:

$$\frac{1}{2} - \frac{1}{4} + \frac{1}{6} - \frac{1}{8} + \frac{1}{10} - \frac{1}{12} + \cdots$$

This is exactly half of what we started with! So this series converges to $\frac{1}{2} \ln 2$. In fact the situation is even worse. Bernhard Riemann's *rearrangement theorem* says that this series can be rearranged to converge to any number you choose (or diverge to $\pm \infty$).

The same is true for any *conditionally convergent* series, that is, one which does *not* converge when all its terms are made positive. When the terms of this series are made positive, it produces the divergent harmonic series. Happily, other *absolutely convergent* series are better behaved.

Generalized binomial coefficients

The **binomial theorem** tells us how to expand brackets such as $(1 + z)^{17}$ without having to work through sixteen intermediate stages.

The answer is a sum of terms of the form $\binom{17}{r} z^r$, as r takes the values from 0 to 17. Here $\binom{17}{r}$ is a **combination**. Writing it out:

$$\binom{17}{r} = \frac{17 \times 16 \times 15 \times \dots \times (17 - r + 1)}{r!}$$

On the face of it, there is no reason to expect this theorem to have anything to say about $(1 + z)^{-17}$, for example. This is something different and cannot be expressed as a sum of positive powers of z in any obvious way.

Yet, nothing ventured, nothing gained. Isaac Netwon experimented to see what would happen if the exponent was replaced with a negative number. He might have ended up with nonsense; in fact he found a powerful generalization of the theorem. First, he had to define the *generalized binomial coefficient* $\binom{-17}{r}$:

$$\binom{-17}{r} = \frac{(-17) \times (-18) \times (-19) \times \dots \times (-17 - r + 1)}{r!}$$

In general, if a is any complex number, and r is a natural number, then:

$$\binom{a}{r} = \frac{a \times (a - 1) \times (a - 2) \times \dots \times (a - r + 1)}{r!}$$

Generalized binomial theorem
Mimicking the original binomial theorem, Newton's generalized version is intended to express objects such as $(1 + z)^{-17}$ as a sum of terms of the form $\binom{-17}{r} z^r$, as r assumes the values 0, 1, 2, 3, ... So, if the theorem is to be true, it will produce an infinite series:

$$(1 + z)^{-17} = \sum_{r=0}^{\infty} \binom{-17}{r} z^r$$

In full generality, the statement is that, for any complex number a:

$$(1 + z)^a = \sum_{r=0}^{\infty} \binom{a}{r} z^r$$

However, the issue of convergence is delicate here. For most values of z, this series will not converge, and the theorem will fail. But whenever z is a complex number with $|z| < 1$ this series will converge, and this *generalized binomial theorem* holds.

CONTINUITY

Achilles and the tortoise

Around 450 BC, the philosopher Zeno of Elea assembled a list of paradoxes. The most famous purport to prove that physical motion is impossible.

The first paradox involves the mythical hero Achilles. When this great warrior sets out to race a tortoise, he encounters an unexpected problem. Agreeing to give the tortoise a head-start, he has no trouble running to where the tortoise started. But when he gets there, he finds the tortoise has moved on slightly. He can run to the tortoise's new position, but again once he arrives it has moved ahead, and so on. He can never reach the tortoise without first arriving at its previous position, but every time he does, it moves on slightly. Therefore he can never catch the tortoise.

Achilles and the tortoise have further paradoxical adventures in Lewis Carroll's logical dialogue *What the tortoise said to Achilles*, and then in one of the great books of the 20th century, Douglas Hofstadter's *Gödel, Escher, Bach*.

It is possible that Zeno genuinely believed that all motion and change is illusory. However, the enduring importance of his paradoxes, from a mathematical perspective, is in illustrating the subtleties in the relationship between discrete and continuous systems.

Zeno's dichotomy paradox

Zeno's *dichotomy paradox* makes a similar point to the one of Achilles and the tortoise, but even more forcefully. Suppose Achilles is training for his big race by running a mile, alone this time. Before he can cross the finish line, he must first reach the half-way mark. But, before he can reach that, he must reach the point half-way to that, the quarter mark. Before he can get there, though, he must pass the eighth mark, and so on, *ad infinitum*. Suddenly he is faced with an infinity of tasks to do, and no first step. This time Achilles cannot even get started.

Zeno's paradoxes

Zeno's paradoxes prefigure modern mathematical analysis, and come to us through the works of Aristotle. It was he who made an important observation about the dichotomy paradox: as the distances become ever smaller, so too does the time required to run them. Indeed, as the distances and times both become minuscule, the distance divided by the time approaches a limit, which defines Achilles' speed at that moment. This is the very definition of the **derivative**.

In the case of Achilles' race against the tortoise, if each phase of the race took the same length of time, then it really would be true that he could never catch the creature. But this is not the case; successive phases take ever smaller quantities of time, which together form a *convergent series*. The limit of this series will be the moment Achilles moves into the lead. What is misleading about the paradox is that the time taken to *say* each step remains the same, and this forms a divergent series.

The rational numbers have gaps

Take a piece of paper with a horizontal line across the middle of the page. Put one mark in the bottom half of the page, and another in the top. It seems obvious that any curve you can draw between the two marks, without your pen leaving the paper, must cross the dividing line at some point (leaving aside tricks such as going around the back of the page).

This is not a mathematical scenario, of course, but it is a model that any sensible theory of **continuous functions** should follow. Unfortunately, when working in the **rational numbers**, this does not happen. Take the function $f(x) = x^2 - 2$. This seems an innocuous, continuous function. If we plot the graph of $y = f(x)$ then at $x = 0$, y is negative, and when $x = 2$, y is positive. So the graph should cross the horizontal line $y = 0$ somewhere. But, working in the rational numbers, it does not: there is no rational number such that $f(x) = 0$. There is a gap in the rational line at $\sqrt{2}$, and the graph has sneaked through.

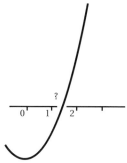

The intermediate value theorem

The beauty of the **real numbers** is that all the gaps in the rational numbers have been filled in. The *intermediate value theorem* asserts this once and for all. If you have a continuous function f on the real numbers, which is negative at some point (a) and positive somewhere else (b), then there is guaranteed to be an intermediate point (c) where f crosses the horizontal line, that is to say $f(c) = 0$. This theorem shows that the real numbers capture our intuition about points, curves and continuity, and form the right setting for geometry and analysis.

Discreteness and continuity

The whole numbers are *discrete*, meaning that they come in separate packages, with gaps in between. We jump from 1 to 2 and then from 2 to 3. The real numbers, in contrast, are the classic example of a *continuous* set. Here we glide smoothly between 1 and 2, covering an infinite number of intermediate points. The intermediate value theorem shows that, even on the finest imaginable scale, there are no gaps.

Discreteness and continuity are the north and south poles of mathematics. Each provides the setting for a great deal of fascinating work. **Topology** and **analysis** are concerned with continuous functions, whereas **combinatorics** and **graph theory** are wholly discrete in nature. These two areas have very different feels, and the tension between them poses many technical and conceptual challenges, of which Zeno's paradoxes were an early taste. When discrete and continuous situations collide, the fireworks can be spectacular. Examples include the number theoretical heights of **Diophantine geometry** and the mysterious quantum phenomenon of **wave–particle duality**.

Continuous functions

A *real function* is a **function** which takes real numbers as inputs and gives real numbers as outputs. These are hugely useful for modelling all sorts of physical processes. For example, walking down a road can be modelled by a real function which takes numbers representing time as input, and gives numbers representing distance as output. In this and in many applications, the function we get is *continuous*, meaning that it doesn't contain any gaps or jumps. (A discontinuous function would be needed to model someone walking down the road and then suddenly teleporting 10 feet forwards.)

Being continuous need not imply that the curve is *smooth*, however. It may be very jagged, as long as it doesn't contain any gaps. So every differentiable function (see **differentiability**) is certainly continuous, but the reverse does not hold (an example being the **Koch snowflake**).

Weiertrass' *epsilon–delta* definition of continuity confuses everyone on first meeting. But it illustrates the level of rigour that Augustin Cauchy and Karl Weierstrass introduced into mathematical analysis in the 19th century, replacing the old fallacious reasoning about **infinitesimals**. It is remarkable that such an intuitive notion as continuity should require such a technical formulation.

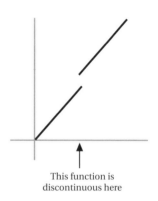

This function is discontinuous here

Epsilon and delta

Suppose that f is a real function, where $f(0) = 1$. To say that f is *continuous* at this point intuitively means that it does not have a gap at this point. The idea is simple enough: as x gets closer to 0, then $f(x)$ gets closer to 1. To formalize this idea of 'getting closer', Weierstrass rephrased it as follows: for any number ε (no matter how small), whenever x is close enough to 0, then $f(x)$ will be within ε of 1.

This version still contains the undefined notion of being 'close enough' however. This can be removed as follows: for any number ε, there exists another positive number δ, so that, whenever x is within δ of 0, then $f(x)$ will be within ε of 1. This is now the modern definition of continuity. It can be written out more quickly using **logical quantifiers**:

$$(\forall \varepsilon > 0)(\exists \delta > 0)(|x| < \delta \rightarrow |f(x) - 1| < \varepsilon)$$

More generally, a function $f : \mathbf{R} \rightarrow \mathbf{R}$ is continuous at a point a, if:

$$(\forall \varepsilon > 0)(\exists \delta > 0)(|x - a| < \delta \rightarrow |f(x) - f(a)| < \varepsilon)$$

If we want to say that f is continuous everywhere, we need even more:

$$(\forall a)\, (\forall \varepsilon > 0)(\exists \delta > 0)(|x - a| < \delta \rightarrow |f(x) - f(a)| < \varepsilon)$$

DIFFERENTIAL CALCULUS

Differentiability

The illustrated curve has a *cusp*, a sharp point. Everywhere else the curve is *smooth*, but not at this particular point. How can we define what it means to be smooth? Differentiation provides an answer. This is a subtle process, but what it amounts to is finding the **tangent** to the curve. The problem is that you cannot always do this. At its cusp, the curve does not have a unique tangent. Any line you can draw is as good as any other, so the tangent is undefined, and the curve is *not differentiable* at this point.

Differentiability encapsulates our understanding of smoothness. It is a stronger requirement than *continuity*. A function is continuous if it does not contain any jumps or gaps. This curve fits this definition, even at its cusp. On the other hand, every differentiable function is automatically continuous.

Of course, the pictured curve is smooth everywhere except at its cusp. For many years, mathematicians thought this was typical: every continuous function should be differentiable, except possibly at a few points. However, in 1872 Karl Weierstrass shocked the establishment by producing a function which is continuous everywhere, but not differentiable *anywhere*. The **Koch snowflake** is another such curve. The theory of **smooth functions** works much more cleanly in the complex numbers.

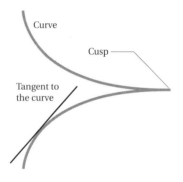

Curve

Cusp

Tangent to the curve

The gradient of a tangent

Gradient is a measure of steepness. To measure it for a straight line is easy: calculate the change in height, divided by the change in base length, over some range. Usually this can be read straight from the equation. The line $y = 4x + 1$ has a gradient of 4, for example. This is very important in many practical applications, as the gradient of the line represents the *rate of change of y*, written as $\frac{dy}{dx}$. Velocity, for example, can be modelled as the **rate of change of position**.

When we have a curve instead of a straight line, this process is trickier. To start with, it is not obvious what the *gradient* of a curve should mean. The agreed meaning is that the gradient is that of the *tangent* to the curve, the straight line which touches it at exactly one point.

Choosing this point makes a huge difference. Unlike straight lines, the gradient of a curve varies from place to place. Dating back to Archimedes, the basic method for calculating this gradient is as follows: locate the chosen point on the curve, draw a tangent by hand, and calculate its gradient. The problem with this is that it relies on being able to draw curves and straight lines with perfect accuracy. In reality, this method will only ever produce an estimate of the true value. For an exact procedure mathematicians had to wait for the development of the **derivative**.

Approximating tangents with secants

Suppose we want to calculate the gradient of the curve $y = x^2$ at the point $(1, 1)$. Archimedes' method would be to draw the tangent at this point, and calculate it. Ideally, however, the accuracy of our mathematics should not be hostage to our artistic skill. To draw an exact tangent at a point is difficult. But to draw an approximate tangent is easy: pick another nearby point on the curve and join the two with a straight line (called a *secant*). We could pick $(2, 2^2)$, that is, $(2, 4)$, as our second point. We can join these with a straight line, and easily calculate its gradient: $\frac{4-1}{2-1} = 3$.

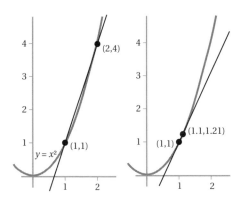

If we had chosen a second point nearer $(1, 1)$ we would get a better approximation. The point $(1.5, 2.25)$ would give a gradient of $\frac{2.25 - 1}{1.5 - 1} = 2.5$. If we choose $(1.1, 1.21)$, we get a gradient of $\frac{1.21 - 1}{1.1 - 1} = 2.1$. Similarly, $(1.01, 1.0201)$ gives a gradient of 2.01. As the second point moves ever closer to $(1, 1)$, it seems the gradient is getting closer to 2. So we could guess that this is what we really wanted, the gradient of the tangent at $(1, 1)$. To prove this rigorously we need a little algebra.

Differentiating from first principles

To calculate the gradient of $y = x^2$ at the point $(1, 1)$ we need to pick a second point on the curve, near $(1, 1)$. Choose a small number h, and set the second point to have x-coordinate $(1 + h)$. From the equation of the curve, the coordinates of our second point are $((1 + h), (1 + h)^2)$. Expanding this bracket, we get $((1 + h), (1 + 2h + h^2))$.

We want to know the gradient of the line passing through this and $(1, 1)$. As always this is the change in height divided by the change in base: $\frac{(1 + 2h + h^2) - 1}{(1 + h) - 1}$. Simplifying, this becomes $\frac{2h + h^2}{h}$, which is $2 + h$. This is the gradient of the secant.

It is now clear what happens as the second point gets closer to $(1, 1)$. This corresponds to h getting smaller and smaller, meaning that the gradient $2 + h$ gets closer and closer to 2.

If we had chosen the point $(3, 9)$ we would have found that the curve has gradient $2 \times 3 = 6$. Similarly at $(-4, 16)$ it has gradient $2 \times -4 = -8$. In general, at the point (x, x^2), the curve has gradient $2x$. This shows that the *derivative* of the function $y = x^2$ is $y = 2x$.

Derivative

Suppose we have a function f which has real numbers as its inputs and outputs. The *derivative* of f is a new function (written f') which describes the rate of change of f. In graphical terms, if we have the graph of $y = f(x)$, then we can look at its gradient at the point a. This is the number given by $f'(a)$. How is this defined?

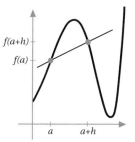

An approximate gradient would be given by $\frac{f(a + h) - f(a)}{h}$ for some small number h. So, to calculate the exact gradient, we take the limit of this as h gets smaller and smaller:

$$f'(a) = \lim_{h \to 0} \frac{f(a + h) - f(a)}{h}$$

This is the definition of the derivative. (It is an important caveat that this limit must exist uniquely, otherwise we have a non-differentiable function on our hands.)

When y is another variable related to x by $y = f(x)$, we write the result as:

$$\frac{dy}{dx} = f'(x)$$

When working with derivatives in practical applications, this technical definition usually remains firmly in the background. In practice, the derivative of a great many functions can be deduced from a few standard derivatives.

The derivative song

Tom Lehrer is best known for his satirical songs such as 'Poisoning Pigeons in the Park'. He is also a mathematician, and taught at Harvard University. In one of his lesser known works, Lehrer set the definition of a derivative to music. Very appropriately, the tune was 'There'll Be Some Changes Made' by W. Benton Overstreet (with original lyrics by Billy Higgins).

You take a function of x, and you call it y,
Take any x_0 that you care to try,
You make a little change and call it δx,
The corresponding change in y is what you find nex',
And then you take the quotient, and now carefully
Send δx to zero and I think you'll see
That what the limit gives us, if our work all checks,
Is what we call $\frac{dy}{dx}$. It's just $\frac{dy}{dx}$!

Here x_0 ('x nought') is the point at which we are calculating the derivative (a in the example above). Similarly δx ('delta x') is the small change denoted by h above. So the 'corresponding change in y' is $f(x_0 + \delta x) - f(x_0)$.

Standard derivatives

The definition of the derivative can be tricky to apply directly. Happily, there are several standard facts we can use instead:

y	$\dfrac{dy}{dx}$	y	$\dfrac{dy}{dx}$
x^n	nx^{n-1}	C^x	$C^x \ln C$ (valid for $x > 0$)
C (any constant number)	0	$\log_c x$	$\dfrac{1}{x \ln c}$ (valid for $\lvert c \rvert \neq 0, 1$)
e^x	e^x	$\ln x$	$\dfrac{1}{x}$
$\sin x$	$\cos x$	$\sinh x$	$\cosh x$
$\cos x$	$-\sin x$	$\cosh x$	$\sinh x$
$\tan x$	$\sec^2 x$	$\tanh x$	$\operatorname{sech}^2 x$
$\operatorname{cosec} x$	$-\operatorname{cosec} x \cot x$	$\operatorname{cosech} x$	$-\operatorname{cosech} x \coth x$
$\sec x$	$\sec x \tan x$	$\operatorname{sech} x$	$-\operatorname{sech} x \tanh x$
$\cot x$	$-\operatorname{cosec}^2 x$	$\coth x$	$-\operatorname{cosech}^2 x$
$\sin^{-1} x$	$\dfrac{1}{\sqrt{1-x^2}}$ (for $-1 < x < 1$)	$\sinh^{-1} x$	$\dfrac{1}{\sqrt{1+x^2}}$
$\cos^{-1} x$	$\dfrac{-1}{\sqrt{1-x^2}}$ (for $-1 < x < 1$)	$\cosh^{-1} x$	$\dfrac{1}{\sqrt{-1+x^2}}$ (for $\lvert x \rvert > 1$)
$\tan^{-1} x$	$\dfrac{1}{1+x^2}$	$\tanh^{-1} x$	$\dfrac{1}{1-x^2}$ (for $-1 < x < 1$)

Together with the product rule, quotient rule, and chain rule, these standard derivatives are enough to differentiate a wealth of common functions, without having to worry about tricky limiting processes.

Product rule

The table of standard derivatives tells us how to differentiate $y = x^3$ and $y = \sin x$. But how do we differentiate $y = x^3 \sin x$? This is an instance of the general question of how to differentiate the product of two functions: $y = f(x)g(x)$.

A common mistake is to differentiate the first two separately and multiply the results together. This is wrong (as can be seen by taking $f(x) = g(x) = x$, for example). The correct answer, known as the *product rule* or *Leibniz' law*, is:

$$\frac{dy}{dx} = f(x)g'(x) + f'(x)g(x)$$

So in the above example, where $y = x^3 \sin x$, we get $\frac{dy}{dx} = x^3 \cos x + 3x^2 \sin x$.

Proof of the product rule

Why should the product rule be true? The reason comes from manipulating the definition of the derivative of $f(x)g(x)$. It must be the limit of:

$$\frac{f(x + h)\, g(x + h) - f(x)g(x)}{h}$$

By adding and subtracting $f(x + h)\, g(x)$ to the top row, we can rewrite this as:

$$\frac{f(x + h)\, g(x + h) - f(x + h)g(x) + f(x + h)g(x) - f(x)g(x)}{h}$$

This equals

$$\frac{f(x + h)\, (g(x + h) - g(x)) + (f(x + h) - f(x))g(x)}{h}$$

which is

$$f(x + h)\left(\frac{g(x + h) - g(x)}{h}\right) + \left(\frac{f(x + h) - f(x)}{h}\right)g(x)$$

As $h \to 0$, $f(x + h) \to f(x)$ and so the whole thing approaches $f(x)g'(x) + f'(x)g(x)$.

The chain rule

Using the table of standard derivatives we can differentiate $y = x^3$ and $y = \sin x$, and thanks to the product rule, we can also differentiate $y = x^3 \sin x$. But what if we combine these functions in a different way, such as $y = \sin(x^3)$? The general question is to find $\frac{dy}{dx}$ when $y = f(g(x))$.

The answer, known as the *chain rule*, is:

$$\frac{dy}{dx} = f'(g(x)) \times g'(x)$$

So in the above example $\frac{dy}{dx} = \cos(x^3) \times 3x^2$. Similarly, if $y = e^{\sin x}$, then $\frac{dy}{dx} = e^{\sin x} \times \cos x$.

Iterated applications of the chain rule can allow even more complicated functions to be differentiated. For instance, if $y = e^{\sin(x^3)}$, x then $\frac{dy}{dx} = e^{\sin(x^3)} \times \cos(x^3) \times 3x^2$.

Proof of the chain rule

The chain rule is extremely useful, as it enormously increases the number of functions we can differentiate. To see why it should be true requires getting our hands dirty with the technical definition of the derivative.

If $y = f(g(x))$ then $\frac{dy}{dx}$ must be the limit of:

$$\frac{f(g(x + h)) - f(g(x))}{h}$$

The top line almost looks like $f(g(x) + h) - f(g(x))$, in which case the whole thing would approach $f'(g(x))$. This isn't quite right though.

We can manoeuvre ourselves into this position by introducing a new small number j, defined as $j = g(x + h) - g(x)$. As h becomes very small, so too does j (since g is continuous). Now $f(g(x + h)) = f(g(x) + j)$. So $\frac{dy}{dx}$ is the limit of:

$$\frac{f(g(x) + j) - f(g(x))}{h}$$

To arrive at $f'(g(x))$, we would like to have j on the bottom row, instead of h. We can arrange this, by multiplying the whole thing by $\frac{j}{h}$. Now $\frac{dy}{dx}$ is the limit of:

$$\frac{f(g(x) + j) - f(g(x))}{j} \times \frac{j}{h}$$

The first fraction does indeed approach $f'(g(x))$. So we need to understand the second, $\frac{j}{h}$. From the definition of j, we know that $\frac{j}{h} = \frac{g(x + h) - g(x)}{h}$, and as $h \to 0$ this approaches $g'(x)$.

The quotient rule
If the product rule allows us to differentiate $x^3 \ln x$, and the chain rule $\ln (x^3)$ then how do we differentiate $y = \frac{x^3}{\ln x}$? The general problem here is to differentiate $y = \frac{f(x)}{g(x)}$.

The answer is given by the *quotient rule*, which says that:

$$\frac{dy}{dx} = \frac{g(x)f'(x) - g'(x)f(x)}{g(x)^2}$$

So, in the above example, if $y = \frac{x^3}{\ln x}$, then

$$\frac{dy}{dx} = \frac{\ln x \times 3x^2 - \frac{1}{x} x^3}{(\ln x)^2}$$

which is

$$\frac{3x^2}{\ln x} - \frac{x^2}{(\ln x)^2}$$

Rather than being a rule in its own right, the quotient rule follows from applying the chain rule to differentiate $(g(x))^{-1}$ to get $\frac{-g'(x)}{g(x)^2}$, and then the product rule to $y = f(x)g(x)^{-1}$.

Implicit differentiation
A typical question about differentiation might present us with a formula of the form $y = f(x)$, where f is some function. By applying the rules above to f, we can (hopefully) differentiate it to arrive at $\frac{dy}{dx} = f'(x)$. However, there is no law that says we have to begin with y alone on one side of the equation. If we have $2y + 5x = x^3$, we can differentiate this immediately to get $2\frac{dy}{dx} + 5 = 3x^2$.

This process of *implicit* differentiation works in exactly the same way as before. We just have to remember that whenever we differentiate y, we get $\frac{dy}{dx}$ (just as whenever we differentiate $\sin x$

we get cos x). So, if we have $yx^2 = \sin x$, we can differentiate the left-hand side using the product rule, to get $\frac{dy}{dx}x^2 + y(2x) = \cos x$. Similarly, if we want to differentiate $y^2 = \ln x$, we can use the chain rule on the left-hand side, to get $2y\frac{dy}{dx} = \frac{1}{x}$.

Maxima and minima

Looking at the graph of $y = x^3 - 3x$, there are two places where the graph is perfectly horizontal. Let's find them exactly. A horizontal graph corresponds to a gradient of 0, which means that at these two points $\frac{dy}{dx} = 0$. So we start by differentiating: $\frac{dy}{dx} = 3x^2 - 3$.

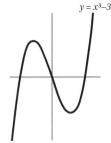

$y = x^3 - 3x$

If $\frac{dy}{dx} = 0$, that means $3x^2 - 3 = 0$, and we can solve this to find $x = 1$ or $x = -1$. Substituting these into the original equation to find the y-coordinates, we get the two points we wanted: $(1, -2)$ and $(-1, 2)$.

In some sense these points represent maximum and minimum values of y. But $(-1, 2)$ is not really the maximum value of y (when $x = 10$, $y = 970$, for example). It is a *local* maximum, however. There is no point in its immediate vicinity which is greater.

Local maxima and minima represent a curve's *turning points*, and they can be found by finding the points where $\frac{dy}{dx} = 0$.

Care is needed, however: if we try this with the graph $y = x^3$ we find $\frac{dy}{dx} = 0$ at the origin. Although the graph is flat here, this is not a turning point, it is a *stationary inflection point*.

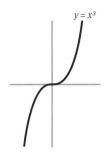

$y = x^3$

Maxima, minima, and stationary inflection points can be distinguished by the second derivative test.

The second derivative

If we start with a function f and differentiate it, we arrive at the derivative f' of f. For example, if $f(x) = \sin x$, then $f'(x) = \cos x$. This function f' describes the rate of change of f. What happens if we differentiate again? We get the *second derivative f''*. In this case $f''(x) = -\sin x$. When the variable y is defined by $y = f(x)$, we write $\frac{dy}{dx} = f'(x)$, and $\frac{d^2y}{dx^2} = f''(x)$. ($\frac{d^2y}{dx^2}$ is pronounced 'd two y by d x squared'.)

The derivative measures the gradient of the graph $y = f(x)$, so what does the second derivative mean? Of course, it must be the 'rate of change of the gradient', but what geometric meaning does this have? If $\frac{d^2y}{dx^2}$ is positive, that means the curve is getting steeper and appears convex from the left. If you draw a tangent, it will appear below the curve. If $\frac{d^2y}{dx^2}$ is negative, then the gradient is decreasing and appears concave from the left. If you draw a tangent it will be above the curve. If $\frac{d^2y}{dx^2} = 0$, then the gradient is unchanging and the tangent runs concurrent to the curve.

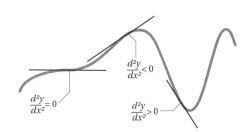

$\frac{d^2y}{dx^2} = 0$ $\frac{d^2y}{dx^2} < 0$ $\frac{d^2y}{dx^2} > 0$

The second derivative test

The second derivative provides a useful test for identifying maxima, minima, and stationary inflection points. At a maximum, the gradient is decreasing, meaning that $\frac{d^2y}{dx^2} \leq 0$. At a minimum, the gradient is increasing, and $\frac{d^2y}{dx^2} \geq 0$. At a stationary inflection point, the gradient is unchanging, meaning that $\frac{d^2y}{dx^2} = 0$.

So if $\frac{d^2y}{dx^2} < 0$, then our stationary point is a maximum. If $\frac{d^2y}{dx^2} > 0$, it is a minimum. However, if $\frac{d^2y}{dx^2} = 0$, then we cannot be sure. It might be a stationary point of inflection, but it might not. For example, $y = x^4$ has a minimum at 0, but $\frac{dy}{dx} = \frac{d^2y}{dx^2} = 0$ there.

Rates of change of position

'Rates of change of rates of change' can certainly be difficult to get one's head around. Luckily, there is a ready-made example: a moving object. Suppose a bicycle is moving along a straight road, and the function s gives its displacement from a particular house after t seconds. Then $\frac{ds}{dt}$ is its rate of change of distance, that is to say, its *velocity*. $\frac{d^2s}{dt^2}$ is the rate of change of velocity, namely its *acceleration*.

Displacement and velocity

Why do we mathematicians say 'displacement' instead of 'distance'? The answer is that two people can be the same distance from the house, but not be in the same place: for example, if one is 5 metres up the road, and the other is 5 metres down the road. The first would be said to have a displacement of -5 metres, and the second a displacement of $+5$ metres. Similarly a bicycle coming up the road might have a velocity of $+6$ m/s, and one going down the road a velocity of -6 m/s. These are different velocities, even though they have the same speed. In more complex situations displacement and velocity are each given by **vectors**.

Partial derivatives

Ordinary differentiation takes place where we have a relationship between two variables, such as x and y. For example, the formula $y = x^2$ describes a curve, which we can differentiate with respect to x, to get $\frac{dy}{dx} = 2x$.

Often more than two variables are involved. For example, the formula $z = 2xy$ describes a surface (specifically a **hyperbolic paraboloid**) in three dimensions. We can think of this as a function which takes a pair of numbers (x and y) as input, and produces a single number (z) as output. We can return to a familiar scenario by fixing a value of y, such as $y = 5$. Geometrically, this corresponds to taking a slice through the surface, to get a curve with equation $z = 10x$. We can now differentiate this as usual, to get $\frac{dz}{dx} = 10$. Of course, if we had fixed a different value of y, such as $y = -2$, we would have found a different curve, $z = -4x$, with a different derivative, $\frac{dz}{dx} = -4$.

In fact, whatever value of y we fix, we will get twice that number as the derivative. This says that $\frac{\partial z}{\partial x} = 2y$. This is the *partial derivative* of x with respect to y. (In partial differentiation, it is traditional to use a curly d, ∂. This is not a Greek delta, δ.)

Partial differentiation

The basic laws for partial differentiation are exactly the same as for ordinary differentiation. For example, suppose we have a formula $z = x^2 + 3xy + \sin y$, which we want to differentiate with respect to x. The only new rule is that y is treated in exactly the same way as a constant term is (as if we have fixed it at some value). So we get $\frac{\partial z}{\partial x} = 2x + 3y$. (The $\sin y$ term disappears as it represents a constant.) Starting again, we might wish to differentiate $z = x^2 + 3xy + \sin y$ with respect to y instead. Now x must be treated as a constant, and we get $\frac{\partial z}{\partial y} = 3x + \cos y$.

Similarly if $y = 4t^2x$, then $\frac{\partial y}{\partial x} = 4t^2$ and $\frac{\partial y}{\partial t} = 8tx$.

Tangent spaces

In a curve, such as $y = x^2$, the derivative $\frac{dy}{dx} = 2x$ has a nice geometrical interpretation, as the gradient of the curve. So, at the point $x = 4$, the curve has gradient 8. More precisely, the derivative is the gradient of the *tangent* to the curve. This is the straight line which touches the curve exactly once, at the point (x, y).

On a surface, the tangent is no longer a line, but a 2-dimensional *tangent plane*. We would expect differentiation to analyse the slant of this plane. But this cannot be captured by a single number. Instead, it is described by a pair of numbers: the plane's slope in the x–z plane, and its slope in the y–z plane. These are given by the partial derivatives $\frac{\partial z}{\partial x}$ and $\frac{\partial z}{\partial y}$ respectively.

Higher-dimensional **manifolds** have higher-dimensional *tangent spaces*, described by correspondingly more partial derivatives.

Second partial derivatives

With ordinary differentiation, once we have differentiated y to get $\frac{dy}{dx}$, we then differentiate again to arrive at $\frac{d^2y}{dx^2}$, the second derivative. We can do the same with partial differentiation, but there is more choice. If we have $z = x^2 + 3xy + \sin y$, we can differentiate with respect to x to get $\frac{\partial z}{\partial x} = 2x + 3y$, and then differentiate that with respect to x again, to get $\frac{\partial^2 z}{\partial x^2} = 2$. Similarly, we can differentiate with respect to y twice, to get $\frac{\partial z}{\partial y} = 3x + \cos y$ and $\frac{\partial^2 z}{\partial y^2} = -\sin y$.

There are also *mixed derivatives*. If we differentiate $\frac{\partial z}{\partial x} = 2x + 3y$ with respect to y, we get $\frac{\partial^2 z}{\partial x\, \partial y} = 3$. Similarly, we may differentiate $\frac{\partial z}{\partial y} = 3x + \cos y$ with respect to x to get $\frac{\partial^2 z}{\partial y\, \partial x} = 3$. It is no coincidence that these two come out the same. *Clairaut's theorem* guarantees that $\frac{\partial^2 z}{\partial x\, \partial y} = \frac{\partial^2 z}{\partial y\, \partial x}$ always holds. So the order of variables by which we differentiate does not matter. We can continue to give higher partial derivatives, such as $\frac{\partial^3 z}{\partial y^3} = -\cos y$, $\frac{\partial^3 z}{\partial y^2 \partial x} = 0$, and so on.

INTEGRAL CALCULUS

Integration

The integration symbol $\int$ was first employed by Gottfried Leibniz. It is a deformed 'S', standing for 'summatorius', Latin for 'sum'. In a series, such as $1 + \frac{1}{4} + \frac{1}{9} + \frac{1}{16} + \cdots$ we sum up a sequence of terms as we go along. This is easy to understand, as the terms are *discrete*. But integration applies to *continuous* functions, such as $y = x^2$. What might it mean to 'sum' this function as we travel along it? The answer is that integration gives the *area* under this curve.

$$\int_1^4 x^2 \, dx$$

represents the area enclosed by the curve $y = x^2$ and the x-axis, between the points 1 and 4.

The term dx indicates that integration is being performed with respect to the variable x. In a subtle way, this tells us how the area is being measured. In this case, the area we want is that swept out as x moves smoothly between its two limits, 1 and 4. The fundamental theorem of calculus provides the method for calculating this area.

Step functions

It is not obvious how to evaluate the area under a curve. Some functions are easier to manage though. Take the curve which is equal to 2 between 1 and 4, and 0 elsewhere. We can write this as:

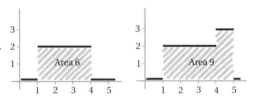

$$f(x) = \begin{cases} 2 & \text{if } 1 \le x \le 4 \\ 0 & \text{otherwise} \end{cases}$$

The area beneath this is a rectangle of width 3 and height 2, and therefore with area $3 \times 2 = 6$.
Similarly the function

$$f(x) = \begin{cases} 2 & \text{if } 1 \le x < 4 \\ 3 & \text{if } 4 \le x \le 5 \\ 0 & \text{otherwise} \end{cases}$$

has an area comprising two rectangles, giving a total area of 9. These are two examples of *step functions*, which are easy to integrate.

The definition of integration for more general curves is technical, and was first fully worked out by Bernhard Riemann and Henri Lebesgue in the late 19th century. But the idea is simple enough: it

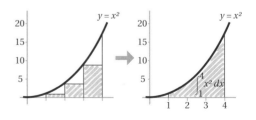

comes from approximating the target function ever more closely by step functions. Almost every function can be approximated this way, and is therefore integrable.

Happily, awkward sequences of step functions remain firmly in the background for most practical problems. If we want to integrate a curve such as $f(x) = x^2$, the fundamental theorem of calculus provides a much easier method.

Definite integrals

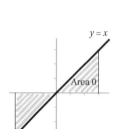

Integrals come in two forms: *definite*, in which the endpoints are specified, and *indefinite*, without endpoints. The integral

$$\int_1^4 x^2 \, dx$$

is definite, because it has endpoints 1 and 4. The result of a definite integral should be a specific number (in this case 21), which gives the area under the curve.

When the curve passes below the x-axis, the area comes out as a negative number. For example, integrating $y = -x$ between 0 and 4:

$$\int_0^4 -x \, dx = -8$$

If we integrate $y = x$ between -3 and 3 the positive and negative areas cancel each other out:

$$\int_{-3}^3 x \, dx = 0$$

This total 'area' of 0 can give a misleading picture!

Indefinite integrals

Definite integrals produce numbers, representing areas. But how can we calculate these numbers? There is a function which gives them, called the *indefinite integral*. For example, consider the definite integral:

$$\int_1^4 x^2 \, dx$$

The corresponding indefinite integral is

$$\int x^2 \, dx = \frac{x^3}{3}$$

(using the fundamental theorem of calculus, and omitting the **constant of integration**). This function now allows any definite integral to be calculated. If the indefinite integral of f is F, then any definite integral is calculated according to the rule:

$$\int_a^b f(x) \, dx = F(b) - F(a)$$

In the above example, this comes out as $\int_1^4 x^2 \, dx = \frac{4^3}{3} - \frac{1^3}{3} = 21$.

The next question is how to evaluate indefinite integrals. The fundamental theorem of calculus provides the answer.

The fundamental theorem of calculus

The subject of calculus has two components: differentiation and integration. The *fundamental theorem of calculus* relates the two. The answer is that differentiation and integration are *inverse* procedures, one is the other one 'done backwards'.

If we differentiate a function f to get f', and then integrate that, we arrive back at f (along with a **constant of integration**). Writing this out formally, we get:

$$\int f'(x)\, dx = f(x) + C$$

Going the other way, if $\int f(x)\, dx = F(x)$ then $F'(x) = f(x)$.

A sketch of the proof of the fundamental theorem of calculus

Although the fundamental theorem of calculus is one of the first to be introduced in any calculus course, it is not straightforward. Differentiation is first defined in terms of gradients of curves, and integration as areas under curves. It is not obvious that these two should be closely related. Suppose we have a curve whose equation is $y = f(x)$. Let's say that the area under the curve f between 0 and x is given by the function $F(x)$. According to the fundamental theorem of calculus, it should hold that $F'(x) = f(x)$. Why should this be true?

If we increase the value of x by a small amount, h, the area increases from $F(x)$ to $F(x + h)$. Geometrically, a small strip has been added, which is h wide and approximately $f(x)$ high. This strip has an area of approximately $h \times f(x)$. (Of course the strip is not really a perfect rectangle; this is why this is only a sketch of a proof!) Therefore the total new area should satisfy $F(x + h) \approx F(x) + h \times f(x)$.

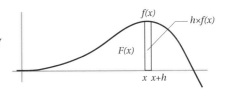

Rearranging this, we get:

$$f(x) \approx \frac{F(x + h) - F(x)}{h}$$

This formula looks like the definition of the derivative of F; as h tends to zero, the right-hand side approaches $F'(x)$, as required.

Evaluating indefinite integrals

The fundamental theorem of calculus gives a method for evaluating indefinite integrals. To find $\int f\, dx$, the question becomes: what function differentiates to f?

So to find $\int x^2\, dx$, we need a function that differentiates to give x^2. A first guess might be x^3, but this differentiates to $3x^2$. To get x^2 alone, we can try $\frac{x^3}{3}$. Differentiating this, we get the right answer: $\frac{3x^2}{3} = x^2$. So $\int x^2\, dx = \frac{x^3}{3} + C$, where C is a constant of integration.

More generally:

$$\int x^n \, dx = \frac{x^{n+1}}{n+1} + C$$

Other standard integrals can be found by working backwards from the table of **standard derivatives**. For example:

$$\int \cos x \, dx = \sin x + C$$

Constants of integration

Constant functions such as $y = 5$ and $y = -2$ have a gradient of zero. So when we differentiate any such function, we always get $\frac{dy}{dx} = 0$.

This is highly relevant when integrating. Suppose we want to evaluate:

$$\int \cos x \, dx$$

The fundamental theorem of calculus tells us to search for a function y that differentiates to give $\frac{dy}{dx} = \cos x$. The obvious answer is $y = \sin x$. Certainly in this case $\frac{dy}{dx} = \cos x$.

But this is not the only answer. If $y = \sin x + 5$, then also $\frac{dy}{dx} = \cos x$, as the constant term disappears. Similarly $y = \sin x - 2$ has derivative $\cos x$, as does $y = \sin x + C$ for any number C. So there is not just one possible answer: $y = \sin x + C$ is a valid solution, for every possible value of C.

It is a good idea to present the most general answer possible, so the best way to write it is as $\sin x + C$. The unknown number C is called a *constant of integration*.

In a definite integral, the constant disappears. In an indefinite integral, if we are also given a **boundary condition**, we may be able to pin C down more precisely.

Evaluating definite integrals

A definite integral should produce a numerical value for an area. For example, $\int_1^4 x^2 \, dx$ is the area bounded by the curve $y = x^2$ and the axis, between the points 1 and 4. The first step to calculating this is to evaluate the corresponding indefinite integral. In this case $\int x^2 \, dx = \frac{x^3}{3}$. (We may ignore the constant of integration, as it would cancel out in the next step anyway.) The second step is to substitute the two limit values for x (in this case 1 and 4) into this function. So we get $\frac{4^3}{3} = \frac{64}{3}$ and $\frac{1^3}{3} = \frac{1}{3}$. Finally, we subtract the value for the higher limit (4) from that for the lower (1) to give our answer. In this case $\frac{64}{3} - \frac{1}{3} = \frac{63}{3} = 21$.

Square brackets with indices $[\ ]_1^4$ are useful notation to put all this together. So:

$$\int_1^4 x^2 \, dx = \left[\frac{x^3}{3} \right]_1^4 = \frac{4^3}{3} - \frac{1^3}{3} = \frac{64}{3} - \frac{1}{3} = \frac{63}{3} = 21$$

Similarly:

$$\int_0^{\frac{\pi}{2}} \cos x \, dx = \left[\sin x \right]_0^{\frac{\pi}{2}} = \sin \frac{\pi}{2} - \sin 0 = 1 - 0 = 1$$

Integration by parts

According to the product rule for differentiation, if we have two functions of x, say $f(x)$ and $g(x)$, then when we differentiate $f \times g$, we get $(f \times g)' = f' \times g + f \times g'$. So $f \times g' = (f \times g)' - f' \times g$. If we reintegrate this, we get

$$\int f \times g' \, \mathrm{d}x = f \times g - \int f' \times g \, \mathrm{d}x$$

This technique of *integrating by parts* is often useful where no other method seems available. For example, if we want to work out

$$\int x \cos x \, \mathrm{d}x$$

we might spot that this is in the correct form, with $f(x) = x$ and $g(x) = \sin x$. So, applying the formula above, this integral becomes

$$x \sin x - \int 1 \sin x \, \mathrm{d}x$$

which comes out as

$$x \sin x + \cos x + C$$

Integration by substitution

The chain rule for differentiation says that the derivative with respect to x of $f(u)$ is $f'(u) \frac{\mathrm{d}u}{\mathrm{d}x}$. This provides a method for integrating anything of the form $f'(u) \frac{\mathrm{d}u}{\mathrm{d}x}$. Namely:

$$\int f'(u) \frac{\mathrm{d}u}{\mathrm{d}x} \, \mathrm{d}x = \int f'(u) \, \mathrm{d}u = f(u) + C$$

We can think of the two $\mathrm{d}x$ terms on the left as cancelling each other out (though we should not take this too literally). This is a useful technique for integration; the skill is in spotting a suitable *substitution* to make for u.

For example, suppose we want to evaluate:

$$\int \cos(x^2) \, 2x \, \mathrm{d}x$$

We can make the substitution $u = x^2$, in which case $\frac{\mathrm{d}u}{\mathrm{d}x} = 2x$. So, our integral becomes:

$$\int \cos u \frac{\mathrm{d}u}{\mathrm{d}x} \, \mathrm{d}x = \int \cos u \, \mathrm{d}u = \sin u + C$$

Finally we substitute x back in to get an answer of $\sin(x^2) + C$.

Integrability

It is certainly not true that every function is differentiable. Integration, however, is a much more robust procedure. It does not require the function to be smooth, or even continuous, just that it should be approximable by step functions. It is *almost* correct to say that every function satisfies this condition. Certainly, any function you can write down will be integrable.

Ultimately, this question closely depends on the logical structure underlying the real numbers. At the root of this structure is the prickly problem of the **axiom of choice**. If we take this to be true then the existence of a *non-measurable* function automatically follows. Such a function cannot be integrated, and there is no satisfactory way it can even be written down. (This existence of non-measurable functions is also the cause of the **Banach–Tarski paradox**.)

This is not a problem which impacts people's lives very often, since every function encountered in a practical situation will certainly be integrable. Far more troublesome is the problem of non-elementary integrals.

Liouville's non-elementary integrals

Students of integration are often presented with problems of the form $\int f(x)\, dx$. An example is:

$$\int e^{-x^2}\, dx$$

Thanks to the fundamental theorem of calculus, it is enough to look for a function which differentiates to e^{-x^2}. In this case, however, no matter how hard we try, we will never find one, as Joseph Liouville showed in the 19th century. The function e^{-x^2} is certainly integrable; that is not the obstacle. The problem is that there is no easy formula which gives the integral. More precisely, it cannot be expressed in terms of *elementary functions*. So no combination of polynomials, exponentials, logarithms, trigonometric or hyperbolic functions will ever work.

This is rather inconvenient, since evaluating $\int e^{-x^2}\, dx$ is the key to working with the **normal distribution**, amongst other things. Nor is this an isolated case. The *logarithmic integral function* which occurs in the **prime number theorem** is also always expressed as an integral. It too cannot be expressed by any combination of elementary functions. Similarly $\frac{\sin x}{x}$, though integrable, has no elementarily expressible integral.

Numerical analysis

Non-elementary integrals are one area where the comforting certainty provided by a single mathematical formula is lost. But this problem is much more widespread. As we investigate more complicated **differential equations**, this phenomenon recurs, or seems to, all too often. A notable instance is the **Navier–Stokes problem**. This is a moment when the interests of pure mathematicians diverge from those of engineers, for example. While mathematicians worry about the theoretical **existence and uniqueness** of exact solutions, most applications require something which works well enough for practical purposes.

Numerical analysis is the solution of equations by approximate methods. It is a large industry, with its own flavour and array of tools. The areas of science which rely on numerical analysts are too numerous to list, ranging from astronomy and climatology to architecture and economics.

Catastrophe theory

According to geologists, at some moment within the next few thousand years, the earth's magnetic poles will spontaneously switch. This phenomenon of geomagnetic reversal is an example of what mathematicians call a *catastrophe*. This is not to suggest that the consequences for humanity will be disastrous, rather it is an example of a process which carries along perfectly smoothly, and then changes abruptly and without warning.

Closely related to **chaos**, catastrophes are widespread in both mathematics and the natural world, and are even invoked by sports psychologists to explain sudden slumps in performance. Such occurrences cause a headache for numerical analysts, as even slight errors in calculation risk putting them on the wrong side of a catastrophe, invalidating their answer.

COMPLEX ANALYSIS

Real and imaginary parts

What do **complex numbers** look like and how can we perform arithmetic with them? The first example of a complex number is i (the square root of -1). Others are $-2 + 3i$, and $\frac{1}{3} + 1001i$. Every complex number can be written in the form $a + bi$, where a and b are real numbers (indeed this is the definition of a complex number).

The numbers a and b are called, respectively, the *real part* and the *imaginary part* of z, written as $\text{Re}(z) = a$ and $\text{Im}(z) = b$. For example, if $z = -2 + 3i$, then $\text{Re}(z) = -2$ and $\text{Im}(z) = 3$. (Notice that, somewhat confusingly, the 'imaginary part' of z is in a fact a *real* number.)

It is very easy to add or subtract complex numbers presented in this form: simply add or subtract their real and imaginary parts. So:

$$(1 + 2i) + (3 + i) = 4 + 3i \quad \text{and} \quad (1 + 2i) - (3 + i) = -2 + i$$

Multiplying complex numbers

There is nothing special about multiplying complex numbers; the procedure is the same as expanding brackets of real numbers. The only extra consideration is that we have to watch for two i's which multiply together to produce -1. To calculate $(1 + 2i)(3 + i)$:

$$1 \times 3 + 1 \times i + 2i \times 3 + 2i \times i = 3 + i + 6i - 2 = 1 + 7i$$

In general $(a + bi)(c + di) = (ac - bd) + (bc + ad)i$.

Numbers written in **modulus and argument** form are even easier to multiply:

$$3e^{\frac{\pi}{2}i} \times 4e^{\frac{\pi}{3}i} = 3 \times 4e^{(\frac{\pi}{2} + \frac{\pi}{3})i} = 12e^{(\frac{\pi}{2} + \frac{\pi}{3})i} = 12e^{\frac{5\pi}{6}i}$$

Dividing complex numbers

Every complex number can be written in the form $a + b$i, where a and b are real numbers. But complex numbers can also be divided, so $\frac{2 + 3i}{1 + 2i}$ is a complex number. How can we write it in the form $a + b$i?

The procedure is essentially the same as **rationalizing the denominator**. We will multiply top and bottom by $1 - 2$i:

$$\frac{(2 + 3i)\,(1 - 2i)}{(1 + 2i)\,(1 - 2i)} = \frac{2 + 3i - 4i + 6}{1 + 2i - 2i + 4} = \frac{8 - i}{1 + 4} = \frac{8}{5} - \frac{1}{5}i$$

This is now in the usual format, with its real and imaginary parts clearly visible.

Modulus and argument

The complex number z $= \sqrt{3} + $ i is presented in the ordinary way, via its real and imaginary parts. If we think of the complex numbers as a plane (called the *Argand diagram*), then the real and imaginary parts correspond to the **Cartesian coordinates** of the point z, namely $(\sqrt{3}, 1)$.

There is an alternative: we could use **polar coordinates** instead. This means that we need to measure the distance of the point $(\sqrt{3}, 1)$ from 0, and find its angle from the real axis (measured in radians). Pythagoras' theorem and a little trigonometry reveal these to be 2 and $\frac{\pi}{6}$ respectively. These are called the *modulus* and *argument* of the complex number, written $|z| = 2$ and Arg $z = \frac{\pi}{6}$.

Putting these together, the second way of writing the number z is as $2e^{i\frac{\pi}{6}}$. In general the complex number with modulus r and argument θ can be written as $re^{i\theta}$.

The appearance of the number e here is due to **Euler's trigonometric formula**. This theorem, along with a little triangular geometry, is all that is needed to switch between the modulus–argument representation $re^{i\theta}$ and the real–imaginary representation $a + i b$. According to Euler's trigonometric formula, $re^{i\theta} = r(\cos\theta + i\sin\theta) = r\cos\theta + ir\sin\theta$, so $a = r\cos\theta$ and $b = r\sin\theta$.

Complex analysis

At the end of the 18th century, Carl Friedrich Gauss established the profound importance of the complex numbers, by proving the **fundamental theorem of algebra**. In the early 19th century, Augustin-Louis Cauchy set about investigating the spectacular feats which this opened up. In particular, he pursued the ideas of **calculus**, and discovered how elegantly they work in this setting, more smoothly in fact than in the real numbers. This realization marked the dawn of *complex analysis*, perhaps the moment that mathematics entered full-blown adulthood. The central objects of study are complex functions.

Complex functions

The fundamental theorem of algebra tells us that the theory of **polynomials** works much more smoothly at the level of the complex numbers than in the real numbers. The same holds true when we look at the more general theory of **functions**. Complex functions take complex numbers as inputs, and give complex numbers as outputs,

$f: \mathbf{C} \rightarrow \mathbf{C}$. The difficulty is that while we can 'see' a real-valued function as a 2-dimensional graph, our human minds are incapable of visualizing the 4-dimensional graph that a complex function demands. Despite this inconvenience, we can learn a great deal about them. The most important complex functions are the smooth functions.

Smooth functions

In analysis, differentiable functions (see **differentiability**) are particularly important, as they capture the intuitive notion of *smoothness*. However, when we are working in the real numbers, we need to decide *how smooth* we want our functions to be. A differentiable function f is quite smooth, but its derivative f' may not be at all smooth. Smoother are the *doubly differentiable* functions, which can be differentiated twice to get f''. Smoother still are the triply differentiable functions, and so on. These classes are all distinct. There are functions which can be differentiated 1000 times, but not 1001, for example.

At the end of this sequence are the *infinitely differentiable* functions, the smoothest of all. A small subsection of these are the *analytic functions*, essentially those which can be written as **power series**. These are the functions which are easiest to work with.

All told, it is a messy picture, and it is easy to get lost in this confusing hierarchy. Augustin-Louis Cauchy's theory of analytic functions shows that, in the complex numbers, the situation is infinitely simpler.

Analytic functions

One of Cauchy's most important theorems was a highly unexpected result about smooth functions in the complex numbers: every differentiable function is automatically *infinitely* differentiable. Knowing that you can differentiate f once guarantees that you can differentiate it twice, three times, and as many times as you like. This is emphatically not true in the real numbers. It shows what a hospitable world the complex numbers are, once you get to know your way around.

This is a consequence of an even stronger theorem, also proved by Cauchy. In the complex numbers, just being differentiable guarantees that a function is analytic, that is, it can be expressed by power series. More precisely, f is *analytic* if the complex plane can be divided into overlapping discs, and the function is given by a power series in each region. This sounds as if there is a large potential for patching together different functions. The analytic continuation theorem shows that this is not the case.

Analytic continuation theorem

When we are working with the real numbers, smooth functions can easily be chopped up and glued together. For example, we can take the curve $y = x^2$, cut it in two at its trough, and insert a portion of straight line 3 units long between the two halves. This hybrid curve may not have a concise formula to define it, but it is perfectly valid, and indeed differentiable.

The *analytic continuation theorem* says that the situation is utterly different in the complex numbers. If we know the values of an analytic function on a patch of plane, then there is exactly one way to extend it to the whole complex plane. This is two theorems in one, and they are two good ones! Firstly, if we are only provided with a function on a tiny patch of the plane, we know that it automatically extends to cover the whole plane. Secondly, there is only ever one way to do this. If two analytic functions f and g coincide, even on a tiny patch of the plane, then they must be equal everywhere. (A 'patch' cannot just be a smattering of individual points, it must have positive area, however small.)

This shows that differentiable functions in the complex numbers are far more rigid than their slippery counterparts in the real numbers. A particularly famous application of this fact is to the **Riemann zeta function**.

Picard's theorem
The theory of analytic functions, and the analytic continuation theorem in particular, tell us that we cannot carry over our intuition about the real numbers to understand complex functions. Consider the function $f(x) = x^2 + 2$. Viewed as a function with real numbers as inputs, what are its outputs? The answer is that every real number from 2 onwards appears as an output, those below 2 do not. The real function $f(x) = \sin x$ is even more restricted, producing outputs only between -1 and 1. The simplest of all are the constant functions, such as $f(x) = 3$. For every input, this produces the same output: 3.

In the late 19th century, Charles Émile Picard showed that yet again the situation is dramatically different in the complex numbers. If f is a complex analytic function and is non-constant, Picard proved that every single complex number must appear as an output of f, with possibly one solitary exception. For example, viewed as a complex function, $f(z) = z^2 + 2$ produces every complex number as an output. The exponential function, $f(z) = e^z$ misses one number: 0.

POWER SERIES

Summing powers
Ordinary series consist only of numbers. But if we include a variable z, we can end up with a *function*. An important way to do this is to add up successive powers of z. For example:

$$\sum_{r=1}^{\infty} z^r = 1 + z + z^2 + z^3 + \cdots$$

Care is needed for this series to converge. Substituting $z = 2$, for example, gives no chance of the series approaching a finite limit. But it will converge when $|z| < 1$. In this region, it produces the function $(1-z)^{-1}$, as it is a geometric progression. Another power series is given by:

$$\sum_{r=1}^{\infty} \frac{z^r}{r} = z + \frac{z^2}{2} + \frac{z^3}{3} + \frac{z^4}{4} + \cdots$$

This also turns out to converge for $|z| < 1$. Less obviously, this converges to the function $-\ln(1 - z)$. These are both examples of *power series*.

Power series

Generally, a *power series* is of the form

$$a_0 + a_1 z + a_2 z^2 + a_3 z^3 + \cdots = \sum_{r=0}^{\infty} a_r z^r$$

where $a_0, a_1, a_2, a_3 \ldots$ is some sequence of numbers.

Many important functions are built in this fashion: the **exponential function**, the **trigonometric functions**, and of course all **polynomials** are power series where most of the a_i are 0. This is a very powerful unifying framework for analysis.

In fact, according to the theory of analytic functions, it is almost true to say that *all* important functions arise this way. Certainly all well-behaved functions do, though pathological examples such as the **Koch snowflake** do not.

Calculus of power series

In analysis, we often want to *differentiate* and *integrate* functions. Power series make this easy. We know how to differentiate polynomials: the derivative of z^n is nz^{n-1}. So to differentiate a power series $a_0 + a_1 z + a_2 z^2 + a_3 z^3 + a_4 z^4 + \cdots$ we proceed, term by term, to get $a_1 + 2a_2 z + 3a_3 z^2 + 4a_4 z^3 + \cdots$

Suppose we start with the series $z + \frac{z^2}{2} + \frac{z^3}{3} + \frac{z^4}{4} + \cdots$. If we differentiate this term by term, we get $1 + z + z^2 + z^3 + \cdots$. There are some technical arguments suppressed here. There are two processes going on: the differentiation of a function, and the limit of a series. I have assumed that it is legitimate to swap the order of these. Happily it is valid, but Riemann's rearrangement theorem is a warning not to take these types of fact for granted.

Functions as power series

Power series may initially seem as if they have been dreamt up to torment the unsuspecting student. But, with a little familiarity, they really do provide an excellent language for analysis. It is a remarkable fact that any reasonable function can be written as a power series: this is formalized as *Taylor's theorem*.

First we *assume* that a function, such as sin, can be written as a power series:

$$\sin z = a_0 + a_1 z + a_2 z^2 + a_3 z^3 + a_4 z^4 + \cdots$$

We want to find the numbers $a_0, a_1, a_2, a_3, a_4 \ldots$ But how can we start? a_0 is easy. If we set $z = 0$, then all the other terms disappear. So $a_0 = \sin 0$, which of course is 0. To find a_1, notice what happens when we differentiate the equation above. We get:

$$\cos z = a_1 + 2a_2 z + 3a_3 z^2 + 4a_4 z^3 + 5a_5 z^4 + \cdots$$

If we set $z = 0$ in this new equation, we get $a_1 = \cos 0 = 1$. To find a_2, we can differentiate again:

$$-\sin z = 2a_2 + 6a_3 z + 12a_4 z^2 + 20a_5 z^3 + \cdots$$

Now setting $z = 0$, we get $2a_2 = -\sin 0 = 0$, so $a_2 = 0$.

Differentiating one more time:

$$-\cos z = 6a_3 + 24a_4 z + 60a_5 z^2 + \cdots$$

Setting $z = 0$ again we get $a_3 = -\frac{1}{6}$.

The general pattern might become clear now: $a_n = 0$ whenever n is even. When n is odd, $a_n = \pm \frac{1}{n!}$, the sign alternating. Putting this altogether, we find that:

$$\sin z = z - \frac{z^3}{3!} + \frac{z^5}{5!} - \frac{z^7}{7!} + \cdots$$

Taylor's theorem

Taylor's theorem guarantees that the above method for writing functions as power series actually works. The resulting series really does converge to $\sin z$. In this example the series is valid for all z, although this is not true for every function. In general, Taylor's theorem says that if f is a complex function which is differentiable, then the complex plane can be divided into discs, and f can be expressed as a power series on each disc. If D is one such disc centred at a, say, then f will be expressed as a power series, not in terms of z, but in terms of $(z - a)$.

The sine, cosine and exponential functions are particularly nice, because only one region is needed. These functions are given by the same series everywhere.

EXPONENTIATION

The exponential function

Power series are central to modern mathematics, and the *exponential function* is the most important of all:

$$1 + x + \frac{x^2}{2} + \frac{x^3}{6} + \frac{x^4}{24} + \cdots = \exp x$$

Factorials make the pattern clearer:

$$\frac{x^0}{0!} + \frac{x^1}{1!} + \frac{x^2}{2!} + \frac{x^3}{3!} + \frac{x^4}{4!} + \cdots = \exp x$$

There are several properties that make this function very special. Firstly, by multiplying the series for $\exp x$ and $\exp y$, we find that:

$$\exp x \times \exp y = \exp (x + y)$$

Taken together with the fact that $\exp 0 = 1$, this makes the exponential function perfect for building **complex exponentiation**.

A second crucial property is what happens when we differentiate this function. The $\frac{x^4}{4!}$ term, for example, produces $\frac{4x^3}{4!}$, which is $\frac{x^3}{3!}$. This shows the general pattern. The result is the same series as before. In short:

$$\frac{\mathrm{d}}{\mathrm{d}x}(\exp x) = \exp x$$

This says that the exponential function describes its own rate of change. Indeed, the function can be defined by this property. This is one reason that this function is so widespread within mathematics, because it has a tendency to appear whenever calculus is used (and mathematicians are always using calculus). **Radioactive decay** is one example.

When we feed the value 1 into the exponential function, exp 1, we get the definition of the important number e. The function is also commonly written as e^x.

e

If the imaginary number i is the cornerstone on which the complex numbers are built, then e is the key to the front door. Whereas π is a number with remarkable properties, e is more than a number: it is the public face of the exponential function. This function carries great power; e, as its most visible part, takes all the plaudits.

The number e is defined as exp 1, that is to say as the limit of the series:

$$1 + 1 + \tfrac{1}{2!} + \tfrac{1}{3!} + \tfrac{1}{4!} + \cdots$$

This converges to a numerical value of approximately 2.7182818285. Being **irrational** (and indeed **transcendental**), this sequence of digits continues for ever without repetition.

e can equivalently be defined as the base of the **natural logarithm**, or the limit of the **continuous interest** sequence. The exponential function is the parent of the trigonometric functions sin and cos, and the basis of the **modulus and argument** approach to complex numbers. In fact, it is almost impossible to do mathematics without encountering this number at every turn.

Complex exponentiation

What might 4^i mean? '4 multiplied by itself i times' is not an acceptable answer!

To answer this question, we need to extend 'taking powers' or *exponentiation* to the complex numbers. But we want to do this in a way that preserves its basic properties. Luckily, there is a complex function which can help: the exponential function. It is time for it to live up to its name! We begin just by deciding how to raise the number e to any complex power z. The answer is to take $e^z = \exp z$.

The first criteria that we should demand of complex exponentiation are that $a^0 = 1$ and $a^1 = a$ for any number a. Since $\exp 0 = 1$ and $\exp 1 = e$, this translates as $e^0 = 1$ and $e^1 = e$, as required. This is a good start.

A more significant rule that we want complex exponentiation to follow is the first law of powers: $a^b \times a^c = a^{b+c}$, for any a, b, c. The exponential function also satisfies this: $e^x \times e^y = e^{x+y}$. We can then extend this from e to other complex powers. The most famous example of complex exponentiation in action is **Euler's formula**.

Complex powers

The exponential function allows us to perform complex exponentiation when the number e is the base, such as e^{2i}. The natural logarithm also allows us to take logarithms to base e. How can we extend this to have other complex numbers as the base, as in 4^i?

To define the general exponent a^b, it would be helpful first to write a as a power of e. We can do this: $a = e^{\ln a}$. Then we can raise this to the power b, to define:

$$a^b = e^{b \ln a}$$

So in particular, $4^i = e^{i \ln 4}$, which is around $0.18 + i0.98$.

Compound interest

If you deposit $100 in a bank account which produces 5% interest, one year later you will have the original $100, plus $5 of interest, making $105. An elementary, but common, mistake is to believe that after two years you should have $110. The error is that the account does not add $5 each year; it adds 5% of the amount in the beginning of the year. At the start of the second year this is $105, and 5% of this is $105 \times 0.05 = \$5.25$. So at the end of the second year, there is $110.25 in the account.

How much money will be in the account after 25 years? Rather than working through 24 intermediate calculations, we want a short cut. Each year, the account grows by 5%, which is equivalent to multiplying the total by 1.05. At the end of the first year there is 1.05×100. At the end of the second, $1.05 \times 1.05 \times 100$, that is $1.05^2 \times 100$. At the end of the third there is $1.05^3 \times 100$, and so on. In general, at the end of the nth year, there will be $1.05^n \times 100$. Taking $n = 25$, after 25 years the account will contain $1.05^{25} \times 100 = \$338.64$.

In general, if you put $\$n$ into an account that pays $m\%$ each period, then after k periods the amount of money there will be $(1 + \frac{m}{100})^k \times n$.

Continuous interest

In 1689, Jacob Bernoulli discovered that beneath the arithmetic of compound interest is some intriguing mathematics. Suppose I put $1 into a bank account which pays 100% interest each year. After 1 year, the account will have grown to $(1 + 1)^1 = 2$. Suppose instead that the account pays 50% every 6 months, or $\frac{1}{2}$ every $\frac{1}{2}$ year. Now, after one year it will contain $(1 + \frac{1}{2})^2 = 2.25$. If the account pays $\frac{1}{3}$ every $\frac{1}{3}$ year, then after one year it will contain $(1 + \frac{1}{3})^3 = 2.37$, to the nearest cent. Next, $\frac{1}{4}$ every $\frac{1}{4}$ year produces $(1 + \frac{1}{4})^4 = 2.44$.

The interesting question is what happens if we continue this line of thought. If we split the year into tiny pieces, seconds perhaps, can we manufacture a huge yearly total? Or is there some bound that it can never exceed? The question is: what happens to the sequence $(1 + \frac{1}{n})^n$? Bernoulli's answer is that this sequence gets ever closer to the number e. Dividing the year into hours produces $\$e$ (to the nearest cent). In fact, this is often used as an alternative definition: $e = \lim_{n \to \infty} (1 + \frac{1}{n})^n$.

In the limiting system, we no longer have *discrete* interest being awarded at regular periods; instead the money increases *continuously*, (see **discreteness and continuity**) growing at every single moment, with e determining the rate of increase.

Euler's trigonometric formula

The exponential function is defined as a power series:

$$e^z = 1 + z + \frac{z^2}{2!} + \frac{z^3}{3!} + \frac{z^4}{4!} + \cdots$$

Taylor's theorem allows us to form the power series of other functions, including:

$$\sin z = z - \frac{z^3}{3!} + \frac{z^5}{5!} - \frac{z^7}{7!} + \cdots \qquad \cos z = 1 - \frac{z^2}{2!} + \frac{z^4}{4!} - \frac{z^6}{6!} + \cdots$$

Leonhard Euler noticed that these three series seem to be very closely related. In fact, it almost looks as if $\sin z$ and $\cos z$ should combine together to make e^z. But simply adding them does not quite work.

If we replace z with iz (where $i - \sqrt{-1}$), we get a new series:

$$e^{iz} = 1 + iz - \frac{z^2}{2!} - \frac{iz^3}{3!} + \frac{z^4}{4!} + \frac{iz^5}{5!} - \frac{z^6}{6!} - \frac{iz^7}{7!} + \cdots$$

The odd terms here are the series for $\cos z$. The remaining ones are i multiplied by the series for $\sin z$. Putting this together, we get *Euler's trigonometric formula*:

$$e^{iz} = \cos z + i \sin z$$

The power series above also produce other important formulas:

$$\cos z = \frac{e^{iz} + e^{-iz}}{2} \qquad \sin z = \frac{e^{iz} - e^{-iz}}{2i}$$

De Moivre's theorem

Abraham de Moivre provided powerful evidence of the value of working with complex numbers, even if we are fundamentally interested in the more familiar real numbers. The starting point is the observation that $(e^{i\theta})^n = e^{in\theta}$. De Moivre realized that, when taken with Euler's trigonometric formula, this dull equation took on a surprising and useful new appearance:

$$(\cos \theta + i \sin \theta)^n = \cos n\theta + i \sin n\theta$$

This is *de Moivre's theorem* and it is excellent at generating trigonometric formulas. Examples are the double-angle formulae for $\cos 2\theta$ and $\sin 2\theta$. To find them we take $n = 2$. De Moivre's theorem then tells us that:

$$(\cos \theta + i \sin \theta)^2 = \cos 2\theta + i \sin 2\theta$$

Now we expand this bracket:

$$\cos^2 \theta - \sin^2 \theta + 2i \sin \theta \cos \theta = \cos 2\theta + i \sin 2\theta$$

Equating the real parts of this equation produces:

$$\cos 2\theta = \cos^2 \theta - \sin^2 \theta$$

Equating the imaginary part gives:

$$\sin 2\theta = 2 \sin \theta \cos \theta$$

These are very useful formulas. Taking $n = 3, 4, \ldots$ in de Moivre's theorem can produce triple, quadruple angle trigonometric formulas, and so on.

Euler's formula

Leonhard Euler's work appears in this book more often than that of any other mathematician. His mathematics was extensive and decisive. He also made major contributions to other areas of science, including astronomy and optics. But he may be best known for something he probably never actually wrote down (although it is an immediate consequence of his work).

Euler's formula is an equation which beautifully unites the five fundamental constants of mathematics, and hints at awesome, unimagined depths to the complex world:

$$e^{i\pi} + 1 = 0$$

The Nobel-prize-winning physicist Richard Feynman has called this 'the most remarkable formula in mathematics'. Why should this exquisite equality hold true? It follows directly from Euler's trigonometric formula. According to that, $e^{i\pi} = \cos \pi + i \sin \pi$. Since $\cos \pi = -1$ and $\sin \pi = 0$, Euler's formula follows.

Natural logarithm

The *natural logarithm* is the inverse of the exponential function: if $\exp x = y$, then $\ln y = x$. It is written 'ln' but pronounced 'log'. Equivalently, the natural logarithm is the **logarithm** to base e. The natural logarithm was one of the first glimpses that mathematicians had of the exponential function, and was first tabulated by John Speidell in 1619.

Calculus of the natural logarithm

Integrating powers is routine work: the integral of x^2 is $\frac{x^3}{3} + C$, where C is a **constant of integration**. Similarly, the integral of x^{-5} is $\frac{x^{-4}}{-4} + C$, and in general the integral of x^n is $\frac{x^{n+1}}{n+1} + C$. There is, however, an exception. If we try to apply this to x^{-1}, we seem to get $\frac{x^0}{0} + C$, which is meaningless, as we cannot divide by zero.

So what is the integral of x^{-1}, or equivalently $\frac{1}{x}$? The answer is the natural logarithm:

$$\int \frac{1}{x} \, dx = \ln x + C$$

Indeed, this is what makes the natural logarithm 'natural'. To see why this should be true, we approach the problem from the other side, and show that if $y = \ln x$, then $\frac{dy}{dx} = \frac{1}{x}$. If $y = \ln x$, then by definition $e^y = x$. Now we can differentiate this with respect to x, using the chain rule: $e^y \frac{dy}{dx} = 1$. So $\frac{dy}{dx} = \frac{1}{e^y}$, which says that $\frac{dy}{dx} = \frac{1}{x}$.

Hyperbolic trigonometry

To make Euler's trigonometric formula work, we had to introduce complex numbers. A way to avoid this is by using the following *hyperbolic functions*:

$$\cosh z = 1 + \frac{z^2}{2!} + \frac{z^4}{4!} + \frac{z^6}{6!} + \cdots \qquad \sinh z = z + \frac{z^3}{3!} + \frac{z^5}{5!} + \frac{z^7}{7!} + \cdots$$

(The pronunciation of 'sinh' is a matter of disagreement. Some people say 'sinch', others 'shine'.) Equivalently:

$$\cosh z = \frac{e^z + e^{-z}}{2} \qquad \text{and} \qquad \sinh z = \frac{e^z - e^{-z}}{2}$$

Straight from this definition, we get $\cosh x + \sinh x = e^x$, the hyperbolic equivalent of Euler's trigonometric theorem.

All standard trigonometric facts and formulas have hyperbolic counterparts. For instance, instead of $\cos^2 x + \sin^2 x = 1$, we have $\cosh^2 x - \sinh^2 x = 1$. This fact provides a clue to their name too. In the real numbers, if you plot a graph of the points $(\cos\theta, \sin\theta)$ as θ varies, the result is a circle. Plotting $(\cosh\theta, \sinh\theta)$ produces a **hyperbola**.

Hyperbolic functions are excellent examples of the complex numbers at work. In the geometry of the real numbers, the graphs of sinh and cosh had been encountered, in the Bernoullis' study of **catenaries**. They do not resemble the sine and cosine waves, however, being non-periodic. Nevertheless, the complex numbers reveal them as the close cousins of the usual trigonometric functions: $\cosh z = \cos iz$ and $\sinh z = -i \sin iz$.

FRACTALS

Self-similarity

The word *fractal* does not have a formal definition. But these fascinating shapes share a property of *self-similarity*: a scale-defying tendency to look the same, however far you zoom in. For example, the *Sierpinksi triangle* is obtained by taking an equilateral triangle, dividing it into four smaller triangles, and removing the central one. Then the same process is repeated with each of the three remaining triangles, and so on. Once this is completed the set that remains is *self-similar*: if you shrink its width by a half, then it exactly fits into one of the corners of the original.

Sierpinksi's triangle is a fractal directly constructed as a geometrical object. Other examples of this type include the **pentaflakes** and **Cantor dust**. Other fractals emerge more gradually, from the study of **dynamical systems**, including the most famous fractal of all, the **Mandelbrot set**, and the closely related **Julia sets**, which show how fractals appear as **strange attractors** in chaos theory.

Pentaflakes

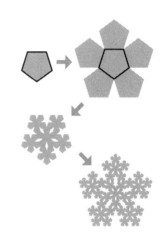

The only regular polygons which *tessellate* (see **tessellations**) are equilateral triangles, squares and hexagons; pentagons do not. Unperturbed by this, in his 1525 book *Instruction in Measurement* the German artist and polymath Albrecht Dürer proceeded to lay six regular pentagons edge to edge creating a new shape, resembling a pentagon, but with a notch cut out of each side. Then he laid six of these notched pentagons edge to edge, to obtain a more complicated figure. The beautiful shape which results from repeating this process is now known as a *pentaflake*, and has a strong claim to being history's first fractal.

Koch snowflake

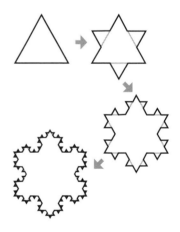

Fractal patterns often appear when familiar objects are subjected to minor changes, repeated infinitely often. One such was the *Koch snowflake*, discovered by Helge von Koch in 1906. The idea is simple: first draw an equilateral triangle. Then take each side in turn, divide it into three pieces, and build an equilateral triangle on the middle section. Now, delete the original portion of line, just leaving the two new sides. At this stage, we have a six-pointed star. Now repeat this process on every straight section of the new shape, and keep repeating. The Koch snowflake is defined to be the curve generated by this process.

The resulting curve is infinitely long, but encloses a finite area (in fact $\frac{8}{5}$ times the area of the original triangle). The **fractal dimension** of this curve is $\frac{\log 4}{\log 3}$. As well as being aesthetically appealing, mathematically this curve aroused interest as an example of something *continuous* (it doesn't have any jumps or gaps), but not *differentiable*: it is not smooth anywhere; you cannot draw a tangent to it at any point.

Cantor dust

Start with a segment of line, 1 unit long. Divide it into three equal parts, and throw away the middle section. Now repeat the process with each of two remaining pieces: chop each into thirds and throw away the middle section. Keep repeating this process. The collection of points which never get thrown out, and still remain after infinitely many steps, is called *Cantor dust*.

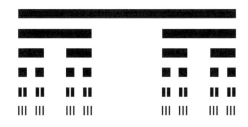

Georg Cantor himself constructed this fractal object to illustrate that the connection between his new infinities (see **countable and uncountable infinities**) and geometry is not at all straightforward. From a set-theoretic point of view, Cantor dust contains a lot of points. In fact it is in **one to one correspondence** with the entire set of real numbers. But, from a geometric point of view, there is barely anything there. At the start, the segment has length 1. After the first step, it has length $\frac{2}{3}$, and then $\frac{4}{9}$, $\frac{8}{27}$, and so on. After infinitely many steps, it has reached a total length of 0.

Dimension

Suppose we have a segment of straight line, 1 m long. Covering it in smaller lines, each $\frac{1}{3}$ as long, will require three of them. Beginning instead with a square 1m × 1m, and covering it in smaller squares each $\frac{1}{3}$ as long, we will need 9 ($= 3^2$) of them. Starting with a cube 1 m × 1 m × 1 m, filling it with smaller cubes each $\frac{1}{3}$ as long will require a supply of 27 ($= 3^3$) of them.

Each time, the *dimension* of the original shape (1 for the line, 2 for the square, 3 for the cube) appears as the *exponent* (see **powers**) in the required number of smaller shapes: 3^1, 3^2 and 3^3. (The base number 3 is arbitrary; any other number would do as well.) The difference with fractal dimension is that the answer may not come out as a whole number. But we can still calculate it in the same way, using a fractal's property of self-similarity.

Fractal dimension

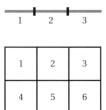

Looking at the top part of the Koch curve, if we want to cover it with smaller copies of itself, $\frac{1}{3}$ as long, we need four of them. So, if its dimension is D, this should satisfy $4 = 3^D$. Anyone adept at **logarithms** can solve this as $D = \frac{\log 4}{\log 3}$, approximately 1.26. With fractal dimension between 1 and 2, Koch's curve can be thought of as intermediate between a line and a surface. Cantor dust has fractal dimension $\frac{\log 2}{\log 3}$, which is around 0.63, lying between a point and a line.

Fractal dimension was first discovered by Lewis Fry Richardson, in his investigation of the coastline problem.

The coastline problem

In the early 20th century, the wide-ranging British scientist Lewis Fry Richardson was gathering data on how the lengths of national boundaries affect the likelihood of war, when he hit an inconvenient obstacle: the Spanish measured their border with Portugal as being 987 km, but the Portuguese put it at 1214 km. Contemplating this, Richardson realized that whereas the length of a straight line is unambiguously defined, for a very wiggly one the answer would depend on the scale you used to measure it. This work was rediscovered by Benoît Mandelbrot, who in 1967 wrote an article: 'How Long is the Coast of Britain?'

To try to answer this question, you could first take a map from a schoolbook and measure the outline in straight 100 km sections. But using a higher-resolution map and a shorter scale of say 10 km, taking into account all the extra kinks and wiggles, will produce a larger result. In the extreme case, you could set out yourself with a ruler, and try to measure the exact line of high tide by hand. The extra distance could push the estimate up to millions of times your original answer; this is the so-called *Richardson effect*.

Following Richardson, Mandelbrot wrote that 'Geographical curves are so involved in their detail that their lengths are often infinite or, rather, undefinable'. Richardson had invoked a number (*D*) to quantify the wiggliness of different coastlines. In Mandelbrot's hands this developed into fractal dimension, and he defended its application to geographical coastlines, arguing that they possess enough 'statistical self-similarity' to qualify as approximate fractals.

Richardson had found that 'At one extreme, $D = 1.00$ for a frontier that looks straight on the map. For the other extreme, the west coast of Britain was selected because it looks like one of the most irregular in the world; it was found to give $D = 1.25$'.

Peano's space-filling curve

An extreme example of the coastline phenomenon is the extraordinary curve discovered by Giuseppe Peano in 1890. We can consider a *curve* as a function which takes every real number from 0 to 1 and assign it to a point in space. It must do this in a way which is *continuous*: without jumps or gaps. Common examples are circles or polygons in two dimensions, or **knots** in three dimensions. Fractal curves also exist, including pentaflakes and the Koch snowflake.

Peano constructed a curve which completely fills a square in the plane. That is, his curve passes through every single point inside the square. It was a consequence of Cantor's work on **set theory** that the square and the line should contain the same number of points (that is, they are in one to one correspondence). This realization prompted Cantor to remark, 'I see it, but I don't believe it!' That this can be achieved by a *continuous* curve is doubly surprising.

Because Peano's curve covers the whole square, it has fractal dimension 2. Subsequently generalizations of Peano's curve have been found which can fill a 3-dimensional cube, or any *n*-dimensional **hypercube**.

Kakeya's needle

Imagine you have a needle lying on a table-top, and you want to rotate it a full 360° by sliding it around (you are not allowed to pick it up). But there's a catch: the needle is covered in ink, and as you push it around it leaves its path painted out behind it. The question Soichi Kakeya asked in 1917 is: what is the minimum possible area for the resulting shape?

The surprising answer, supplied by Abram Besicovitch in 1928, is that you can make the area as small as you like. There is no way to do it leaving a shape with zero area, but it can be done by leaving an area of 0.1 square units, or 0.001, or as small an area as required.

Kakeya's conjecture

The sets left by Kakeya's needle have an interesting property: they contain a line 1 unit long pointing in every single direction. Shapes like this are known as *Kakeya sets*. The needle problem concerns such sets on a 2-dimensional plane, but the same definition makes sense in higher dimensions.

Kakeya's conjecture concerns the fractal dimension of Kakeya sets. It says that a Kakeya set in n-dimensional space must have the maximum possible fractal dimension, that is to say n. This means that the sets are substantial, unlike Cantor dust. At time of writing, Kakeya's conjecture remains open.

DYNAMICAL SYSTEMS

Dynamical systems

One of the joys of mathematics is the way fabulous beauty and flabbergasting complexity can arise from seemingly simple situations. Often, all that is needed is the right angle of approach.

The formula $z^2 + 0.1$ is hardly a wonder of the mathematical world. If we substitute the value $z = 0$ into it we get $0^2 + 0.1 = 0.1$. What happens if we put this value back into the formula? We get $0.1^2 + 0.1 = 0.11$. Putting this back in again, we get $0.11^2 + 0.1 = 0.1121$. If we keep doing this, after about 13 iterations the result settles down to a number close to 0.112701665.

A *dynamical system* arises when the output of a function is repeatedly fed back in as its input. This example is a *quadratic system*, as it is based on the quadratic function $z \mapsto z^2 + c$.

Quadratic systems

When $c = 0.1$, the quadratic dynamical system $z \mapsto z^2 + c$ converges to a single number, close to 0.112701665. If we slightly change the system to $z \mapsto z^2 + 0.5$, and begin by substituting in $z = 0$, then after 13 iterations the result is so huge that most pocket calculators will give an error message. So, for some values of c (such as 0.5), the quadratic system $z \mapsto z^2 + c$ produces a sequence which grows uncontrollably large. But for others, like 0.1, it stays within finite bounds.

The Mandelbrot set

The *Mandelbrot set* is the collection of values of c for which the quadratic system $z \mapsto z^2 + c$ produces a bounded sequence. This turns out to be the real numbers between -2 and 0.25. This does not seem very exciting, until we also consider complex values of c, when the spectacular fractal discovered by Benoît Mandelbrot in 1980 is fully revealed.

The different bulbs of the Mandelbrot set correspond to different types of attracting cycles of the function $z \mapsto z^2 + c$. The central heart-shaped region (called the main cardioid) comprises the set of values of c for which the system has a unique attracting fixed point. The largest circular disc then represents those where there is an attracting 2-cycle (see **attracting cycles**). The smaller bulbs correspond to attracting cycles of different lengths.

Julia sets

The number -1 lies in the Mandelbrot set, because the sequence we get by repeatedly applying $z \mapsto z^2 - 1$, starting at $z = 0$, remains bounded (oscillating for ever between -1 and 0). But what happens if we start the sequence at somewhere other than 0, say at $z = 2$? This time the sequence does rush off to infinity. This means that 0 is in the *Julia set* of $z^2 - 1$, but 2 is not.

We can replace -1 with any other complex number c, and look at the Julia set for $z^2 + c$.

Julia sets typically form fantastically intricate patterns. In fact for some values of c the Julia set of $z^2 + c$ resembles the Mandelbrot set itself (such as $1 - \phi$, where ϕ is the **golden ratio**). The Mandelbrot set serves as a map of all these patterns, with every point in it producing its own Julia set (and those outside producing Cantor dust).

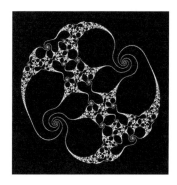

More generally, we can replace $z^2 + c$ with any function f, and ask which starting numbers cause the system obtained by repeatedly applying f to remain bounded. Many extraordinarily organic patterns can be formed this way. This idea was discovered by Gaston Julia in 1915, before the term 'fractal' had been coined, and without the aid of modern computers to display them in their full glory.

The logistic map

The Mandelbrot set is a picture of the simplest non-linear system, the quadratic system, $x \mapsto x^2 + c$. Only slightly more complicated is the *logistic map*:

$$x \mapsto r \times x(1 - x)$$

Again, just one number needs to specify this system, this time known as r. With the logistic map, the characteristic properties of a dynamical system can be seen without leaving the real numbers.

We form a sequence by repeatedly applying this function, and watch how it evolves. The starting value of the sequence does not matter too much as long as it is between 0 and 1. For convenience, we will begin at $x = \frac{1}{3}$. The evolution of the sequence depends on r. If $r = 0$, then the sequence quickly settles down to a single value: $\frac{1}{3}, 0, 0, 0, 0, 0, \ldots$ If $r = 5$, the sequence grows and grows in magnitude. To two decimal places, the sequence runs: $0.33, 1.11, -0.62, -4.99, -149.54, -112557.70, \ldots$ This sequence quickly diverges to $-\infty$. Between the values of $r = 0$ and $r = 4$ interesting action happens: these produce a sequence of bifurcations, followed by chaos.

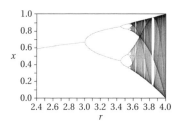

Attracting cycles

In the logistic map $x \mapsto r \times x(1 - x)$, if we take $r = 0$, the sequence starting at $x = \frac{1}{3}$ (or any other randomly selected number between 0 and 1) quickly settles down to a single value of 0. If $r = 2$, this sequence proceeds: 0.33333, 0.44444, 0.49383, 0.49992, ... to five decimal places. In this case the sequence never arrives at a fixed value, but it does get ever closer to the number $\frac{1}{2}$. (The same occurs with any other starting point between 0 and 1.) So $\frac{1}{2}$ is called an *attracting point* of the system.

When $r = 3.2$, the sequence does not get close to a single value, but flickers alternately between values near 0.51304 and 0.79946. This is called an *attracting 2-cycle*. The values of the 2-cycle depend closely on r, but *not* on the starting value. The sequence beginning at $\frac{1}{3}$ settles on the same 2-cycle as the one starting at $\frac{9}{10}$ (and almost every other point between 0 and 1).

Similarly when $r = 3.5$, the system has an *attracting 4-cycle* at approximately: 0.38282, 0.82694, 0.500088, 0.87500. Not all attractors are finite cycles like this. In higher-dimensional situations, the sequence may instead be drawn towards a **strange attractor**.

Bifurcations

When r is between 0 and 3, the logistic map has a single attracting point. From 3 to around 3.45, it has an attracting 2-cycle. Between approximately 3.45 and 3.54, it has an attracting 4-cycle, and between approximately 3.54 and 3.56, an attracting 8-cycle. As r gets closer to 3.57, the attracting cycle keeps doubling in length: 16, 32, and so on. These *bifurcations* are illustrated in the diagram overleaf. When r is bigger than 3.57, we see a new form of behaviour, namely chaos. The threshold of chaos is called the *Feigenbaum number*, and is approximately 3.5699.

Chaos

The logistic map is one of the most commonly studied dynamical systems, as it illustrates the basic aspects of *chaos theory*. Between $r = 3$ and around 3.57, the logistic map exhibits a sequence of bifurcations. At this point, chaos takes over. A sequence, starting at $\frac{1}{3}$, with $r = 3.58$, begins as follows: 0.333, 0.796, 0.582, 0.871, 0.403, 0.861, 0.428, 0.876, 0.388, ... One might hope that after some time this sequence would settle down into a discernible pattern. But the essence of chaos is that this is not so. After $r = 3.57$ most sequences will not be drawn into an attracting cycle, but will jump around for ever, apparently at random.

There are islands of stability amid the chaos, however. At $r = 1 + \sqrt{8}$ (around 3.82843), the system has an attracting 3-cycle.

Chaos versus randomness

It is tempting to use words such as 'random' or 'unpredictable' in relation to chaotic behaviour, such as that of the logistic map. But this is technically inaccurate. Once the value of r has been decided, along with the starting point of the sequence, the entire remainder of the sequence is completely predetermined. Its behaviour undoubtedly *appears* random, however. To an observer who didn't know the rule being used, it would be difficult to distinguish it from a sequence generated by a genuinely random process. Indeed chaotic systems can be well modelled by **Markov chains**, which do assume randomness.

Of course in the real world, it is impossible to know the value of *r* with infinite precision. So when sequences such as this appear in practical applications, such as population dynamics, the result will be indistinguishable from random behaviour. This phenomenon is typified by the *butterfly effect*.

Three-cycle theorem

The bifurcations of the logistic map suggest that attracting cycles tend to come with lengths that are a power of 2. This system does also have a 3-cycle, however, the first occurring at $r = 1 + \sqrt{8}$. The logistic map is an example of a *1-dimensional system*, since it only involves one variable *x*. In 1975, Tien-Yien Li and James Yorke proved that for any such system a 3-cycle is, paradoxically, a sure indicator of chaos. In their paper 'Period Three Implies Chaos', they proved that if a 1-dimensional system has an attracting 3-cycle anywhere, then it must also have attracting cycles of length 2, 4, 5, and every other finite length, as well as chaotic cycles.

The butterfly effect

A crucial characteristic of chaos is *sensitivity to initial conditions*. In the logistic map, for example, even minuscule changes in the value of *r* can produce completely different eventual outcomes of the sequence. Similarly tiny changes in the quadratic system can completely alter the type of attracting cycle, as illustrated by the intricacy of the Mandelbrot set.

The phrase coined by Edward Lorenz to describe this was the *butterfly effect*. The equations which describe the Earth's weather are also believed to be chaotic. So even a tiny change in air flow, such as a butterfly fluttering its wings in Brazil can trigger dramatic changes in weather patterns, eventually leading to tornadoes in Texas.

Chaotic systems

The logistic map was the first simple, chaotic process studied in depth. In the 1940s, John von Neumann first saw the potential of the system $x \mapsto 4\,x(1 - x)$ as a pseudo-random number generator. In the 1970s, the system was revisited by the biologist Robert May. The logistic map was intended as a simple model of the changes of a population of fish from year to year. Of course, the dynamics of the fish population turned out to be rather less simple than had been hoped.

Many chaotic systems have been identified throughout science. The origins of the subject date back to Newtonian physics and Henri Poincaré's work on the chaotic **three body problem**. Subsequently, chaotic systems have been used to analyse phenomena from epileptic seizures to stock-market crashes. A lava lamp illustrates the potential for chaos in fluid dynamics.

Strange attractors

The quadratic system and the logistic map are both examples of 1-dimensional dynamical systems, based on functions which take one number as input and one as output.

Higher-dimensional systems are also possible. For example, the *Henon phase* is a 2-dimensional function which takes as input a pair of numbers, x and y, and applies the following rule:

$$x \rightarrow x \cos \theta - (y - x^2) \sin \theta$$
$$y \rightarrow x \sin \theta - (y - x^2) \cos \theta$$

Here θ is any fixed angle. In the resulting dynamical system, points are gradually drawn towards a particular subset of the plane, called the system's *attractor*. The exact geometry of this attractor will depend on the choice of θ. As the Henon phase illustrates, attractors of dynamical systems are a rich source of beautiful imagery. If the fractal dimension of the attractor is not a whole number, it is called a *strange attractor*.

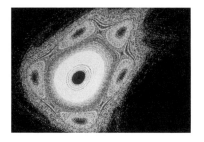

DIFFERENTIAL EQUATIONS

Differential equations

When we first learn to differentiate, we start with a function such as $y = 3x$ and then calculate its derivative: $\frac{dy}{dx} = 3$. Often, however, we need to go the other way. We may have some information about $\frac{dy}{dx}$, such as $\frac{dy}{dx} = 2x$, from which we want to deduce a formula for y.

According to the fundamental theorem of calculus, evaluating $\int 2x \, dx$ is the same as solving the *differential equation* $\frac{dy}{dx} = 2x$. So in this case we can integrate to find the solution $y = x^2 + C$ (where C is a constant of integration). This is the simplest form of differential equation. It is *first order*, because it involves only $\frac{dy}{dx}$ and not higher derivatives such as $\frac{d^2y}{dx^2}$. And, since it only involves $\frac{dy}{dx}$ and x, it can be solved directly by integration. When y is additionally involved, we may try **separating the variables**.

Differential equations appear in innumerable practical applications. **Newton's laws** (first and second), for example, are expressed in terms of rates of changes of position.

Solution spaces

When we solve polynomial equations such as $x^2 + 3x + 2 = 0$, we expect a certain number of fixed solutions, in this case two. When we solve differential equations such as $\frac{dy}{dx} = 6x$ we are looking for something subtler: an expression for y which explicitly describes how it relates to x. A solution to $\frac{dy}{dx} = 6x$ is $y = 3x^2$. But this is not the only solution. $y = 3x^2 + 5$ also satisfies the original equation, as does $y = 3x^2 - \frac{2}{5}$ and indeed $y = 3x^2 + C$ for any constant C. So, instead of getting a single solution, we get a *space of solutions* described by $y = 3x^2 + C$. This space is 1-dimensional, meaning that it is governed by one arbitrary constant.

For second-order differential equations such as $\frac{d^2y}{dx^2} = 6x$, we expect a 2-dimensional solution space: $y = x^3 + Cx + D$ is a solution for any constants C and D. Similarly a third-order differential equation generally has a 3-dimensional solution space, and so on. With the introduction of some *boundary conditions*, these constants may be pinned down to a unique solution (or smaller space of solutions).

Boundary conditions

Suppose a boy is cycling down a hill with constant acceleration of $2\,\text{m/s}^2$. This corresponds to a differential equation $\frac{d^2s}{dt^2} = 2$, where s is the distance travelled down the hill, and t is the time since he started travelling.

We can solve this by integrating. First we get:

$$\text{(i)} \quad \frac{ds}{dt} = 2t + C$$

Integrating again, we arrive at our 2-dimensional solution space:

$$\text{(ii)} \quad s = t^2 + Ct + D$$

Some extra data can specify a solution more precisely. If we are additionally told that the boy's speed at the start is $4\,\text{m/s}$, this translates as saying that, at $t = 0$, $\frac{ds}{dt} = 4$. Substituting this into equation (i), we find that $C = 4$. So our solution space has been narrowed to a 1-dimensional space:

$$\text{(iii)} \quad s = t^2 + 4t + D$$

Suppose we have a second bit of data, at time $t = 0$ the boy is at the top of the hill, that is, $s = 0$. Then we can put this into equation (iii) and find that $D = 0$, so now we have a unique solution:

$$\text{(iv)} \quad s = t^2 + 4t$$

The pieces of data which provide the values of s and $\frac{ds}{dt}$ at some specified value of t are called *boundary conditions*.

Separating the variables

Some differential equations do not describe $\frac{dy}{dx}$ straightforwardly, but set up a relationship between $\frac{dy}{dx}$ and y. For example, consider:

$$2y\frac{dy}{dx} = \cos x$$

Our aim is to deduce a formula relating y and x only. If we understand **implicit differentiation** and the chain rule, then we notice that the term on the left-hand side, $2y\frac{dy}{dx}$, is the derivative of y^2. Therefore we may integrate the left-hand side to get y^2. As the right-hand side is the derivative of $\sin x$, we can integrate to get:

$$y^2 = \sin x + C$$

More generally, if we have a differential equation $f'(y)\frac{dy}{dx} = g'(x)$, we recognize the left-hand side as the derivative with respect to x of $f(y)$, and solve it to get:

$$f(y) = g(x) + C$$

It is very tempting to think of this as 'multiplying both sides by dx', and then integrating: $\int f'(y)dy = \int g'(x)dx$. But some caution is needed, as is it is by no means clear what, if anything, 'multiplying both sides by dx' actually means. Ultimately, this procedure relies on spotting applications of the chain rule. But this short-hand of *separating the variables* is useful.

Another example is $e^y\frac{dy}{dx} = 4x$. Then $\int e^y\,dy = \int 4x\,dx$ and $e^y = 2x^2 + C$.

Radioactive decay
A lump of radioactive material decreases in mass as it decays. This is one of innumerable examples of a physical process which can be well modelled by a differential equation. The underlying physics tells us that the rate at which the lump decreases is proportional to the amount of mass remaining. In the simplest case (or by rescaling and working in different units) the two numbers will be equal.

If we let y be the mass of the lump and let t stand for the time since the clock started, then the rate of change of mass is $\frac{dy}{dt}$. Since y is shrinking, we expect $\frac{dy}{dt}$ to be negative. So the physics is modelled by the following differential equation:

$$\frac{dy}{dt} = -y$$

By separating the variables (or by treating this as a homogeneous equation; see **higher-order differential equations**) we end up with a formula $y = Ce^{-t}$. If we additionally have a boundary condition, say at $t = 0$ the lump had an initial mass of 2 kg, then we can solve the equation fully: $y = 2e^{-t}$. This formula tells us the mass of the lump after any length of time.

Higher-order differential equations
Second-order differential equations, which additionally involve $\frac{d^2y}{dx^2}$, can be very awkward to solve. The simplest type are those like $\frac{d^2y}{dx^2} - 5\frac{dy}{dx} + 6y = 0$, in which the left-hand side consists only of multiples of y, $\frac{dy}{dx}$, and $\frac{d^2y}{dx^2}$ added together, and the right-hand side is zero. These *homogeneous* equations are more manageable.

Their solutions can be found by considering related polynomials, called *auxiliary equations*. In this case the auxiliary equation is $A^2 - 5A + 6 = 0$, which illustrates the general pattern. This can then be solved to get $A = 2$ and $A = 3$. The theory of homogeneous equations then tells us that $y = e^{2x}$ and $y = e^{3x}$ are solutions to the original equation, and the general solution is $y = Be^{2x} + Ce^{3x}$ (for any constants B and C).

This procedure works just as well for higher-order homogeneous equations, and other forms of second and higher-order equations have different methods of solution. However, many cannot be solved exactly at all, and rely on **numerical analysis** for approximate solutions.

Partial differential equations

Partial differential equations are equations involving partial derivatives. For example, we might want a function z of x and y which satisfies $\frac{\partial z}{\partial x} = 0$. Partial differential equations are of great importance in physics, where they are often expressed in terms of **vector calculus**. Many physical phenomena are described by such equations. Examples include the **heat equation**, **Maxwell equations** for electromagnetic fields, the **Navier–Stokes equations** for fluid flow, and **Schrödinger's equation**, which governs quantum mechanics.

Solutions of partial differential equations

When solving an ordinary differential equation such as $\frac{dy}{dx} = 6x$, we usually get infinitely many solutions: $y = 3x^2 + 4$ is one solution, $y = 3x^2 - 1$ is another. At least, however, we can give a complete description of the solution space: $y = 3x^2 + C$, for any constant number C.

The situation is worse for partial differential equations. If we want a function z of x and y which satisfies $\frac{\partial z}{\partial x} = 0$, then $z = 4$ works, as does $z = 3y$, and $z = \sin y$. In fact, $z = f(y)$ is a solution for any function f. When differentiating with respect to x, any term involving y alone is treated as a constant and therefore disappears.

This illustrates that partial differential equations may have much bigger families of solutions than ordinary differential equations. Finding precise descriptions of these families can pose a major challenge. In other cases it is not clear whether the equation has any solutions at all. A notable example is the **Navier–Stokes problem**.

FOURIER ANALYSIS

Sine waves

To a mathematician's ear, the purest sound is that carried by a *sine wave* or *sinusoid*. For a sine wave at concert pitch A (a **frequency** of 440 Hz), and volume around that of ordinary conversation (an **amplitude** of around $\frac{1}{2,000,000}$ metres), the equation of the wave will be $y = \frac{\sin(440t)}{2,000,000}$, where y is in metres and t in seconds. But it is convenient to rescale this to:

$$y = \sin t$$

The *cosine wave* ($y = \cos t$) is identical, just shunted along by $\frac{\pi}{2}$. One of the advantages of working with sine and cosine waves is that their **harmonics** have easy equations. If $y = \sin t$ is the first harmonic, then $y = \sin 2t$ is the second, $y = \sin 3t$ the third, and so on.

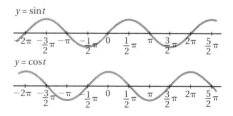

Musical instruments produce much more complex *waveforms* than the plain sine wave. In 1807, Joseph Fourier made the magnificent discovery that all such waves can be built from sine waves. This is the subject of Fourier analysis, and is of incalculable importance in modern technology.

Building waveforms

A *waveform* is a mathematical function with a special property: it is *periodic*, meaning that it repeats itself. Its *period* is another name for its *wavelength*, the length of one cycle. By rescaling, we can assume that any waveform has the same period as a sine wave, namely 2π.

The basic waveforms we have are the sine wave, with its equation $y = \sin t$, and its family of harmonics: $y = \sin 2t$, $y = \sin 3t$, $y = \sin 4t$, and so on. We can form many more interesting waveforms by adding together or *superposing* multiples of these: $y = \sin t + \sin 3t$ or $y = \sin t + \frac{3}{2}\sin 2t - \frac{1}{2}\sin 4t$.

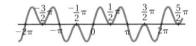
$y = \sin t + \sin 3t$

Two superposed waves will reinforce each other at some places, and cancel each other out at others; this is known as *interference*, and has the result that the superposition of the two waves will have a much more complicated appearance than either of the original waves.

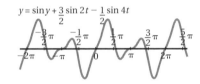
$y = \sin y + \frac{3}{2}\sin 2t - \frac{1}{2}\sin 4t$

For more complex waveforms still, we need Fourier series.

Fourier series

Waves come in all shapes and sizes. A *saw-tooth wave* (pictured top) looks radically different to the mathematician's favourite, the sine wave. It sounds different too. It also looks different from the smooth waves that can be built from sine waves by simple addition. We can get something quite close to it though:

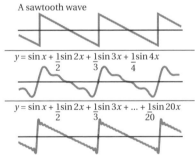
A sawtooth wave

$y = \sin x + \dfrac{\sin 2x}{2} + \dfrac{\sin 3x}{3} + \dfrac{\sin 4x}{4}$

$y = \sin x + \frac{1}{2}\sin 2x + \frac{1}{3}\sin 3x + \frac{1}{4}\sin 4x$

$y = \sin x + \frac{1}{2}\sin 2x + \frac{1}{3}\sin 3x + ... + \frac{1}{20}\sin 20x$

Once we have spotted the pattern, it is clear what needs to happen: next we need to add on $\frac{\sin 5x}{5}$ and then $\frac{\sin 6x}{6}$. As we add more and more such terms, we get closer to the saw-tooth wave, without ever quite getting there.

In a practical application, such as a musical synthesizer, we would settle for some number of terms as being an adequate approximation. But in the mathematical realm, there is another piece of technology we can exploit, that of an infinite series.

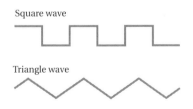
Square wave

Triangle wave

By taking the infinite limit of these terms, we can produce an exact formula for the saw-tooth wave, as a *Fourier series*:

$$\sin x + \frac{\sin 2x}{2} + \frac{\sin 3x}{3} + \frac{\sin 4x}{4} + \cdots = \sum_{n=1}^{\infty} \frac{\sin nx}{n}$$

Similar tricks work for other waveforms such as the square wave

$$\sin x + \frac{\sin 3x}{3} + \frac{\sin 5x}{5} + \frac{\sin 7x}{7} + \cdots = \sum_{n=1}^{\infty} \frac{\sin(2n-1)x}{2n-1}$$

and the triangular wave:

$$\sin x + \frac{\sin 3x}{9} + \frac{\sin 5x}{25} + \frac{\sin 7x}{49} + \cdots = \sum_{n=1}^{\infty} \frac{\sin(2n-1)x}{(2n-1)^2}$$

These series are all of the form $\sum a_n \sin nx$, where the a_n are carefully chosen numbers. In general a Fourier series can also include harmonics of cosine waves, giving the general formula:

$$\sum a_n \sin nx + \sum b_n \cos nx$$

Fourier's theorem

Objects similar to Fourier series had been studied before Joseph Fourier turned his mind to them in 1807. His great contribution was to realize that *every* reasonable periodic function can be given by a Fourier series. This was *Fourier's theorem*. It is certainly a powerful idea, but how is such a series to be found?

Suppose we have a periodic function f (perhaps the waveform of a French horn). If Fourier is correct, then there should be a Fourier series which is equal to f. That is,

$$f(x) = \sum a_n \sin nx + \sum b_n \cos nx$$

where a_n and b_n are some numbers. The challenge is to find the correct values of a_n and b_n. Other than blindly experimenting, how can we proceed? In a piece of mathematical magic, Fourier was able to provide a complete answer to this question, in Fourier's formulas.

By evaluating these formulas any reasonable waveform can be modelled as a Fourier series, and thus as a superposition of basic sine and cosine waves. This amazing fact has been of tremendous value in mathematics, physics and technology. For example, it is the basis on which computers can sample and recreate the sound of musical instruments.

Fourier's formulas

To express a waveform f as a Fourier series

$$f(x) = \sum a_n \sin nx + \sum b_n \cos nx$$

we need to find the right values for a_n and b_n. Fourier proved that the answers are:

$$a_n = \frac{1}{\pi} \int_{-\pi}^{\pi} f(t) \sin nt \, dt \qquad b_n = \frac{1}{\pi} \int_{-\pi}^{\pi} f(t) \cos nt \, dt$$

To split any reasonable waveform into sine and cosine waves, all that needs to be done is to evaluate these two integrals. In some cases this may be more easily said than done, of course, and can require sophisticated techniques of numerical analysis.

Complex Fourier series

A Fourier series breaks a waveform into trigonometric functions $\sin x$ and $\cos x$. According to **Euler's trigonometric formula**, these two are the offspring of the exponential function. This observation lured Fourier to take his series into the realm of complex numbers. By trading in the old constants a_n and b_n for some new complex constants c_n and c_{-n}, we can rewrite:

$$a_n \sin nx + b_n \cos nx = c_n e^{inx} + c_{-n} e^{-inx}$$

Now we get a much slicker expression for the complex Fourier series:

$$f(x) = \sum_{-\infty}^{\infty} c_n e^{inx}$$

To write a function $f(x)$ in this form, we have to evaluate the complex version of Fourier's formula:

$$c_n = \frac{1}{2\pi} \int_{-\pi}^{\pi} f(t) e^{-int} \, dt$$

Although more abstract than Fourier's original series, this discovery has been no less important.

The Fourier transform

Complex Fourier series express a function f as a series of exponential functions. A series is of course *discrete* (see **discreteness and continuity**): although there are infinitely many functions $c_n e^{inx}$, they are separate from each other. The Fourier transform is a continuous version of the complex series. The key to the Fourier series is the sequence c_n. We can think of this sequence as a new function $\hat{f}: \mathbf{Z} \to \mathbf{C}$, defined by $\hat{f}(n) = c_n$. The function f determines $\hat{f}$ and vice versa.

In the continuous case, $\hat{f}$ is a function on the real numbers, $\hat{f}: \mathbf{R} \to \mathbf{C}$, and the series becomes an integral:

$$f(x) = \int_{-\infty}^{\infty} \hat{f}(t) e^{itx} \, dt$$

This is undoubtedly rather abstract, but we can think of it as representing f as a weighted average of the functions e^{itx}. The function $\hat{f}$ defines the weighting and is called the *Fourier transform* of f.

Starting with f, how do we get at $\hat{f}$? The answer is again given by Fourier's formula, slightly tweaked:

$$\hat{f}(t) = \frac{1}{2\pi} \int_{-\infty}^{\infty} f(x) e^{-itx} \, \mathrm{d}x$$

These two formulas look surprisingly similar, revealing a deep symmetry between f and $\hat{f}$ which was not observable before. Exploiting this symmetry allows many difficult problems to be rephrased in ways that make them manageable; important examples are tricky partial differential equations. The Fourier transform is a powerful weapon in the modern mathematician's arsenal, and has wide applications, from **representation theory** to **quantum mechanics**.

LOGIC

THE LANGUAGE OF LOGIC pervades mathematics. To prove a theorem, you need to show that its conclusion can be deduced logically from its premises. But logic is also an object of study in its own right. An early milestone was Aristotle's classification of categorical syllogisms. This work remained the state of the art, until George Boole picked up the subject in the 19th century and began the rigorous analysis of formal logical systems.

Shortly after Boole's work, logic was unexpectedly propelled into the limelight, by Georg Cantor's work on set theory. His astonishing revelation that there are different levels of infinity precipitated what Hermann Weyl called a 'new foundational crisis' in mathematics. Were there firm logical foundations on which the rest of mathematics could depend? Russell's paradox brought special urgency to this question.

The challenge was articulated most clearly by David Hilbert, who set out the requirements he thought a logical framework for mathematics should obey. *Hilbert's program* stands as a monument to the fact that great thinkers can be profoundly wrong, and that mistaken ideas can be hugely important. During the 20th century, Hilbert's dream was shattered first by the *incompleteness theorems* of Kurt Gödel, and then by Alonzo Church and Alan Turing's work on the *Entscheidungsproblem*.

Although the mathematical realm turned out to be infinitely more complex than Hilbert expected, from the wreckage of his program sprang a whole new world. Most importantly, *Turing machines* formed the theoretical basis on which physical computers would later be built. Beyond this, new branches of mathematical logic blossomed, bringing powerful new tools. These include proof theory, model theory, complexity theory and computability theory.

BASIC LOGIC

Necessary and sufficient

'Socrates is mortal' is a *necessary* condition for the statement 'Socrates is human', because he must be mortal in order to be human; there is no other way around it. On the other hand, it is not a *sufficient* condition; being mortal does not on its own guarantee being human. Socrates could be a dog, a duck or a demigod. However, 'Socrates is a man' is a sufficient condition for 'Socrates is human'. If Socrates is a man, then he is definitely human. But it is not a necessary condition. Socrates could be a woman or a baby, and still be human.

Necessity and sufficiency are two sides of the same coin: if statement P implies statement Q, then P is a sufficient condition for Q, and Q a necessary condition for P. When P is both a necessary and sufficient condition for Q to hold, the two are *logically equivalent*, and mathematicians often use the phrase 'if and only if' to express this. For example, Socrates is a bachelor if and only if he is an unmarried man.

Contrapositive

Statements of the form 'P implies Q' or 'If …, then …' are *implications*. Here are two examples:

1 If Socrates is human, then he is a mammal.

2 If Socrates is not a mammal, then he is not human.

These two implications have essentially the same meaning: that humans form a subset of mammals. Statement 2 is called the *contrapositive* of 1. In general the contrapositive of 'P implies Q' is '*not Q* implies *not P*'. The contrapositive is a rephrasing of the original implication, a different way of expressing the same thing. The same is not true of the *converse*.

Converse

Not be confused with the contrapositive, the *converse* of 'P implies Q' is 'Q implies P'. The converse is not equivalent to the original, as illustrated by the converse to statement 1 above:

3 If Socrates is a mammal then he is human.

Or, equivalently,

4 If Socrates is not human, then he is not a mammal.

This is false (Socrates might be a dog, for example). If both 'P implies Q' and its converse 'Q implies P' hold, then P and Q are logically equivalent: P is true *if and only if* Q is true.

De Morgan's laws

Three fundamental particles of logic are NOT, AND and OR, sometimes written as $\neg$, $\wedge$, and $\vee$ respectively. In broader mathematics, as well as within logic, 'OR' is interpreted inclusively. That is, 'a OR b' always means 'a or b or both'. (When the exclusive OR is needed, it is written XOR.)

While the earliest formal logic, **propositional calculus**, was being developed by George Boole in the 19th century, Augustus de Morgan formulated two laws relating NOT, AND and OR.

1 'NOT (*a* AND *b*)' is equivalent to '(NOT *a*) OR (NOT *b*)'

2 'NOT (*a* OR *b*)' is equivalent to '(NOT *a*) AND (NOT *b*)'

A little thought shows that these hold in everyday language, and they can easily be verified using **truth tables**.

∀ and ∃

These two symbols are called *quantifiers*. ∀ stands for 'for all' (or 'for every' or 'for each') and ∃ for 'there exists' (or 'there is'). These phrases feature heavily in any self-respecting mathematician's vocabulary. To illustrate how to use them, consider the statement 'Everyone has a parent'. To avoid ambiguity, what is really meant is 'For every person in the world (living or dead), there is another person who is/was their biological parent'.

First we bring in some symbols, 'For every person x, there exists y, such that y is a parent of x'. Now we introduce a **predicate**: $P(y, x)$, to stand for 'y is a parent of x'. Then the sentence becomes:

$$\forall x\, \exists y\, P(y, x)$$

The order of the quantifiers is important here. If we swap them round, $\exists y\, \forall x\, P(y, x)$ we get the false assertion that there exists some person who is the parent of everyone.

Another delicate issue is that we have to specify the domain we're working in. The statement is not true (or at least not meaningful) if x is the Galaxy Andromeda, or the number π. So everything has to be restricted to the collection of people (living and dead). This is the *range of quantification*. It becomes increasingly difficult to discern the meaning of a statement, the more alternating quantifiers it contains. (See the **epsilon–delta** definition of continuity as an example. This is implicitly of the form $\forall \varepsilon\, \exists \delta\, \forall x\, Q(\varepsilon, \delta, x)$). More than three alternating quantifiers are very difficult to unravel.

THE SCIENCE OF DEDUCTION

Syllogisms

Logic received its first thorough, formal analysis in the fourth century BC, at the hands of Plato's student Aristotle. His six-volume *Organon* laid out the basic rules of logical deduction. *Syllogisms* played a central role. A syllogism is an argument which proceeds from two premises to reach a conclusion. The most famous is not due to Aristotle, but to the later thinker Sextus Empiricus:

1 All men are mortal.   (premise)

2 Socrates is a man.   (premise)

Therefore

3 Socrates is mortal.   (conclusion)

Sextus was a sceptic who thought this type of reasoning useless, arguing that, unless we already know Socrates to be mortal, we are in no position to assert that all men are mortal.

Categorical sentences

The first premise of the 'All men are mortal' syllogism is of the form:

(a) Every *X* is *Y*.

(With a slight mental contortion, the second and third can also be understood in this form.) The opposite form to (a) is known as (o):

(o) Some *X* is not *Y*.

The remaining possibilities are known as (i) and (e):

(i) Some *X* is *Y*.

and its opposite

(e) No *X* is *Y*.

Aristotle held that these were the four possible forms of *categorical sentence*, relating the categories *X* and *Y*. He used them as a basis for his classification of categorical syllogisms.

Aristotle's classification of categorical syllogisms

Aristotle's analysis of syllogisms proceeded by placing the four sentence forms a, o, i, e in different combinations at positions 1, 2 and 3 in the argument's scaffolding. This produces a list of 64 different syllogisms. Then he made a further refinement. In the sentence 'All men are mortal', 'men' is the *subject*, and 'mortal' the *predicate*. There are three elements in Sextus' syllogism: men, mortals and Socrates. Considering the various arrangements of these three as subject and predicate in the three sentences increases the number of possible syllogisms to 256.

Aristotle's analysis concluded that exactly 15 of these 256 possible syllogisms are valid, and a further four are valid so long as they do not apply to an empty category (that is, make statements about *X*, when there is no such thing as an *X*). The medieval names for the 19 valid syllogisms are: Barbara, Celarent, Darii, Ferio, Cesare, Camestres, Festino, Baroco, Darapti, Disamis, Datisi, Felapton, Bocardo, Ferison, Bramantip, Camenes, Dimaris, Fesapo and Fresison. These are mnemonics: the order of the vowels in each name gives the order of the categorical sentences a, o, i, e.

Dodgson's soriteses

The 19th century British mathematician and novelist Charles Dodgson, better known as Lewis Carroll, considered *soriteses*: longer arguments resembling Aristotle's categorical syllogisms, but containing more than two premises.

In his book *Symbolic Logic*, he challenged the reader to come up with the logically strongest conclusion from a set of premises such as this:

1 No one takes in *The Times*, unless he is well-educated.

2 No hedgehogs can read.

3 Those who cannot read are not well-educated.

A solution to this would involve breaking this down into a sequence of valid syllogisms. The most spectacular of Dodgson's sorites was Froggy's problem.

Froggy's problem
Dodgson set his readers the task of finding the strongest possible conclusion which can legitimately be deduced from this set of premises:

1 When the day is fine, I tell Froggy 'You're quite the dandy, old chap!';

2 Whenever I let Froggy forget that 10 dollars he owes me, and he begins to strut about like a peacock, his mother declares 'He shall *not* go out a-wooing!';

3 Now that Froggy's hair is out of curl, he has put away his gorgeous waistcoat;

4 Whenever I go out on the roof to enjoy a quiet cigar, I'm sure to discover that my purse is empty;

5 When my tailor calls with his little bill, and I remind Froggy of that 10 dollars he owes me, he does *not* grin like a hyena;

6 When it is very hot, the thermometer is high;

7 When the day is fine, and I'm not in the humour for a cigar, and Froggy is grinning like a hyena, I never venture to hint that he's quite the dandy;

8 When my tailor calls with his little bill and finds me with an empty pocket, I remind Froggy of that 10 dollars he owes me;

9 My railway shares are going up like crazy!

10 When my purse is empty, and when, noticing that Froggy has got his gorgeous waistcoat on, I venture to remind him of that 10 dollars he owes me, things are apt to get rather warm;

11 Now that it looks like rain, and Froggy is grinning like a hyena, I can do without my cigar;

12 When the thermometer is high, you need not trouble yourself to take an umbrella;

13 When Froggy has his gorgeous waistcoat on, but is *not* strutting about like a peacock, I betake myself to a quiet cigar;

14 When I tell Froggy that he's quite a dandy, he grins like a hyena;

15 When my purse is tolerably full, and Froggy's hair is one mass of curls, and when he is *not* strutting about like a peacock, I go out on the roof;

16 When my railway shares are going up, and when it's chilly and looks like rain, I have a quiet cigar;

17 When Froggy's mother lets him go a-wooing, he seems nearly mad with joy, and puts on a waistcoat that is gorgeous beyond words;

18 When it is going to rain, and I am having a quiet cigar, and Froggy is *not* intending to go a-wooing, you had better take an umbrella;

19 When my railway shares are going up, and Froggy seems nearly mad with joy, *that* is the time my tailor always chooses for calling with his little bill;

20 When the day is cool and the thermometer low, and I say nothing to Froggy about his being quite the dandy, and there's not the ghost of a grin on his face, I haven't the heart for my cigar!

Sadly Lewis Carroll died before publishing Froggy's solution. He did, however, hint that the problem 'contains a beautiful "trap".'

Formal systems

Longer logical deductions than syllogisms, such as the solutions to Dodgson's sorites, can be built up as longer sequences of categorical sentences. We would like an analysis of these compound arguments. However, given that there is no limit to their possible length, how is this possible? Gottfried Leibniz had the first idea in the 1680s, but his work was not picked up, and the subject did not come to fruition until the 19th century, when George Boole, Augustus de Morgan and others developed the first *formal system* for logic.

Their philosophy was to excise from logic all human subjectivity and intuition, and build up logical arguments from the ground. As Boole put it, 'the validity of the processes of analysis does not depend upon the interpretation of the symbols which are employed, but solely upon the laws of combination'.

Any formal system has three ingredients:

1 *a language*, which is a list of permitted symbols, together with a grammar which says how to combine these into legitimate formulas

2 *some axioms*, which are specified formulas, taken as the starting points for logical deduction

3 *laws of deduction*, which say how to deduce a formula from previous ones.

Then, a formula is judged *valid*, and called (rather grandiosely) a *theorem* purely if it can be deduced from the axioms, by a sequence of applications of the laws of deduction.

The first formal system is propositional calculus.

Modus ponens

In a formal system, *formulas* are permissible arrangements of symbols. Some formulas are *assumed* to be logically valid: these are the *axioms*. To derive more valid formulas from these, we need laws of deduction. The most universal such law is *modus ponens* (from Latin, meaning the *affirming mode*). This says that from any formulas 'P' and '$P \rightarrow Q$', we may deduce the formula 'Q'.

Propositional calculus

The first formal system, propositional calculus, was intended to provide a complete framework for logical deduction, such as longer sequences of categorical sentences.

1 There are variables p, q, r, etc. (We think of each of these as standing for a categorical sentence, such as 'all men are mortal'.)

The language also includes brackets '(' and ')' and the *connective* symbols $\wedge$, $\vee$, $\rightarrow$ (which we think of as *and*, *or* and *implies*, respectively), as well as $\neg$ (which we think of as *not*).

The grammar says how to combine these symbols into legitimate formulas, intended to mirror meaningful (though not necessarily true) statements. So we get '$(p \wedge q) \rightarrow r$' but not gibberish like '$\rightarrow \rightarrow p \neg \wedge q$'.

2 The *axioms* are a collection of specific formulas. For every formula P, the formula $P \rightarrow P$ is an axiom, as is $((\neg Q) \rightarrow (\neg P)) \rightarrow (P \rightarrow Q)$ (see **contrapositive**). Other axioms are needed to encode the intended meaning of the logical symbols; these include de Morgan's laws, as well as general logical principles deriving from Aristotle's laws of thought. When written down, each has an obvious feel to it.

3 For propositional calculus we only need one law of deduction, *modus ponens*.

The question is what does all this produce? What is the list of formulas that can legitimately be deduced from the axioms?

There is a very neat description of these via the informal method of truth tables. That this provides a complete answer is guaranteed by the **adequacy and soundness theorems**.

Not
In ordinary language, as opposed to a logical formal system, the function of the word 'not' is usually to negate whatever comes next, changing the sentence into its opposite. So 'not P' (or '$\neg P$' as logicians write it) should be true when P is false, and false when P is true.

We can represent this in a little *truth table*:

P	¬P
T	F
F	T

The left-hand column of this truth table shows the possible truth values for P (either True or False), and the right-hand column gives the resulting value for $\neg P$.

Truth tables
We can use a larger truth table to analyse 'and', which is sometimes denoted by the wedge symbol $\wedge$. This time we need two columns on the left to list the possible truth values for P and Q and the right-hand column gives the corresponding truth values for $P \wedge Q$.

P	Q	P∧Q
T	T	T
T	F	F
F	T	F
F	F	F

The same trick works for 'or' ($\vee$). Notice that in mathematics 'or' is inclusive. So, 'P or Q' always means 'P or Q or both'.

P	Q	P ∨ Q
T	T	T
T	F	T
F	T	T
F	F	F

Another important *connective* is 'implies' (written as →). Here we have to make a choice: should the statement 'if the moon is made of cheese, then I am 1001 years old' be considered true or false? What about 'if am 1001 years old, then London is the capital of the UK'? In ordinary life we do not worry about implications where the conditional clause (the 'if…' part) fails. But the logicians of the 19th century had to decide, and they settled on the convention that in both cases the implication should be counted valid. That gives '→' the following truth table:

P	Q	P → Q
T	T	T
T	F	F
F	T	T
F	F	T

Another symbol is '↔' standing for 'if and only if'. It has the truth table:

P	Q	P ↔ Q
T	T	T
T	F	F
F	T	F
F	F	T

Tautology and logical equivalence

We can apply the truth table rules to particular examples. For example, to assess the statement $P \to P$, we use the truth table:

P	P → P
T	T
F	T

Because the column for $P \rightarrow P$ contains only Ts, it is always true no matter what its inputs. Such a statement is known as a *tautology*. In ordinary language, 'If Sophocles is an unmarried man then he is a bachelor' is a tautology and is true whatever Sophocles' actual marital status.

Using this idea we can build up truth tables for more complicated formulas:

P	Q	$P \rightarrow Q$	$\neg P$	$(\neg P) \vee Q$	$(P \rightarrow Q) \leftrightarrow ((\neg P) \vee Q)$
T	T	T	F	T	T
T	F	F	F	F	T
F	T	T	T	T	T
F	F	T	T	T	T

Again the column for $(P \rightarrow Q) \leftrightarrow ((\neg P)) \vee Q$ contains only Ts, so it is a *tautology*. In this case, this means that $P \rightarrow Q$ and $(\neg P) \vee Q$ are *logically equivalent*.

NAND, NOR, XOR and XNOR

The commonest logical connectives are 'and', 'or', 'implies' and 'if and only if'. But there are four others, which are best described by their truth tables.

NAND stands for 'not … and'. So 'P NAND Q' means 'not (P and Q)'. It has truth table:

P	Q	P NAND Q
T	T	F
T	F	T
F	T	T
F	F	T

NOR is 'not … or', as in 'not (P or Q)':

P	Q	P NOR Q
T	T	F
T	F	F
F	T	F
F	F	T

XOR is the exclusive 'or'. So 'P XOR Q' means 'P or Q, but not both':

P	Q	P XOR Q
T	T	F
T	F	T
F	T	T
F	F	F

Finally XNOR is the exclusive 'nor'. So 'P XNOR Q' means 'not (P XOR Q)':

P	Q	P XNOR Q
T	T	T
T	F	F
F	T	F
F	F	T

Logic gates

The relations that logicians put in truth tables are important in electronics, as *logic gates*. A logic gate is a component with inputs (typically two of them) and one output. The output of the component is on or off, according to its two inputs and the rule coming from the truth table. So a NOR gate will emit an output when both its inputs are off, and not otherwise.

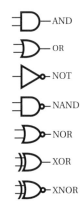

In this context NAND and NOR are the primary gates. In fact, everything can be built up just from NAND. For example, 'NOT *P*', can be obtained as '*P* NAND *P*', and then '*P* AND *Q*' as 'NOT (*P* NAND *Q*)'. Using this idea, more sophisticated rules for the behaviour of a device can be hardwired, as logical deductions.

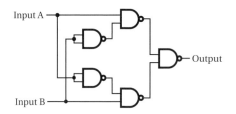

The adequacy and soundness theorems

Truth tables do suffer from one problem, namely that they grow exponentially larger, the more variables are included. The full truth table for Froggy's problem would need 131,072 $(= 2^{17})$ rows. Nevertheless truth tables encapsulate human intuition about logical operations. In particular, *tautology* perfectly captures the concept of logical validity.

If the system of propositional calculus is to do its job properly, every tautology should be a theorem of that formal system. This is indeed true, and is known as the *adequacy theorem*. The converse is also true: every theorem of the formal system is a tautology. This is the *soundness theorem*.

Beyond propositional logic

There are two ways to view logic: George Boole's highly formal propositional calculus and the more intuitive approach of truth tables. The adequacy and soundness theorems say that the two perspectives produce the same result. This effectively renders propositional calculus obsolete. Everything it can express can be arrived at more quickly through common sense and truth tables.

On the other hand, truth tables also vindicate Boole's highly formal approach to logic. Although in the context of propositional calculus the result has not turned out to be especially useful, it does demonstrate that formal systems do work. When turbo-charged and applied in broader settings (**predicate calculus**, **Peano arithmetic** and **axiomatic set theory**), this approach would go on to transform mathematics.

Predicate calculus

Most mathematical arguments do not fit the pattern of an Aristotelian syllogism, or even a long sequence of them. For example:

1 x is greater than y.

2 y is greater than z.

Therefore:

3 x is greater than z.

We would like a logical system to be able to cope with a deduction such as this. But not every argument of this form works:

A x is one more than y.

B y is one more than z.

Therefore:

C x is one more than z.

The overall form of these arguments is not enough to judge whether they are valid. It depends on the detailed behaviour of the relations '… is greater than …' and '… is one more than…'.

Propositional calculus is too simplistic to cope with this. The solution is to introduce subtler *predicates*, which can represent more delicate relationships between objects. The result is *predicate calculus*.

'… is greater than …' can be formalized as a predicate. Others can encode mathematical functions such as addition or multiplication. The choice of which predicates to put in the language depends on the structure under investigation.

Predicates

For a mathematical logician, 'is greater than' is an example of a *predicate*. This is a mathematical adjective for describing an object, or in this case a relationship between two objects. (Philosophers use the term somewhat differently.) Of course, we have a good idea what one number being 'greater than' another *means*. But, in a formal system, all meaning and intuition are stripped away, and only unthinking axioms and deduction apply. For any remnant of the intended meaning to survive, we must axiomatize it. **Hilbert's program** expressed the hope that the whole of mathematics could be incorporated into just one system of predicate calculus.

Axiomatizing predicates

If we want to construct a system which incorporates 'is greater than' we need to introduce a symbol to stand for it, say $G(x, y)$, which we think of as saying 'x is greater than y'. Now we want the system to contain the deduction above (1 and 2 implies 3). So we build it in as an axiom:

$$(G(x, y) \ \& \ G(y, z)) \rightarrow G(x, z)$$

We might also want to say that there is no largest number. We can express that by saying that, for every number, there exists another which is greater. So we also include the axiom:

$$\forall y \, \exists x \, G(x, y)$$

Continuing in this fashion will produce a system of predicate calculus for describing orderings with no maximum element. Every theorem of the system will be a true fact about such a structure. A similar approach allows building formal systems for **number systems**, **groups** and other structures. A key example is Peano arithmetic, which axiomatizes the natural numbers.

Peano arithmetic

Giuseppe Peano's work marked a coming of age of mathematical logic. For the first time, he was able to provide an axiomatization of the most fundamental of all mathematical structures: **N**, the system of **natural numbers**. The aim was to come up with a formal system which precisely captures the behaviour of the number system **N**. Peano's solution revolves around the number 0, and the *successor function*, which should be interpreted as taking any number (such as 5) to the next one (6).

Peano's axioms were a landmark. But, like so many great discoveries, his system opened up as many questions as it answered. First: are there any models of Peano arithmetic? Secondly: can every statement in number theory really be derived within this system? This was the challenge posed by **Hilbert's program**, and later answered by **Gödel's incompleteness theorems**.

Axioms of Peano arithmetic

The first two axioms of Peano's system set out the basic properties of the successor function:

1 No number has 0 as its successor: $\forall x \, S(x) \neq 0$

2 Two different numbers cannot have the same successor:

$$\forall x, y \quad x \neq y \rightarrow S(x) \neq S(y)$$

Peano's language also contains addition, which is axiomatized:

3 Adding 0 to any number does not change it: $\forall x \, x + 0 = x$

4 The successor of the sum of two numbers is the same as adding the first to the successor of the second:

$$\forall x, y \quad x + S(y) = S(x + y)$$

Similarly, multiplication is in the language, and is axiomatized:

5 Any number when multiplied by zero gives zero: $\forall x \, x \times 0 = 0$

6 For any numbers x and y, multiplying x by $y + 1$ gives the same result as multiplying x by y and then adding x:

$$\forall x, y \quad x \times S(y) = (x + y) + x$$

The final axiom is the most complicated. It builds into the system the principle of **mathematical induction**:

7 This says, if ϕ is any property of numbers, such that 0 satisfies ϕ, and whenever x satisfies ϕ, $x + 1$ also does, then every number satisfies ϕ. So, for every formula ϕ, we have an axiom:

$$\big(\phi\,(0)\ \&\ \forall x\,(\phi(x)) \rightarrow \phi(S(x))\big) \rightarrow \forall y\,\phi(y)$$

Models of Peano arithmetic

Peano's axiomatization of the natural numbers was a real breakthrough. But the rulebook alone does not make a game of football. Could the same care and precision which produced the axioms also yield a *model* of them, that is, a structure which obeys them?

Since the dawn of humanity, the primary use of natural numbers has been to count things, that is, to represent the sizes of sets of objects. The task facing logicians, therefore, was first to develop a theory of sets, to abstract the notion of the *size* of a set, and then to show that these sizes do behave as Peano arithmetic says they should. Various approaches were taken, notably *Principia Mathematica* and then **Zermelo–Fraenkel set theory**. Peano himself, however, gave up mathematics, to work on a new international language in which he published the final version of his logical work.

Principia Mathematica

At the begining of the 20th century, Bertrand Russell and Alfred North Whitehead set themselves a monumental task: to deduce the entirety of mathematics from purely logical foundations. Gottlob Frege had already tried this (see **logicism**) but was undone at the last moment, when Russell wrote to him with the news of **Russell's paradox**. Frege had been too permissive regarding what can classify as a *set*, and allowed in paradoxical monsters. In their three-volume masterpiece *Principia Mathematica*, Russell and Whitehead were more cautious, and in the first volume they painstakingly developed type theory.

In the second, they defined numbers within the theory, and proved that their system does indeed conform to Peano's rules. (It takes until page 83 of the second volume to deduce that $1 + 1 = 2$, accompanied by the comment 'The above proposition is occasionally useful'.) The third volume develops higher-level mathematics, including the **real numbers**, **Cantor's infinities** and the rudiments of **analysis**. A fourth volume on geometry was planned, but never finished.

The type-theoretic approach was later largely surpassed by Zermelo–Fraenkel set theory. But the overarching philosophy, and many of the proofs, carry over to that context. Principia Mathematica remains a landmark of mathematical logic, although Gödel's incompleteness theorems later showed that neither it nor any other system can hope to capture the whole of mathematics.

Type theory

Invented by Bertrand Russell, *type theory* is an approach for the foundations of mathematics, intended to avoid **Russell's paradox**. Objects are not of one type, but are stratified. In simple terms, atomic objects are assigned level 0. A set of such objects has level 1, and sets of such sets have level 2, and so on. Any object can only refer to others on lower levels than itself. This prevents any form of Russell's paradox creeping in.

In 1937, Willard Van Orman Quine proposed a new type theory known as *New Foundations*, which greatly simplifies Russell's version, and continues to be studied as an alternative and comparator to Zermelo–Fraenkel set theory. As the subject of computer science flourished, type theory assumed a new importance for describing computational objects. It was advanced in 1970 by Per Martin-Löf, who developed a powerful new framework initially intended to act as a foundation for **constructive mathematics**. This type theory had properties which made it exciting to computer scientists, including the ability to encode logical deductions inside itself. Several programming languages, including **proof checking software**, have subsequently grown out of it.

SET THEORY

Sets

A *set* is a collection of objects, called its *elements*, or *members*. So the set of **natural numbers** (usually denoted **N**) has elements 0, 1, 2, 3, … We might informally write **N** = {0, 1, 2, 3, …}, but set theory involves being extremely careful about the meaning of '…'.

It is to be expected that sets are everywhere in mathematics. Every number system and algebraic structure is a set, with some additional properties. But it might be surprising that much mathematics can be done at such a general level. The origins of the subject lie in the pioneering work of Georg Cantor, who published his first work on infinite sets in 1874. His paper of 1891 famously contained **Cantor's diagonal argument** and **Cantor's theorem**. In the early 20th century, following the discovery of **Russell's paradox**, the search began for a version of **axiomatic set theory** to act as a foundation for the whole of mathematics. The search culminated in 1922 with the discovery of **Zermelo–Fraenkel set theory**, but not without leaving loose ends such as the **axiom of choice** and **continuum hypothesis**.

Set membership

'$a \in B$' means that the object a is a member of the set B. The symbol '$\in$' is a disfigured Greek epsilon.

So '$1 \in$ **N**' says that the object 1 is an element of the set **N**. If x is a person, and Y is the set of all Spanish speakers, then '$x \in Y$' asserts that x is a member of the set of Spanish speakers, that is, x can speak Spanish.

Intersection

Suppose A and B are two sets. The *intersection* of A and B is their overlap: that is, the collection of all objects included in both A and B. This is written as $A \cap B$. So, if A is the collection of English speakers and B is the collection of Spanish speakers, then $A \cap B$ is the set of people who speak both Spanish and English. If A is the collection of even numbers and B is the collection of odd numbers, then $A \cap B = \varnothing$ (the **empty set**). This means that the two sets are *disjoint*, having no elements in common.

Union

The *union* of two sets A and B is the set obtained by taking the two together. It is denoted by $A \cup B$. So $A \cup B$ is the collection of objects in *either A or B* (or both). If A is the collection of English speakers, and B is the collection of Spanish speakers, then $A \cup B$ is the collection of people who speak either English or Spanish (or both). When both sets are finite, there is a formula for the **size of the union**.

If A is the collection of even numbers, and B is the collection of odd numbers, then $A \cup B = \mathbf{Z}$, the set of all integers.

Because A and B cover $\mathbf{Z}$ entirely, and are also *disjoint*, they form a *partition* of $\mathbf{Z}$.

Subsets

If A is a set, then a *subset* of A is any collection of elements from A. The set of even numbers is a subset of set of natural numbers ($\mathbf{N}$). Similarly $\{1, 2, 3, 4, 5\}$ is a subset of $\mathbf{N}$, as are the set of prime numbers, the set of square numbers, and every other possible collection of natural numbers. Of course $\mathbf{N}$ is also a subset of itself. The empty set is a subset of every set.

{A,B} is a subset of {A,B,C}

Functions

Every area of mathematics involves the notion of a *function*. This can be thought of as a process into which you feed inputs, to get outputs. It is common to call a function 'f', to write 'x' for an input and '$f(x)$' for the corresponding output.

The formula '$f(x) = x^2 + 2$' describes a function which accepts any number as an input, and produces an output of 2 more than its square. So $f(2) = 6$, for example. If we wanted to draw this function as a graph, we would set $y = f(x)$, and plot points (x, y) for different values of x.

As in this example, inputs and outputs are often numerical, and the function can be written out explicitly with a formula. But there are more abstract types of function too.

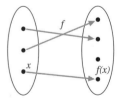

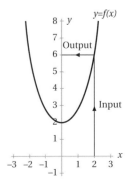

Domain and range

When working with a function, care is always needed to specify its *domain*. This is the set of all allowable inputs. We write $f: A \to B$ to indicate that f is a function from A to B. That is, its domain is the set A, and all the outputs are in the set B. For example, we might consider a function $f: \mathbf{R} \to \mathbf{R}$, given by the formula $f(x) = x^2 + 2$.

The *range* is the set of all outputs of the function. If we have a function $f: A \to B$, the range may not be the whole of B, but can be a subset. In the above example, the range is the set of all real numbers from 2 onwards.

One to one correspondence

A *one to one correspondence*, also known as a *bijection*, between two sets A and B is a way to pair up every element of A with an element of B. These are special functions from A to B which guarantee that every element of B corresponds to one, and only one, element of A.

Bijections are ubiquitous in mathematics. Every permutation (see **permutation groups**) is a bijection, for example, as is every **isomorphism**. In set theory, the importance of bijections stems from the fact that two sets have the same size if and only if there is a one to one correspondence between them. This is obvious for finite sets; indeed this is how we count things. For example, by constructing a one to one correspondence between the set $\{1, 2, 3, 4, 5, 6\}$ and the letters in the word CANTOR, we count the number of letters. It was when Georg Cantor began to apply this principle to infinite sets that the ground beneath 19th-century mathematics began to quake: previously there had been a tacit assumption that all infinite sets were the same size, that is, that any two infinite sets should be in one to one correspondence. Cantor's theorem and diagonal argument shocked many by demolishing that belief.

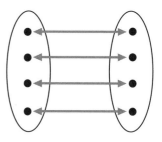

The uncountability of the real numbers

In 1874, Cantor published a demonstration that the natural numbers, **N**, and the **real numbers**, **R**, must have different *cardinalities*: that is, there can never be a one to one correspondence between them. In fact he showed that just between 0 and 1 there are more real numbers than in the whole of **N**. His paper of 1891 contained a new and disarmingly slick proof of the same thing, a classic result now known as Cantor's diagonal argument. Though no longer controversial within mathematics, this result still attracts a great deal of scepticism from those encountering it for the first time. One instance even ended up in court in Wisconsin, USA: the 1996 case of Dilworth versus Dudley.

In 1998 the mathematician and editor of the journal *Bulletin of Symbolic Logic*, Wilfrid Hodges, wrote about various fallacious refutations that he had been sent, and wondered 'why so many people devote so much energy to refuting this harmless little argument – what had it done to make them angry with it?' Hodges went on to suggest that 'This argument is often the first mathematical argument that people meet in which the conclusion bears no relation to anything in their practical experience or their visual imagination'.

Cantor's diagonal argument

If b is a decimal between 0 and 1, we will represent it as $0.b_1 b_2 b_3 ...$, where b_1 is the first decimal place, b_2 is the second, and so on. So if $b = 0.273333...$, then $b_1 = 2$, $b_2 = 7$, and $b_3 = 3 = b_4 = b_5 = \cdots$ (A technical detail is that some numbers have two different decimal expansions: $0.899999... = 0.900000...$ To avoid ambiguity we will always choose the representation ending in zeroes.)

Now, suppose for a contradiction, that there is a one to one correspondence between the natural numbers and the real numbers between 0 and 1. Then we can write out this correspondence in a grid:

1	b^1
2	b^2
3	b^3
4	b^4
$\vdots$	$\vdots$
n	b^n
$\vdots$	$\vdots$

The right-hand column is the supposed complete enumeration of the real numbers between 0 and 1, written out as decimals. (The superscripts are labels to discriminate between the numbers, they do not represent powers.)

Writing out the decimal places gives:

1	$b^1 = 0.b_1^1 b_2^1 b_3^1 ...$
2	$b^2 = 0.b_1^2 b_2^2 b_3^2 ...$
3	$b^3 = 0.b_1^3 b_2^3 b_3^3 ...$
4	$b^4 = 0.b_1^4 b_2^4 b_3^4 ...$
$\vdots$	$\vdots$
n	$b^n = 0.b_1^n b_2^n b_3^n ...$
$\vdots$	$\vdots$

Cantor proceeded to construct a number x which is missing from the list. To begin with, either $b_1^1 = 7$ or it does not. If it does, then let $x_1 = 4$, but if not, then let $x_1 = 7$. Either way, $x_1 \neq b_1^1$. Similarly, if $b_2^2 = 7$, then let $x_2 = 4$. But if $b_2^2 \neq 7$ then let $x_2 = 7$. Either way, $x_2 \neq b_2^2$. Do the same at every stage: if $b_n^n \neq 7$ then let $x_n = 7$. So whatever happens, $x_n \neq b_n^n$. Then we create the number $x = 0.x_1 x_2 x_3 ...$ Now x cannot be equal to b^1 because they differ in the first digit ($x_1 \neq b_1^1$). Similarly, $x \neq b^2$ because $x_2 \neq b_2^2$. In fact, $x \neq b^n$ since they differ in the nth digit ($x_n \neq b_n^n$). So x is a real number between 0 and 1 which is not on the list, and that is a contradiction. Therefore there can be no one to one correspondence between the natural numbers and the real numbers between 0 and 1.

Countable infinities

Cantor's proof of the uncountability of the real numbers cleaved the world of infinity in two types of infinity: *countable* sets which can be put in one to one correspondence with the natural numbers, and *uncountable* sets which cannot. Every infinite subset of the natural numbers, such as the prime numbers, is countable. Write them out in a list: 2, 3, 5, 7, 11, ... Then counting them sets up a correspondence, 1 with 2, 2 with 3, 3 with 5, 4 with 7, 5 with 11, and so on. The integers are also countable, since we can list them like this: $0, 1, -1, 2, -2, 3, -3, \ldots$

More surprisingly, the positive **rational numbers** are countable, as Cantor proved in 1873. These are the positive fractions and they can be set out in a grid as shown. Then, by taking a cleverly snaking path, we can count them all once, though we have to take care to skip any we've already listed. So $\frac{2}{2}$ gets missed out, as 1 has already been counted (as $\frac{1}{1}$). This sets up the correspondence we want.

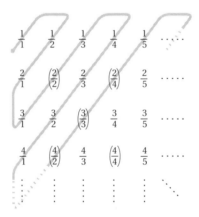

Uncountable infinities

Cantor's dichotomy of countable and uncountable infinite sets had profound consequences for mainstream mathematics. To begin with, he had shown that the real numbers are uncountable and the rational numbers are countable. It must follow that there are uncountably infinitely many irrational numbers in the gaps between the rationals. In a very precise sense then, almost all real numbers are irrational. Cantor also applied this line of thought to the study of the even more enigmatic **transcendental numbers**.

He had not finished yet, however. Cantor's theorem of 1891 subdivided infinity again. In a second shock, he showed that there is not just one level of uncountable infinity, but infinitely many. Many of these have names, given by **cardinal numbers**. The tool Cantor used was the notion of a power set.

Power sets

If A is a set, then its *power set* $\mathbf{P}(A)$ is the collection of all its subsets (making the power set a set of sets). To take an example, the power set of {1, 2, 3} is {∅, {1}, {2}, {3}, {1, 2}, {2, 3}, {1, 3} {1, 2, 3}}. In this example the original set has three elements, and its power set has $2^3 = 8$ elements. This illustrates the general pattern. The power set of a set with n elements contains 2^n elements. Remarkably, this rule extends to infinite sets, as well as the finite ones. This is the content of Cantor's theorem.

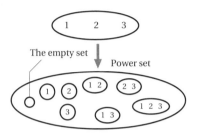

Cantor's theorem

In 1891, armed with just two simple tools, the idea of a power set, and the notion of a one to one correspondence, Georg Cantor tore apart the mathematics of his time. With one short argument (whose form anticipated Russell's paradox) he proved that there can never be a one to one correspondence between any set and its power set.

The consequences were dramatic: start with an infinite set, such as that of the natural numbers $\mathbf{N}$, and take its power set, $\mathbf{P}(\mathbf{N})$. By Cantor's argument, the two can never be the same size. The power set is always bigger. Now start again. Taking the power set of the new infinite set $\mathbf{P}(\mathbf{N})$ reveals a still bigger set $\mathbf{P}(\mathbf{P}(\mathbf{N}))$. By repeatedly taking power sets, infinity opened up before Cantor's eyes. No longer did it seem a monolith, but an infinite hierarchy, with each level infinitely surpassing the last.

Cardinal numbers

Cantor's theorem demonstrated that there are infinitely many levels of infinity, so a way to measure and compare them was needed. Cantor constructed a system of *cardinal numbers*, special sets which could measure any other: every set will be in one to one correspondence with exactly one cardinal, called its *cardinality*.

The first cardinals are the ordinary natural numbers: 0, 1, 2, 3, … But then come the infinite cardinals, the first of which is called $\aleph_0$ (that is 'aleph zero', aleph being the first letter of the Hebrew alphabet). $\aleph_0$ is the cardinality of the set of natural numbers $\mathbf{N}$, and so also the cardinality of every countable set. The next cardinal number is $\aleph_1$, then $\aleph_2$, $\aleph_3$, and so on, as far as $\aleph_{\aleph_0}$, and way beyond. For any cardinal number it always makes sense to talk about the *next* one: the smallest cardinal which is bigger than the one you already have.

Another way to find bigger cardinals is by taking power sets. The cardinality of the power set of $\aleph_0$ is written as $2^{\aleph_0}$, and then we can take the power set of that $2^{2^{\aleph_0}}$, and so on. This sequence is also written as $\beth_0\,(=\aleph_0)$, $\beth_1(=2^{\beth_0})$, $\beth_2(=2^{\beth_1})$, $\beth_3(=2^{\beth_2})$ etc. (where $\beth$ ('beth') is the second letter of the Hebrew alphabet).

The relationship between the sequences $\aleph_0$, $\aleph_1$, $\aleph_2$, $\aleph_3$, … and $\beth_0$, $\beth_1$, $\beth_2$, … is the subject of the continuum hypothesis. Beyond the reach of these sequences is the domain of the **large cardinals**.

Russell's paradox

With a few short words in 1901, the mathematician and philosopher Bertrand Russell seemed to sound the death knell of the set theory that had blossomed in the years following Cantor's theorem. If a *set* is any collection of objects, then the *set of all sets* seems to make perfectly good sense. Furthermore, this special set must contain itself as a member, by definition. This illustrates that some sets contain themselves as members, while others do not. Russell's devastating move was to define a set X: the set of exactly those sets which do *not* contain themselves as members. The paradox is this: is X a member of itself? Either assumption, that it is a member of itself, or that it is not, ends in contradiction.

Although Russell and others struggled to find a way around this paradox, there is no way to resolve it, other than by being far more discriminating about what is classed as a *set*. Russell's paradox killed off the naïve set theory which admitted any collection of objects as a set. In a few years it would be replaced with the more technically demanding and rigorous subject of axiomatic set theory, in which paradoxical monsters such as X and the set of all sets cannot exist.

The barber and librarian paradoxes are real-world analogies for Russell's paradox, and Grelling's paradox transplants the argument into linguistics.

The barber paradox

The *barber paradox* is a translation of Russell's paradox from the domain of set theory to the real world. It was used by Russell in discussion of his work. There is a village, in which there lives a barber, who shaves some of the village men. To be exact, he shaves all the men who do not shave themselves (and only those men). The unanswerable question is: who shaves the barber?

The librarian paradox

The *librarian paradox* is another analogy for Russell's paradox. A librarian is indexing all the books in her library. She compiles two lists, A and B; every book is entered into one or the other (but not both). In A she lists all the books which reference themselves. In B she lists those which do not. Once she has completed the main collection, she has two new books, the books A and B, which need to be indexed too. But where can she list B? If she puts it in A, then it cannot also go in B. This would make it one of those books which do not reference themselves, and so it should be listed in B, not A. But if she does list it in B, then it does reference itself, and so should be listed in A, not B. Again, either way leads to a contradiction.

Axiomatic set theory

'No-one shall expel us from the paradise that Cantor has created', declared the influential German mathematician David Hilbert. But Russell's paradox demonstrated that the set theory of the time contained dangerous contradictions. Would it bring the whole edifice of cardinal numbers crashing down? What was needed was a secure logical grounding for the theory of sets, to replace the informal idea of a set as a 'collection of objects'. With this in place, set theory itself could act as foundation for the whole of mathematics, and safely incorporate Cantor's system of cardinal numbers.

Two main contenders appeared for the role: the type theory of Russell and Whitehead's *Principia Mathematica* (published between 1910 and 1913), and, by the 1920s, the axioms of Zermelo–Fraenkel set theory. Today, several variants of set theory are still studied, including intuitionistic formulations (see **intuitionism**) and Quine's *New Foundations* (see **type theory**). Alternative approaches to the foundations of mathematics come from **category theory**. However, the industry standard remains ZFC, Zermelo–Fraenkel set theory plus the axiom of choice.

Empty set, $\varnothing$

'I got plenty o' nothin' and nothin's plenty for me' — George Gershwin

Written as '$\varnothing$' or sometimes '{ }', the *empty set* is the most trivial object in mathematics. It is simply a set with nothing in it. More accurately, it is *the* set with nothing in it. If two sets have exactly the same elements then they are really the same set. Two empty sets certainly have the same elements (namely none at all). So the set of square prime numbers is the same as the set of regular heptahedra (see **Platonic solids**): both are equal to $\varnothing$.

In axiomatic set theory, the existence of $\varnothing$ is guaranteed by the axioms of Zermelo–Fraenkel set theory (or one of its alternatives). For set theory to underpin the whole of mathematics, the natural numbers must first be encoded in it. This is usually achieved with $\varnothing$ playing the role of 0, $\{\varnothing\}$ that of 1, $\{\varnothing, \{\varnothing\}\}$ that of 2, $\{\varnothing, \{\varnothing\}, \{\varnothing, \{\varnothing\}\}\}$ that of 3, and so on. Not only the natural numbers, but a whole set-theoretic universe, and essentially the whole of mainstream mathematics, can then be built from the empty set in this way.

Zermelo–Fraenkel set theory

'All mathematical theories may be regarded as extensions of the general theory of sets … on these foundations I state that I can build up the whole of the mathematics of the present day …' wrote Nicolas Bourbaki in 1949. (Nicolas Bourbaki was not actually a human being, but the pseudonym for a collective of French mathematicians.)

Before this situation could arise, a proper formalization of set theory was needed, avoiding Russell's paradox. In 1905, Ernst Zermelo had begun the search for such a list of axioms, and in 1922 Thoralf Skolem and Abraham Fraenkel were able to complete the search. The resulting framework, called *Zermelo–Fraenkel Set Theory* (or ZF) posits the existence of the empty set, and axiomatizes the process of taking power sets, as well as unions and intersections. Everything in this universe is a set: they are no longer constructed from more basic objects. So sets may contain other sets, but not every collection of sets itself counts as a set. In particular, the collection of all sets is *not* a set, and no set is permitted to include itself. Thus Russell's paradox is avoided.

The axiom of choice

Suppose you have a collection of sets: $A, B, C, D, \ldots$ Then you can form a new set by taking one element from A, one from B, one from C, and so on. This principle may appear uncontroversial at first sight, but all attempts to prove it from the axioms of Zermelo–Fraenkel set theory (ZF) floundered. In 1940 Kurt Gödel at least managed to prove that this principle did not introduce any contradictions into ZF.

The problem is that the collection of sets $A, B, C, D, \ldots$ may go on for ever, in which case an infinite number of choices have to be made. This is particularly awkward when there is no method for choosing one element from each set. Bertrand Russell's analogy was that if you have infinitely many pairs of shoes, you can apply a general rule: take the left shoe from each pair. But for infinitely many pairs of identical socks, no such rule exists, so infinitely many arbitrary choices are needed.

The status of the axiom was eventually resolved in 1962, when Paul Cohen used **forcing** to show that the axiom of choice is independent of ZF.

Banach–Tarski paradox

In 1924, the analyst Stefan Banach and the logician Alfred Tarski joined forces to prove a very disconcerting fact: if you have a 3-dimensional ball, then the axiom of choice allows you to chop it into six pieces which can be slid around (using only rotations and translations) and reassembled into two balls each identical to the original. This looks very much as if it violates the original sphere's volume. Banach and Tarski dodged this problem by cutting the sphere into *non-measurable sets*, which don't have a volume in any meaningful sense. These ghostly sets are impossible to imagine, and their existence is only guaranteed by the axiom of choice.

This strikingly counterintuitive consequence of the axiom of choice has caused some people to question its validity. It is also the source of a favourite mathematical joke, that the catchiest anagram of 'BANACH–TARSKI' is 'BANACH–TARSKI BANACH–TARSKI'.

The cardinal trichotomy principle

Cantor's system of cardinal numbers works perfectly for measuring the size of sets: every set is in one to one correspondence with a cardinal. However, comparing these cardinals is not always as straightforward as might be hoped. You might expect that, for any two cardinals A and B, one of the following should hold: either $A \leq B$, or $A = B$ or $A \geq B$. Cantor certainly believed in this *trichotomy principle* and was happy to use it without proof in his work in 1878. Later, however, he came to realize that it was not quite as self-evident as it had first seemed. In fact, this principle is logically equivalent to the Axiom of Choice. If that fails, then there will exist cardinal numbers which just cannot be compared to each other.

Continuum hypothesis

At the International Congress of Mathematicians at Paris in 1900, David Hilbert delivered a speech in which he listed 23 mathematical problems for the new century. The first of Hilbert's problems concerned cardinal numbers. Its difficulty was such that it had contributed to Georg Cantor's depression, as he had struggled and failed to resolve it. The lowest infinite cardinal number is $\aleph_0$: the cardinality of the set of natural numbers **N**, or any other countably infinite set. The *continuum* was the name given to another important cardinal: that of the power set of **N**, $2^{\aleph_0}$.

Many important sets have cardinality $2^{\aleph_0}$, including the set of real numbers, and the set of complex numbers.

The question that had so frustrated Cantor was whether or not there is another cardinal number between $\aleph_0$ and $2^{\aleph_0}$. That is, whether $2^{\aleph_0} = \aleph_1$. The statement that the two are indeed equal is known as the *continuum hypothesis*. It was finally settled in 1963 when Paul Cohen used forcing to show that this statement is independent of the axioms of ZFC (Zermelo–Fraenkel set theory plus the axiom of choice).

Forcing

Paul Cohen invented a powerful method, called *forcing*, for creating made-to-measure universes of sets satisfying the Zermelo–Fraenkel axioms (ZF), as well as extra requirements he could choose to impose. Cohen showed off his new technique in the most spectacular fashion. In 1962 he constructed a model of ZF in which the axiom of choice failed. With Gödel's earlier work showing that the axiom of choice is consistent with ZF, the independence of the axiom of choice was established.

The following year Cohen built another model of ZF in which the axiom of choice was true, but the continuum hypothesis was false. Again, when taken with Gödel's work, the independence of the continuum hypothesis from ZFC was shown. Cohen's forcing remains the standard way of constructing new models of set theory, in particular in the study of large cardinals.

New models of set theory

When Cantor and Zermelo laid the foundations of set theory at the end of the 19th century, they had in mind a unified world of sets to be described and axiomatized, and then for these to act as the base for the whole of mathematics. Gödel's incompleteness theorems of 1931 meant that this goal had to be radically reassessed: there would always be statements X for which neither X nor its negation ('not X') could be deduced from ZF.

However, for some time this remained only a theoretical possibility; the hope was any such X would be an arcane curiosity, not of any real mathematical importance. The triumph of forcing dashed that hope. As the logician Andrzej Mostowski said in 1967, 'Such results show that axiomatic set theory is hopelessly incomplete … Of course if there are a multitude of set-theories then none of them can claim the central place in mathematics'. The unavoidable conclusion was that mathematicians would have to treat ZF more like the **group axioms**, not as the axiomatization of one, unique categorical structure, but with an infinite variety of models. Some of these are explored by large cardinal axioms.

Large cardinals

After $\aleph_0$, the most familiar infinite cardinal is $2^{\aleph_0}$. Already, its properties are not clear, as the independence of the continuum hypothesis shows all too well. However, when we look at bigger cardinals, their behaviour becomes ever more opaque. (A rare exception is the extraordinary result by Saharon Shelah, under a minor additional hypothesis, that $2^{\aleph_\omega} < \aleph_{\omega_4}$.) When we look at much larger cardinals, very often even their existence is independent of ZFC. As the set theorist Dana Scott put it, 'if you want more you have to assume more'.

The so-called *inaccessible cardinals*, first envisaged by Felix Hausdorff in 1914, cannot be approached by any sequence of smaller cardinals, just as $\aleph_0$ is out of reach of the finite numbers. In particular, if κ is inaccessible, then $\aleph_\kappa = \kappa$. Hausdorff noted that if such cardinals do exist then 'the least among them has such an exorbitant magnitude that it will hardly ever come into consideration for the usual purposes of set theory'. The existence of an inaccessible cardinal was later shown to be independent of ZFC. Inaccessible cardinals form the lowest denomination in the pantheon of large cardinals, with the higher echelons being occupied by such tremendous beasts as the *totally indescribable*, *superhuge* and *ineffable* cardinals.

HILBERT'S PROGRAM

Hilbert's program

In the early 20th century, mathematics was suffering from what Hermann Weyl called a 'new foundational crisis'. The search had been on for a framework in which the whole of mathematics could systematically be deduced. In the 1920s, David Hilbert set out what he thought was required from such a system. It would be based around Peano arithmetic and should satisfy three criteria:

1 *Consistency*. The system should never produce a contradiction (such as $1 + 1 = 3$). That the foundations of mathematics should be consistent was the second problem Hilbert posed in his 1900 address.

2 *Completeness*. Every true statement about natural numbers should be deducible within the system.

3 *Decidability*. There should be a procedure which can determine whether any given statement about natural numbers is true or false.

Hilbert's program must have seemed like a natural, realistic goal. As it turned out, mathematics was far more slippery than he expected. **Gödel's first and second incompleteness theorems** killed off the possibilities of criteria 1 and 2, and Church and Turing's solution of the **Entscheidungsproblem** similarly demolished criterion 3.

Gödel's first incompleteness theorem

Kurt Gödel's *first incompleteness theorem* shattered Hilbert's dreams of a complete, consistent foundation for mathematics. In a 1931 bombshell, Gödel showed that every logical system whose rules could be written down as axioms, and which was powerful enough to incorporate Peano's axioms for the natural numbers, would automatically be either inconsistent or incomplete. More technically, he showed that no system whose axioms were **computably enumerable**, and which interpreted Peano arithmetic, could be both consistent and complete. Computable enumerability is the technical way to convey that the list can be written down, either by a human or by any conceivable computer.

Gödel's method was to encode the self-referential statement 'this statement has no proof'. If it was false, then the system would be inconsistent; if true, the system would be incomplete.

Gödel's second incompleteness theorem

Gödel's *second incompleteness theorem*, published in the same paper as his first, says that no logical axiomatization for arithmetic can ever be proved consistent under its own rules, unless it is in fact inconsistent. This was the starting point for the subject of **proof theory**. Both Gödel's theorems had a huge impact within mathematics, and their philosophical implications continue to be discussed today.

Algorithms

The ninth-century Persian mathematician Muhammad ibn Mūsā al-Khwārizmī showed that any **quadratic equation** can be solved by following a simple set of predetermined instructions, with no further ingenuity or insight needed. Many other mathematical problems are similar. Once you know the procedure that needs to be followed, the rest of the problem can be solved on auto-pilot. It is from al-Khwārizmī's name that we get the work *algorithm*, which, in modern terms, means any set of instructions that a computer can follow. The study of theoretical algorithms was crucial to the development of the physical computer.

Computability theory is the abstract study of problems which can be solved by algorithms, while **complexity theory** investigates the speed of algorithms.

Church's thesis

There are various technical formulations of the notion of an algorithm, notably in terms of theoretical Turing machines and type theory. On first sight these approaches seem very different. However, a network of theorems by Alonzo Church, Alan Turing, Stephen Kleene and John Barclay Rosser showed that the differences are ultimately superficial. All reasonable approaches to formalizing algorithms produce the same results. This idea, known as *Church's thesis*, means that the notion of an algorithm is mathematically robust.

Turing machines

In 1934, Alan Turing conceived of an abstract machine to carry out a list of instructions. It was never intended to be built, but meant as an exploration of the theoretical limits of the power of **algorithms**. From this perspective, Turing's machines were a spectacular success, leading him to solutions of the halting problem, and, along with Alonzo Church, **the Entscheidungsproblem**.

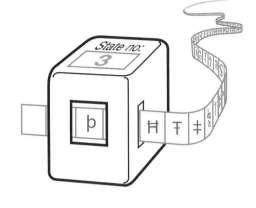

A *Turing machine* involves a tape (as long as required), which serves as its memory. The tape is divided into segments, each of which may contain a symbol. (Only finitely many symbols are allowed; these make up the machine's alphabet.) The machine comes equipped with a finite number of possible *states*. The combination of its current state and the symbol in the current segment determines what the machine does next. It may erase or write on the tape, then move to a neighbouring segment, and finally assume a new state.

This is where the digital computer was born. With a suitable choice of alphabet and states, a Turing machine is theoretically capable of everything a modern machine can do. As *Time* magazine wrote in 1999, '*everyone who taps at a keyboard, opening a spreadsheet or a word-processing program, is working on an incarnation of a Turing machine*'.

The halting problem
Some algorithms will run for ever (or until the universe ends, or someone intervenes with Ctrl-Alt-Delete). Other algorithms stop once they have achieved their goal. The *halting problem* is to decide whether a given algorithm halts, or does not. Dry as it sounds, this is a question of huge mathematical significance. For instance, it is easy to write an algorithm which runs through the natural numbers, stopping if it finds an **odd perfect number**. Knowing whether or not this algorithm halts amounts to knowing whether such a number exists, a major open problem in number theory.

If someone was able to come up with a method to tell in advance whether or not any algorithm will halt, they would settle at a stroke this, and innumerable other questions of mathematics and computer science. However, no-one will find such a method, as none can exist. This seminal result was proved by Alan Turing, in 1936.

The Entscheidungsproblem
Is there a procedure which will determine whether or not any given statement about natural numbers is true? Hilbert's program had posited the existence of such a procedure, later interpreted as an algorithm. Certainly, the history of mathematics does not hold much hope. Different theorems

have required wildly differing techniques, and the individual insight and hard work of many mathematicians, to prove. Was Hilbert right that there should be one overarching perspective which would reduce mathematics to the simple application of one mechanical process?

Closely related to the halting problem, this is the *Entscheidungsproblem* (or *decision problem*). Alonzo Church in 1936 and Alan Turing in 1937 independently proved that it does not have a solution. Echoing and supplementing Gödel's theorems, they showed that every logical system powerful enough to incorporate Peano's axioms for the natural numbers is *undecidable* as well as incomplete. That is, there can be no algorithm to determine whether or not an arbitrary statement is true within the system. A happy consequence is that computers will not be putting mathematicians out of work for the foreseeable future.

Tarski's geometric decidability theorem

Gödel's incompleteness theorems had blasted a hole through mathematics, compounded by Church and Turing's work on the Entscheidungsproblem. What of mathematical systems other than number theory? The oldest of all is Euclidean geometry, with **Euclid's postulates** as its axioms. Indeed, the discovery of the independence of the parallel postulate from the others was the first hint of the incompleteness phenomenon within mathematics.

In 1926, Alfred Tarski set about the task of translating Euclid's postulates into the language of modern logic. The resulting system allows all the usual analysis of points and lines, but not set-theoretic constructions (such as **Cantor dust**). In 1930, Tarski managed to show that his system does not conceal any Gödelian horrors. It is unambiguously *consistent* (it contains no contradictions) and, unlike arithmetic, it is *complete* (every statement in the language will be definitively true or false). Even better, it is *decidable*: there is an algorithm which will take any statement about points on the plane, and produce a result of 'yes' or 'no', depending on whether the statement is true or false. The foundations of geometry, it seems, are more solid than those of number theory.

Proof theory

In 1936, Gerhard Gentzen proved that the ordinary arithmetic of the natural numbers, as laid down in Peano arithmetic, is consistent: following the rules of the system will never produce a contradiction. This seems to fly in the face of Gödel's second incompleteness theorem, where it was shown that such a system can never be proved consistent under its own rules. The resolution to this contradiction is that Gentzen was *not* working within Peano arithmetic. He was working inside another, stronger system. Although the new system could not prove itself consistent, it could prove Peano arithmetic consistent.

Does this mean for certain that no contradictions lurk within Peano arithmetic? Although most mathematicians would subscribe to this conclusion, Gentzen's proof relies on the assumption that his new, stronger system is consistent. Of course, this cannot be taken for granted. *Proof theory* compares the relative strengths of different logical systems (including **intuitionistic** and **modal logic**). Any system can be assigned an *ordinal*, an infinite quantity closely related to a **cardinal number**, which measures its strength.

Reverse mathematics

The standard process of mathematical research is to start with some basic *axioms*, and deduce interesting conclusions from them. It is good practice to be as sparing with the initial assumptions as possible. The worst case would be if your work depended on a conjectural result, such as the **Riemann hypothesis** or, even worse, **the P = NP problem**. Beyond this, some axioms are more controversial than others. If your theorem absolutely requires the axiom of choice (or, worse, the continuum hypothesis) this is worth recording. Otherwise, if you can do without them that would usually be preferable. Similarly, if you can prove your theorem from first principles, instead of relying on monumental results such as the **classification of finite simple groups** or the **four colour theorem**, it will usually be better (and more illuminating) to do so.

Mathematicians generally try to follow this guiding principle (without getting too hung up on it). But the logician Harvey Friedman turned it into a full-blown logical programme. Starting with classical theorems such as the **intermediate value theorem**, Friedman's question is: what are the absolute bare minimum assumptions needed to prove it? This question demands a logical answer, in terms of a minimal proof-theoretic system. In 1999, Stephen Simpson identified the five commonest such systems, in order of strength. Reverse mathematicians analyse which the appropriate foundation is for a given theorem.

Hilbert's 10th problem

For thousands of years people had been investigating equations such as **Fermat's last theorem** and **Catalan's conjecture**, trying to determine whether or not they have any solutions which are whole numbers. David Hilbert believed that this ad hoc approach to **Diophantine equations** was inadequate. In his 1900 address, he called upon the mathematical community '*to devise a process according to which it can be determined by a finite number of operations whether the equation is solvable in rational integers*'. This process should involve inputting an equation, following some predetermined steps, and arriving at an answer: 'yes this equation has a whole number solution', or 'no it does not'. In modern terms, what was required was an algorithm.

The study of Hilbert's 10th problem led to astonishing bridges between number theory and mathematical logic. It became clear that Diophantine equations and Turing machines are really two perspectives of the same underlying subject. Hilbert's hopes were finally dashed by Matiyasevich's theorem.

Matiyasevich's theorem

Different mathematical disciplines give rise to different perspectives on fundamental objects, such as collections of natural numbers. Number theory and logic each have their own ways to describe such collections. In number theory, a set of integers is *Diophantine* if is described by a Diophantine equation. So the set of square numbers is Diophantine, as the equation $y = x^2$ describes it exactly: substituting in whole numbers for 'x' produces the set of squares as the corresponding 'y'-values.

The key concept from logic is that of a collection of natural numbers being ***computably enumerable***, which means that there is an algorithm which lists it. The set of prime numbers is computably enumerable, as there is an algorithm which takes the natural numbers in turn, testing each one for primality, and including or excluding it as appropriate. It is not too difficult to prove that every Diophantine set of integers is computably enumerable. In 1970, building on earlier work by Julia Robinson, Martin Davis and Hilary Putnam, Yuri Matiyasevich proved the deep and counterintuitive result that the opposite is also true: every set of computably enumerable integers is Diophantine. Because we know there are enumerable sets that are not computable, Matiyasevich's theorem immediately implies the existence of **uncomputable equations**.

Uncomputable equations
Church and Turing's solution to **the Entscheidungsproblem** had showed that no algorithm can capture every truth about the natural numbers. Perhaps in the more restricted area of Diophantine equations an algorithm might still exist? This was the question posed in Hilbert's 10th problem. It reduced to a seemingly weaker conjecture: that every Diophantine set is computable, that there is an algorithm which decides whether any number is a member of it or not.

Matiyasevich's theorem killed off this idea, by showing that every computably enumerable set is Diophantine. The critical point is that the class of *computably enumerable sets* is much broader than that of outright *computable* sets. This followed from Turing's work on the halting problem. Therefore, by Matiyasevich's theorem, there are a great many Diophantine sets which are uncomputable.

COMPLEXITY THEORY

Complexity theory
In practical computing applications, not all algorithms are equally efficient. An expert programmer might write a slick piece of code to perform a given task quickly, while an amateur's effort could take hundreds of times longer to do the same thing. This is the art of *algorithm design*. However, not everything comes down to the ingenuity of the programmer. It is likely that even the greatest programmers will never build a quick algorithm for solving the **travelling salesman problem**, because probably no such algorithm can exist. (Whether or not this is true depends on the biggest problem in the subject: $\mathbf{P} = \mathbf{NP}$.)

This is the topic of *complexity theory*, which straddles the boundary of mathematics and theoretical computer science. It studies the inherent difficulty of a task, as measured by the minimum length of time any algorithm will take to solve it.

Complexity classes
Complexity classes measure the innate difficulty of tasks. To complete a certain task, on an input of n pieces of data, any algorithm might have to carry out a minimum of n^2 steps. The polynomial n^2 classifies the task. If the number of steps

is given by a polynomial such as n^2 or n^3, then the task is said to have *polynomial time*. The collection of all such tasks form a class known as P (standing for 'polynomial'). **Cobham's thesis** identifies this with the set of tasks which can be completed fast enough for practical purposes.

On the other hand, some functions grow much faster than polynomials. If the algorithm has to carry out 2^n steps on n pieces of data, then this very quickly explodes out of control. At $n = 100$, even the fastest modern processor will be defeated. Tasks like this are said to run in *exponential time*, and the class of them is known as EXPTIME.

Complexity theorists have identified several hundred different complexity classes. Much of the subject is dedicated to understanding the relationships between them. Of particular interest is the relationship of ordinary classes to **non-deterministic complexity classes**.

Cobham's thesis

Alan Cobham was one of the early post-war workers in complexity theory. He viewed the complexity class P as the best theoretical description of problems that can feasibly be solved on a physical computer. Cobham held that for most purposes 'polynomial time' means 'fast enough'. Problems which are not in P may technically be computable, but any algorithm is likely to be so slow as to be of no practical use.

Cobham's thesis is valuable as a rule of thumb, but is not quite the whole story. In truth, an algorithm which has to carry out n^{1000} operations when n pieces of data are entered is of little use (though technically still *polynomial time*). On the other hand, while most non-polynomial algorithms are of limited practical value, one which grows at the rate $2^{\frac{n}{1000}}$ could well be worth implementing for small sets of data.

Non-deterministic Turing machines

There is – or there seems to be – a large difference between the time needed to solve a problem and the time it takes to check the solution. For example, there is no known algorithm which can factorize large integers such as 10,531,532,731 in polynomial time, suggesting that it is an inherently tough problem. On the other hand, once a solution is provided, it is quick work to check that it is correct. If someone tells me that $101{,}149 \times 104{,}119 = 10{,}531{,}532{,}731$, it is straightforward to verify that she is right.

So there is a short route through the problem; the hard work is in finding it. Although it may not be quickly solvable by an ordinary Turing machine, a machine which is allowed to make guesses during the algorithm could solve it fast, if it got lucky. Such a theoretical device is known as a *non-deterministic Turing machine*.

Non-deterministic complexity classes

The class of problems which can be checked in polynomial time is known as NP, standing for 'non-deterministic polynomial time', meaning that a non-deterministic machine may be able to solve it quickly (if it is lucky). Similarly EXPTIME, and other complexity classes have their non-deterministic equivalents. The relationship of the classes P and NP is the subject of the famous **P = NP problem**. The toughest problems in NP make up the class of **NP-complete problems**.

AKS primality test

How do we decide whether or not a particular number is prime? The simplest method is just to try dividing n by every smaller number (this can be slightly improved by recognizing that we only need to test prime numbers smaller than $\sqrt{n}$). All the same, if n is a large number (say hundreds of digits long), this is an impossibly slow process. The subject of **primality testing** has deep roots, and some better methods have been found. But for a long time it was unclear where the theoretical barriers lie. The **Lucas – Lehmer test** runs in polynomial time, but it only applies to a very special class of inputs, namely **Mersenne numbers**. No-one had been able to construct an efficient primality test which could work for any integer, nor had anyone demonstrated that it could never be done.

In 2002, Manindra Agrawal, Neeraj Kayal and Nitin Saxena at the Indian Instititute of Technology in Kanpur stunned the world with their paper 'Primes is in P', in which they described a new primality test which works for any number, and does run in polynomial time.

Integer factorization problem

If someone gives you a large number, and asks you to break it down into its constituent prime factors, what is the most efficient process to use? As well as being of mathematical interest, this is a problem of great importance in cryptography theory. Modern **public key** systems rely on large integers being very difficult to factorize (which is why RSA Laboratories set up the RSA factoring challenge).

So far, this seems to be the case. Although some special types of integer can be factorized quickly (that is, in polynomial time), the best algorithm for factorizing general large integers (say, longer than 150 digits) is the *number field sieve*, conceived by John Pollard in 1988 (and developed by Menasse, Lenstra and Lenstra), which does not run in polynomial time.

Integer factorization can be checked very quickly just by multiplying the relevant numbers together, so this problem's certainty lies in the complexity class NP. The big question is whether it is in P, that is, whether a polynomial time algorithm might in principle exist. If **P = NP** was proved to have a positive answer then there would have to be a polynomial time algorithm for integer factorization. Such an algorithm could have a devastating impact on internet security.

The RSA factoring challenge

In 1991, the network security company RSA Laboratories published a list of 54 numbers between 100 and 617 digits long. They challenged the world to factorize them, offering prizes of up to $200,000. All of the numbers are *semiprimes*, that is, multiples of exactly two prime numbers, as these have the most significance for cryptography. RSA declared the challenge inactive in 2007 and retracted the remaining prizes. However, the distributed computing project distributed.net continues to work on these challenges, and is offering privately sponsored prizes for successful participation.

At time of writing, the world record for the largest factorization is an RSA number, namely, the 200-digit number known as RSA-200:

2799783391122132787082946763872260162107044678695542853756000992932612840010760934567105295536085606182235191095136578863710595448200657677509858055761357909873495014417886317894629518723786922182398349

This was factorized in 2005 by F. Bahr, M. Boehm, J. Franke and T. Kleinjung into two 100-digit primes:

3532461934402770121272604978198464368671197400197625023649303468776121253679423200058547956528088349

and

7925869954478333033347085841480059687737975857364219960734330341455767872818152135381409304740185467

The effort required around 55 years of computer time, and employed the *number field sieve* algorithm.

The P = NP question

Can every problem which can be checked quickly also be solved quickly? This is, roughly, the meaning of the P = NP problem, one of the outstanding open problems in mathematics today, and carrying a $1,000,000 price tag, courtesy of the Clay Institute. A solution would be worth vastly more, however, because of the potentially seismic repercussions for integer factorization, the travelling salesman problem and numerous other questions of algorithm design.

The class P is the collection of all problems which can be solved by an algorithm in polynomial time, such as telling whether a number is prime. The class NP is those which can be *checked* in polynomial time, such as factorizing a number. It is straightforward to see that everything in P is also in NP, so P $\subseteq$ NP. The million dollar question is whether NP $\subseteq$ P.

Stephen Cook and Leonid Levin independently posed the P versus NP problem in 1971. Although no proof exists either way, the suspicion among mathematicians and computer scientists is largely that P $\neq$ NP. A third possibility is that the question itself could be independent of all our standard mathematical assumptions.

NP-completeness

If you are attempting a particular task, proving that it is in the complexity class P is often the breakthrough you are hoping for. According to Cobham's thesis, this should mean that that your problem is tractable in the real world. In several important cases, this is very difficult to achieve. There are many problems known to be in the complexity class NP, but whose status regarding P remains uncertain. These include integer factorization, the travelling salesman problem and the unknotting problem. A positive answer to the P = NP question would resolve all of these at a stroke.

But what if P ≠ NP, as most people suspect? This seems not to take us any further forward, as these problems could still either be in P or not. In the case of the travelling salesman problem however, there is an extra detail. This problem is NP-*complete*, meaning that, out of all the problems in the class NP, it is among the most difficult to compute by algorithm. If P ≠ NP, then there is some problem in NP which cannot be computed in polynomial time. Being NP-complete, the travelling salesman problem must be at least as difficult as this problem, and so cannot lie in P.

Quantum computing

Traditional algorithms proceed one step at a time, because a Turing machine can only be in one state at a time. According to quantum physics, however, the fundamental particles of nature are not in one fixed state at any moment, but can occupy several states at once, in what is called *superposition*. When disturbed, they *decohere* into one state.

If this could be exploited for computing purposes, it could lead to massively faster machines. All that remains is for a functioning quantum computer to be built. Although this is a daunting challenge, progress has been made. In June 2009, a team at Yale University led by Robert Schoelkopf succeeded in developing a 2-qubit ('quantum bit') processor. It successfully ran Grover's reverse phone book algorithm.

Grover's reverse phone book algorithm

A phone book is organized in alphabetical order. If you want to look up someone's number, it is easy to do. The *reverse phone book problem* is much harder. If you have a phone number and you want to know to whom it belongs, there is no alternative to searching painstakingly through the book, comparing your number to each one in turn.

However, if you happen to have a quantum computer to hand, the problem can be solved faster. In 1996, Lov Grover designed a quantum algorithm, which exploits a quantum computer's ability to adopt different states, and thus check different numbers, simultaneously. If the phone book contains 10,000 entries, the classical algorithm will take approximately 10,000 steps to find the answer. Grover's algorithm reduces this to around 100. (In general, it will take around $\sqrt{N}$ steps, instead of N.) The algorithm was successfully run on a 2-qubit quantum processor in 2009. Of course it is not really useful for checking phone numbers. But for searching for the keys to ciphers, for example, it would be a powerful tool.

Quantum complexity classes

Exactly how quantum complexity classes relate to their classical counterparts is a topic of current research, and largely a mystery. An added complication is that quantum computations are probabilistic; they only *probably* decohere on the right answer. Repeatedly running the algorithm can increase this likelihood to any required level of certainty, but this slows the process down again, partially counteracting its benefits. Significantly, Grover's algorithm was found to be nearly optimal; no other quantum algorithm would solve the problem significantly faster. This shows that, although quantum computers are powerful, they are not omnipotent.

It is unknown whether every problem in NP can be solved in quantum polynomial time (BQP, *bounded-error quantum polynomial time*). However, in 1994 Peter Shor discovered a quantum algorithm for integer factorization which does run in polynomial time, a problem which is unsolved (and maybe impossible) on a classical computer. This result is of potentially huge importance, should a fully functional quantum computer successfully be built.

COMPUTABILITY THEORY

Computability

What are the possible collections of natural numbers? There are the prime numbers, the numbers 1 to 100, the triangular numbers, and all sorts of other collections of endless fascination to mathematicians. But these are vastly outnumbered by a swathe of unstructured, random-seeming sets which are nearly impossible to describe. Is there any way to tell which is which?

The first attempt to split the interesting sets from the morass is the *computability* concept. Suppose that A is a set of numbers, and I want to know whether or not 57 and 1001 are in A. If there is an algorithm which will give yes/no answers to such questions, then A is said to be *computable*. (Whether the corresponding algorithms are fast enough to be practically useful is a matter for complexity theorists to worry about.) This is only the start, however. Through the introduction of Turing **oracles**, computability becomes a relative concept, with some sets being more uncomputable than others. This opens the door to the study of what it means to be truly *random*. Computability theory can also be understood in terms of *computable real numbers*.

Computably enumerable sets

A subtly different notion from being *computable* is the notion of being *computably enumerable*. This means that there is an algorithm which lists the set. The catch is that it may not list it in any comprehensible order.

Every computable set is certainly enumerable, but the reverse does not hold. Suppose B is an enumerable set, and I want to know whether or not 7 is in it. I run the algorithm which lists B, and it begins: 1, 207, 59, 10003, 6, ... If 7 appears in the list, then I know it is in. But if I wait for half an hour, and 7 has not appeared, I cannot conclude that 7 is not in the list. It may yet be listed. I could let the algorithm run all night, or for a million years, but there is no moment when I can be certain that 7 is definitely out.

By Turing's work on the halting problem, we know that there is no way around this obstacle. The class of enumerable sets really is strictly bigger than that of truly computable sets. Computably enumerable sets appear in number theory as the Diophantine sets, courtesy of **Matiyasevich's theorem**.

Encoding sets in binary

Real numbers are commonly written as decimals, but can also be written out in binary notation. An example might look like $r = 0.011010100010100010\ldots$ Binary representation provides an excellent method for encoding sets of natural numbers as real numbers.

We can construct a set of natural numbers from r as follows: list the natural numbers 1, 2, 3, 4, 5, 6, 7, … Now line these up with the bits of r, with 1 meaning in the set, and 0 for meaning out of it:

n	1	2	3	4	5	6	7	8	9	10	11	12	13	14	15	16	17	18
nth bit of r	0	1	1	0	1	0	1	0	0	0	1	0	1	0	0	0	1	0

So in this case r encodes the set {2, 3, 5, 7, 11, 13, 17, …} The reverse procedure produces a real number from a set. So {1, 2, 3, 5, 8, 13, 21, …} defines the real number $0.111010010000100000001\ldots$

Computable real numbers

Computability theory is usually expressed as analysis of sets of integers. But it can equally well be understood in terms of real numbers, once we know how to encode sets in binary. Very simply, a real number is *computable* if the corresponding set is computable. Again there are degrees of uncomputability, measured by **Turing degrees**. This is in some ways a more natural interpretation of the theory, and leads into the study of random real numbers.

Uncomputable real numbers

Uncomputable real numbers infinitely outnumber the computable ones: there are uncountably many uncomputable ones, while the computable ones are countable. However, their ubiquity does not make them easy to find; it is extremely difficult to give any specific examples. The difficulty is that almost any method you might try to use to write down an uncomputable number will constitute an algorithm, and therefore the result will be computable. Even common transcendental numbers such as π and e are computable.

We can at least gesture in the direction of some individual uncomputable real numbers. The best way to do this is to derive them from known uncomputable problems.

The halting number K

The prime example of uncomputability is the **halting problem**. By encoding this problem as a string of binary, we get our first uncomputable real number, known as **K**.

Although **K** is uncomputable, it is exceptional in not being *random*. A more typical uncomputable number is the compressed version of **K**, Chaitin's Ω.

Chaitin's Ω

During the 1980s, the computer scientist Gregory Chaitin travelled as close to an individual random, uncomputable real number as it is possible to go. His number 'Ω' can be understood as the *halting probability*. Every Turing machine will either complete its task and halt, or it will run for ever. Deciding in advance which will happen is the **halting problem**, and is uncomputable.

It makes sense to talk about the probability of a randomly chosen Turing machine halting. This, roughly, is the definition of the number Ω. It is a compressed version of the halting number **K**, with all the non-randomness stripped out. If Ω could be pinned down exactly (which of course it cannot) it would serve as an oracle for the halting problem, and many other uncomputable problems.

Oracles

By definition, the only sets which can be calculated by Turing machines are the computable ones. By cheating, and endowing our machines with superpowers, we can manufacture a much larger range. A *Turing oracle* is a Turing machine with an extra component: an oracle with magical access to some specified set (A). On top of its usual functions, the machine can consult the oracle for information about A. It is almost certain that such a device will never be built (even a quantum computer would not suffice). But, as an abstraction, it throws up some fascinating possibilities. The question is, what sets can be computed now?

Unsurprisingly, the answer depends on A. If A is itself computable, then we get nothing new. If we put an uncomputable set A in the oracle, for a start A itself becomes computable, as does the *complement* of A (the collection of numbers not in A), and many other new sets.

If B is computable on such a machine, then we say that A *computes* B. The relationship this imposes between sets of numbers is astoundingly complicated, and is encapsulated as the Turing degrees.

Turing degrees

With the idea of an oracle, the concept of computability became relative. Even if A and B are uncomputable sets (or equivalently uncomputable real numbers), it makes sense to say that A computes B. This means that there is an oracle with access to A which can compute B. This an ingenious way of capturing the idea that A contains all the information provided by B (and possibly more).

If A and B can each compute each other, then they contain exactly the same information, and are said to have the same

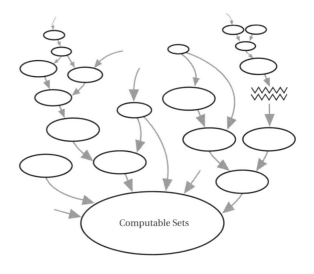

Computable Sets

Turing degree. (Of course *A* and *B* might appear very different indeed; there are many different ways to encode the same information.) If *A* can compute *B*, but not vice versa, we say *A* has a *higher Turing degree* than *B*. This relationship seems natural enough, but the pattern formed by the Turing degrees is enormously complicated. Many pairs of degrees are incomparable, with neither able to compute the other (although there is always a third degree which can compute both).

At the bottom sits the degree of computable sets. Above it, almost anything can happen. There is no top degree which can compute all others; above every degree you can always find another. Similarly, there are infinite descending chains: *A* can compute *B*, which can compute *C*, and so on for ever, but with none of the reverse computations possible. There are degrees which are infinitely nested, in fact almost any configuration you can dream up can be found somewhere within the Turing degrees.

Normal numbers
What does a typical real number look like? Emile Borel, in 1909 formulated the definition of a *normal number*. If you write out the decimal expansion, then each digit 0–9 should appear equally often. This does not have to happen over any short stretch, but over the course of the whole infinite decimal expansion, everything eventually averages out.

More is required, however. There are 100 different possible two-digit combinations: 00 to 99. Each of these should also appear equally often in the long run, and similarly for three-digit combinations, and so on. In general every finite string of digits should appear with the same frequency as every other string of the same length. There is one final requirement for normality. The above definition starts with the number written out in base 10. If you translate it into base 2, 36, or any other base, the same properties should still hold.

Although Borel was able to prove that almost every real number is normal, known examples are rare. The first was provided by Waclaw Sierpinski in 1916. It is conjectured that *e* and π are both normal, but this is not known for sure. Chaitin's Ω is known to be normal. The definition of a *random real* implies normaility.

Random real numbers
The real numbers that we have the easiest access to are the integers, rationals and algebraic numbers. Although some individual transcendental numbers, such as *e* and π are known, these are at least computable. A typical real number will look very different from any of these. Chaitin's Ω is the closest we have to what a *random real* should look like.

Here are two definitions of what it means for a real number to be *random*:

1 An approach based on **Kolmogorov complexity**. If the bits of a real number contain any pattern, this can be exploited to compress the sequence. A random real is one which is incompressible, and therefore completely patternless.

2 Suppose bits of the number were to be revealed one by one: 0.10111001…, and you were to place bets on whether the next digit will be 0 or 1. If the real is computable, you can run an algorithm to predict the next digit exactly, and so win every time. But for a *random* real, there is no strategy you can adopt to win more than 50% of the bets.

Randomness

The two definitions of a random real number above turn out to be equivalent, as do several other variants. The second shows that randomness automatically implies *normality* otherwise you could produce a strategy to win money based on combinations of digits which occur more or less frequently. Randomness, however, is stronger than normality. Random real numbers are scattered throughout the Turing degrees (though of course none are computable). This means that they can encode a lot of information, which has led some researchers to delve further into the question of what randomness is.

Stupidity tests

The definition of random real numbers works well. But it has a somewhat paradoxical consequence. Many random numbers, such as Chaitin's Ω, have high Turing degrees. This means that they encode a great deal of information. In fact, according to the *Kucera–Gacs theorem*, almost any information whatsoever can be encoded within the bits of a random real number. This rather goes against some people's intuition of what randomness should mean.

The computability theorist Denis Hirschfeldt likens randomness to stupidity. There are two ways to pass a test for stupidity: first, if you are really stupid; second, if you are clever enough to predict how a stupid person would answer. Numbers like Ω are the clever sort of random. More stringent tests only allow genuinely stupid numbers through, which contain little information. This idea leads to a whole new hierarchy of randomness.

MODEL THEORY

Model theory

Gödel's incompleteness theorems are widely misunderstood. They do not mean that it is impossible to write down a complete, consistent set of axioms for any mathematical theory. They say that you cannot do this for the natural numbers, where addition and multiplication are too complicated to be captured by one formal system.

A plethora of non-Gödelian theories do exist. These are complete, consistent, and to varying extents well-behaved (though, of course, too weak to describe the arithmetic of the natural numbers). Logical theories of **groups**, **rings**, **fields**, orderings and many other structures exist and are far more amenable to logical analysis than the Gödelian natural numbers.

Lou van den Dries contrasts such 'tame' theories with 'wild' ones which exhibit Gödelian phenomena, and describes *model theory* as the 'geography of tame mathematics'.

An early pioneer in the subject was Alfred Tarski, who provided elegant logical analyses of the real and complex numbers, totally avoiding incompleteness or undecidablility.

Syntax and semantics

In philosophy and linguistics, there is an important distinction between *syntax*, the grammar and internal rules of a language, and *semantics* which is the study of meaning, or how the language relates to the wider world. The same applies in mathematical logic. Formal systems are syntactic frameworks. The semantic question is how well they describe phenomena in broader mathematics.

Gödel's incompleteness theorems show that there can be a chasm between a language and its referents. But, even if we restrict ourselves to non-Gödelian, complete situations, the situation remains far from straightforward. Of course, there are many different logical systems, but the *Löwenheim–Skolem theorem*, applies in many cases. This says that a logical theory can never determine the size of the structure it is talking about. This has dramatic consequences in terms of non-standard models of familiar mathematical structures which can exist with size given by any infinite **cardinal number**.

Beyond this, there is huge variety. Some logical theories can pin down a model exactly (once its size is fixed). This is where the relationship between semantics and syntax is tightest. For others, this breaks down badly, with theory having a huge spectrum of models. Determining which category a given logical theory is in is called *classification theory*.

Non-standard models

Start with a structure, such as the set of real numbers. Look at its *logical theory*, that is, the collection of all true statements about it, in some chosen formal language. (An example of such a statement is 'For any positive number, you can find a smaller one, still not equal to zero.') Now ask what other structures you can find which obey all the same logical rules. Remarkably, you are suddenly faced with a bewildering army of structures, other than the one you first thought of. These are known as *non-standard models*, and were first discovered by the model theorist Abraham Robinson in 1960. In the case of the real numbers, Robinson found a non-standard model he called the *hyperreal numbers*, which include infinitesimals, objects previously discredited as superstition.

Non-standard models are not mere curiosities, but form the setting for **non-standard analysis**. They also raise some philosophical questions. As the logician H. Jerome Keisler wrote, 'we have no way of knowing what a line in physical space is really like. It might be like the hyperreal line, the real line, or neither.'

Infinitesimals

When Archimedes discovered the formula for the volume of a sphere, he did it by cutting the shape into infinitely many infinitely thin slices, measuring the *infinitesimal* volume (that is to say infinitely small) of each, and then adding all these together to get an ordinary, finite number. Mathematicians throughout the ages followed his lead, including Newton and Leibniz, who each relied on infinitesimal numbers in their parallel developments of the **calculus**.

In the 19th century, Karl Weierstrass finally showed how to set calculus on a solid footing, using **limits of sequences** of ordinary numbers, and avoiding infinitesimals. By 1900, infinitesimals had been completely banished from the mathematical repertoire, and for good reason: there are no such things. At least, the set of real numbers, the basic setting for geometry and analysis, does not include any infinitesimals.

Every real number x can be written as a decimal. Either every digit is zero ($x = 0.0000000\ldots$), in which case $x = 0$, or we eventually reach a digit which is not (e.g. $x = 0.00000\ldots000007612415\ldots$). In this case, the number has positive size, and is not infinitely small. It may be unimaginably tiny, of course, but if you zoom in close enough, it will always be a positive, measurable distance away from zero, and therefore not infinitesimal.

The transfer principle
Despite being based on a fiction, the old infinitesimal approach to calculus certainly had an uncanny ability to conjure up the correct answers. An explanation came from a cross-pollination of Weierstrass' analysis with mathematical logic. In the 1960s, the model theorist Abraham Robinson discovered non-standard models of the real numbers which do contain infinitesimal elements. This discovery made the infinitesimal respectable again, and indeed the primary tool of non-standard analysis.

Non-standard models first arose as structures satisfying the same logical rules as the ordinary set of real numbers (in some formal language). Turning this idea around, it provides a useful tool. If we can deduce that something is true in a non-standard model, this is often enough to show that it must also hold true for the ordinary real numbers. This *transfer principle* is the foundation of non-standard analysis.

Non-standard analysis
Also known as *infinitesimal analysis*, non-standard analysis uses exotic non-standard models of the real numbers to study the version we care most about, the original one. The general strategy is to use infinitesimals to study the non-standard model, and then appeal to the transfer principle to carry the results across to the ordinary real numbers.

The subject was begun in 1960 by Abraham Robinson, whom Kurt Gödel described as 'the one mathematical logician who accomplished incomparably more than anybody else in making this science fruitful for mathematics'. Since then, non-standard analysis has successfully been applied to problems in mathematical physics and probability theory, as well as within analysis.

Classification theory
With over 960 papers to his name, the Israeli logician Saharon Shelah is one of the most prolific mathematicians currently working. One of his triumphs is the proof of the *classification problem*.

Some sets of axioms are satisfied by very many structures indeed. At the opposite end, other theories have just one model, once you fix its size (you can always find models of different sizes; this is the Löwenheim–Skolem theorem). The classification problem, roughly speaking, is to tell the difference.

Shelah's analysis proceeded through a host of subtle dichotomies: *stable* theories versus *unstable* ones, *simple* unstable theories versus *non-simple* unstable theories, and so on. This careful sifting led him to a solution of the classification problem, in 1982. The model theorist Wilfrid Hodges wrote 'On any reckoning this is one of the major achievements of mathematical logic since Aristotle'.

Model theoretic algebra

The scraps from the classification theoretic banquet have fuelled many further feasts. Shelah's deep and abstract techniques can provide real insights into structures within mainstream mathematics, such as **rings** and **fields**. This is the subject of *model-theoretic algebra*. Stable **groups**, for example, form the setting for an ongoing attempt to mirror the classification of finite simple groups in an infinite setting.

UNCERTAINTY AND PARADOXES

Aristotle's three laws of thought

Mathematical logic is concerned with deducing true statements from true statements. At the bottom, however, something has to be taken as given, or *axiomatic*. Three laws of thought attributed to Aristotle and posited as non-negotiable axioms are:

  1 The law of identity: anything is equal to itself.
  2 The law of non-contradiction: nothing can be and not be at the same time.
  3 The law of the excluded middle: everything must either be or not be.

The first was so self-evident to Aristotle that he barely mentioned it, other than to opine that 'why a thing is itself is a meaningless inquiry'. It was elevated to a law by later thinkers. Although all three laws persevere, systems have been developed which dispense with 2 and 3, namely **paraconsistent logic** and **intuitionism** respectively.

Law of the excluded middle

The law of the excluded middle is an axiom, not of mathematics, but of logic: the broader framework in which the rest of mathematics takes place. It says that, for any suitably well-posed statement (call it P) either P is true, or the statement 'not P' is true: there is no middle ground. Another way to say this is that

P is logically equivalent to 'not not *P*'. The law of the excluded middle is the foundation of **proof by contradiction**.

As Aristotle noted, it is doubtful whether this rule applies in ordinary language. At 31 years, I am not exactly 'young', but nor am I 'not young'. In mathematics, where terms are defined more precisely, this problem vanishes. However, intuitionist mathematicians and philosophers reject this law. They have a stricter approach to *truth*, equivalent to the *existence of a constructive proof*. So an intuitionist will not allow that either *P* or 'not *P*' is true, until a positive proof for one or the other is supplied.

The liar paradox

'This sentence is false.'

The mother of all logical paradoxes, the liar paradox is attributed to Eubulides in the fourth century BC, also responsible for the **bald man** and other paradoxes. Though not strictly mathematical, it has echoes throughout logic, most notably in Russell's paradox, and the proofs of Gödel's incompleteness theorems and the halting problem. It illustrates the consequences of a language sophisticated enough to refer to itself.

There have been many approaches to resolving the paradox. Bertrand Russell's theory of types sets up a hierarchy whereby objects in the language are only permitted to refer to lower-level objects, so no legitimate sentence may refer to itself. A *paraconsistent* approach simply takes it at face value, and allows that the statement is both true and false.

Grelling's paradox
Some words describe themselves: for example, '*polysyllabic*', '*word*', '*noun*', '*pronounceable*' and '*English*'. We will call these words *autologous*. Other words do not describe themselves: '*monosyllabic*', '*animal*', '*verb*', '*unutterable*' and '*French*'. These words are *heterologous*.

It looks as if every word must be either autologous (if it describes itself) or heterologous (if it does not). What then, about the word '*heterologous*'? If it is autologous, then it describes itself, and so is heterologous (meaning that it does not describe itself). On the other hand, if it is heterologous, that means it does not describe itself, and so is not heterologous after all.

This paradox was cooked up by Kurt Grelling and Leonard Nelson, and closely resembles Russell's paradox, in a linguistic setting.

Uncertain reasoning
Most of the logical systems studied by mathematicians and philosophers are based on rules of deduction, or inference. A classical example is *modus ponens*, which says that from *P* and '*P implies Q*', we may deduce *Q*. Not many people would dispute this. The difficulty is that the human world is rather messier than the domain of pure mathematics, and perfect knowledge is rarely possible.

What happens if we are 90% confident that *P* is true, and believe that *P implies Q*, 75% of the time? This is a problem in *uncertain reasoning*, a skill that human beings are naturally adept at. In a marriage of logic and probability theory, various approaches have been taken to formalize the rules of uncertain systems. A major motivation is artificial intelligence, in particular the development of *expert systems*, such as a machine which can make medical diagnoses. Being able to reason with uncertain information, such as conflicting or unclear symptoms, would be a key component.

Fuzzy sets
Books about mathematics are full of carefully phrased definitions. To the non-mathematician these can seem painfully pedantic, but concepts must be given with the utmost precision for the subject to work. This is particularly true of mathematical sets: any object must be either in or out. We cannot allow set membership to be ambiguous.

In the human world, however, this is not true. The *set of tall people* is certainly meaningful, but not precisely defined. At 6 feet, do I qualify? Maybe. Fuzzy set theory was devised by Lofti Zadeh, in an attempt to extend set-theoretic reasoning to cope with sets which have fuzzy edges. Instead of being given by a crisp in/out, set membership is assigned a number between 0 and 1 (corresponding to definitely out and definitely in, respectively).

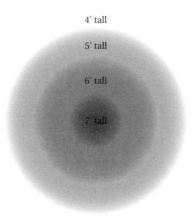

4' tall

5' tall

6' tall

7' tall

Fuzzy logic
An extension of fuzzy set theory is *fuzzy logic*. This is related to uncertain reasoning and deals with logical inference involving similarly fuzzy concepts. An example of Lofti Zadeh's is: *If P is small, and Q is approximately equal to P, then Q is more or less small*. Just as numbers can be defined within traditional set theory, so *fuzzy numbers* have been defined from fuzzy sets. Several parts of algebra have subsequently been fuzzified. Although this approach is not without its sceptics, fuzzy set theory has found applications in the social sciences and information processing.

Multivalued logic

In classical logic, a statement can be assigned one of two truth values: 'true' or 'false'. *Multivalued logics* extend this to three, four, or infinitely many possible truth values. *Intuitionistic* logic can be formalized as a 3-valued logic, by adding the value 'unknown' or 'undetermined'.

Some databases work on a 4-valued *relevance logic*. To a query 'is X true?' there are four possible outcomes: that the database contains 'no relevant information', 'information suggests X is true', 'information suggests X is false' or 'conflicting information' (see also **paraconsistent logic**).

Fuzzy logic, and approaches to uncertain reasoning usually involve infinitely many truth values, with the truth of a statement being measured on a scale between 0 (false) and 1 (true). Other systems of *continuous logic* have been developed which also allow truth values between 0 and 1, to measure how close two functions are to being identical.

Intuitionism

Some mathematicians viewed with suspicion the debate about foundations of mathematics which had developed since the late 19th century, in works such as *Principia Mathematica*, and the movement spawned by Hilbert's program. These *intuitionists*, such as L.E.J. Brouwer, saw mathematics as an activity of the human mind, rather than the unthinking logical consequences of some formal system. If the formalization of mathematical language led to weird and counterintuitive constructs such as Cantor's hierarchy of infinities, this should be taken as a hint that a mistake had been made somewhere, and the language of contemporary logic was clouding, rather than clarifying, the underlying mathematics.

In 1908, Brouwer investigated more closely, trying to isolate the part of logic responsible for the damage. In *The Unreliability of the Logical Principles*, he targeted the law of the excluded middle as the basis on which mathematicians had been welcoming in strange constructs for which there was no direct evidence.

Intuitionistic logic is logic without the law of the excluded middle. The project begun by Brouwer to rebuild mathematics from this base is called **constructive mathematics**.

Explosion

What happens to a logical system when a contradiction creeps in, that is to say, there is some statement P where both P and '*not P*' are deemed to be true? The usual answer is that this acts as a pin to a balloon, and the whole set-up collapses into inconsistency and meaninglessness. Such an event is known as an *explosion*.

To see why this happens, suppose that Q is any statement whatsoever. Now, the two statements 'P implies Q' and '(*not P*) or Q' are logically equivalent (see **tautology and logical equivalence**). But '*not P*' is assumed to be true, and so '(*not P*) or Q' is certainly true. Therefore so is 'P implies Q'. But P is also assumed true, and so it follows that Q is true. But Q was a completely arbitrary statement. Once a contradiction enters the system, then, anything and everything can be deduced from it.

Paraconsistent logic

A wild cousin of intuitionistic logic, the term *paraconsistency* was coined in 1976 by Francisco Miró Quesada. While intuitionism allows the possibility of a statement *P* where neither *P* nor 'not *P*' holds, paraconsistent logic does the opposite: it permits both *P* and 'not *P*' to hold simultaneously.

This would be the kiss of death for any ordinary system, which would immediately explode with the introduction of any such contradiction. The rules of a paraconsistent system, however, are weakened to tolerate some *local* inconsistency. Not every proposition can be both simultaneously true and false, but a limited number may, without bringing the system crashing down. A philosophical motivation is *dialetheism*, which holds that some statements are both true and false, necessitating some sort of paraconsistent logic. (The liar paradox is an example of a statement which some might believe both true and false.) Aside from ideology, paraconsistency has been used to provide methods for large software systems to handle inconsistencies in their data.

Modal logics

After Aristotle's conquest of the **syllogism**, he turned his attention to a thornier problem. In real life, we may not simply assert that a fact *X* is true. There are many ways we may qualify this statement: *X* is necessarily true, possibly true, believed to be true, or will eventually be true. Modal logics are systems which incorporate different ways of being true. The standard modal logic introduces two new symbols, $\square$ and $\diamond$. '$\square$A' is interpreted as '*A* is necessarily true' and '$\diamond$A' for '*A* is possibly true'. They are related, with '*A* is possibly true' meaning '*A* is not necessarily untrue'.

However, there are many variants of modal logic. Kurt Gödel began the study of *provability logic*, which includes the modality 'provably true'. Meanwhile *doxastic logics* attempt to formalize the logic of belief systems, and *temporal logics* are designed to cope with changes of tense. These have to contend with the past being a domain of solid fact, and the future a realm of uncertain possibility. (This caused Aristotle himself to question whether the law of the excluded middle is valid, when applied to statements about the future.)

Possible Worlds

In the 1960s Saul Kripke and others elevated the status of modal logics by demonstrating how to build mathematical models of them. The important fact about these structures is they included different *worlds*. Here *world* is a technical term, corresponding to a part of an abstract mathematical structure. Nevertheless these different domains are indeed intended to mirror parallel universes, or possible worlds, in which different truths may hold. A statement is then necessarily true if it holds in every possible world, and possibly true if it holds in at least one possible world.

METAMAT

MATHEMATICS HAS ALWAYS POSED a dilemma for philosophers. On the one hand, arithmetic seems to be a domain where near perfect knowledge is attainable. Statements such as $1 + 1 = 2$ are freer of caveats and assumptions than almost any other kind. On the other hand, modern mathematics considers a large range of exotic, highly abstract structures. To what extent can these really be said to *exist*?

This question was considered by the ancient Greek philosopher Plato, whose metaphysical answer was provided, in metaphorical form, in *Plato's cave*. But in the late 19th century, this question returned to prominence, principally as a result of Georg Cantor's spectacular work on infinite sets. Gottlob Frege's

EXISTENCE • UNIQUENESS • PROOF BY
• COUNTEREXAMPLES • INDEPENDENCE
ION • CLASSIFICATIONS • HILBERT'S
IUM PROBLEMS • THE FIELDS MEDAL •
S • MATHEMATICS IN THE INFORMATION
RIBUTED MATHEMATICS PROJECTS •
ATH PROJECTS • INSPIRATION AND
ARE • AUTOMATED THEOREM PROVING
• THEORY AND EXPERIMENT • PLATO'S
MATHEMATICS AS LOGIC • FORMALISM •
TRUCTIVE MATHEMATICS • FINITISM •
ATHEMATICS • EXISTENCE • UNIQUENESS

HEMATICS

logicism was an attempt to deduce the whole of mathematics from pure logic. Meanwhile David Hilbert adopted a *formalist* approach, in which mathematics was no more than a set of rules for manipulating symbols on a page. Other thinkers reacted against Cantor's work on set theory. To the *constructivist* school, it was strong evidence that mathematics had gone badly wrong.

In more recent times, any discussion of the nature of mathematics must also consider its symbiotic relationship with technology. Mathematics was fundamental in the development of the computer, and later the internet. In return, these inventions have fundamentally changed the way that mathematics is studied and applied.

WHAT MATHEMATICIANS DO

Proof

Proof is the ultimate goal of mathematics, on which the whole thing stands or falls. A *proof* is a logically watertight argument, starting with some initial assumptions (and underlying *axioms*), and ending with the theorem to be proved. It is the possibility of proof that sets mathematics apart from all other subjects.

Proofs come in many different forms, and mathematicians have numerous strategies available: **proof by contradiction** and **proof by induction** are two important examples. Proofs included in this book are the **irrationality of $\sqrt{2}$**, the **infinity of the primes** and **Cantor's diagonal argument**.

The language of mathematics

Mathematicians have a tendency to name every interesting phenomenon they encounter. This is a good strategy, as it automatically leads towards abstract, axiomatic thinking. On the other hand, it does mean that the subject is completely full of jargon. No doubt this is one reason why outsiders often find it indecipherable and even intimidating.

Aside from all these technical terms, mathematicians have a lexicon of words to describe types of questions they consider.

A *conjecture* is a statement that someone believes to be true, but cannot provide a proof for. A famous example is the **Poincaré conjecture**, which was subsequently proved by Perelman, turning it into a *theorem*. The **Riemann hypothesis**, in contrast, is a conjecture which has not yet been proved, meaning that it remains an *open problem*.

A *lemma* is a minor mathematical result, not of particular of interest in its own right (except when misnamed), but a stepping stone towards a bigger result.

Existence

A common question within mathematics is whether an object of a particular type *exists*. A famous example is whether there exists an **odd perfect number**. Brouwer's fixed point theorem is also a statement that a certain type of object exists.

Euclid's **infinity of the primes** has at its heart a proof of existence. Given some threshold number N, Euclid proved that there exists a prime which is bigger than N. In this case there exist many such primes (rather the point of the proof). The famous **Navier–Stokes problem** and **Yang–Mills problem** ask whether particular equations have solutions; these are also existence questions.

Uniqueness

In some cases an existence theorem comes with something extra: uniqueness. This says that there is *only one* object of the specified type. A famous example is the **fundamental theorem of arithmetic**, which says that every number can be broken down into primes (the existence part), and moreover this can only be done in one way (the uniqueness part). The **analytic continuation theorem** is another celebrated existence and uniqueness theorem. Solving **differential equations** relies on a uniqueness theorem, which specifies the size of the solution space. This means that once we have found a suitable family of solutions, we can be confident that there are no others. Setters of **Sudoku** also have to be sure that their puzzles have unique solutions.

It is possible to prove the uniqueness of a type of object without proving its existence. This amounts to showing that there is *at most one* such object.

Proof by contradiction

Suppose I want to prove that some statement (call it X) is true. The most straightforward method is to start from things I already know, and attempt to deduce X directly. An alternative is to turn the whole process on its head, and begin by supposing that X is *not* true. If I can demonstrate that this leads inescapably to the conclusion that $1 = 2$, then it must be that the assumption was false. So X has to be true after all.

Examples of this technique are Euclid's infinity of the primes, and the irrationality of $\sqrt{2}$. Proof by contradiction rests upon the **law of the excluded middle** and, although it is a standard piece in the mathematician's repertoire, intuitionist and constructivist mathematicians restrict themselves to working without it.

Also known by its Latin title of *reductio ad absurdum*, this technique has a curious by-product: mathematicians can spend much of their time studying things that don't exist.

Things that don't exist

The extent to which mathematical objects truly *exist* is a matter of philosophical debate. Leaving this aside, mathematicians have undoubtedly devoted a great deal of time to studying things whose non-existence is a matter of indisputable fact.

For example, there is a well-known conjecture which says that there is no odd perfect number. A natural way to try to prove this would be by contradiction. Such an attempt would begin by supposing that it is not true, and so assume that there is an odd perfect number, x. In the hunt for an ultimate contradiction, it would be necessary to study the properties of x in great depth. In the course of this work, you might become the world-expert on odd perfect numbers, even though, in all likelihood, there is no such thing.

Counterexamples

If someone conjectures that all grommets are crooked, a *counterexample* is a grommet which is not crooked. Counterexamples are powerful: it only takes one to kill a conjecture stone dead. For example, if someone found a non-trivial zero of Riemann's zeta function not lying on the critical line, that would mark the end of the **Riemann hypothesis**.

Often, when contemplating a question, mathematicians will adopt twin approaches of trying to prove it, and searching for a counterexample. Obviously, only one approach can ultimately succeed, but the obstacles encountered in one attempt can be turned to advantage in the other.

Independence results

This book contains many *theorems:* mathematical statements for which someone has provided a watertight proof. But mathematics is not alchemy; you cannot conjure something out of nothing. At the base of every proof are some initial assumptions, either made implicitly, or stated explicitly as *axioms*. Depending on the field of mathematics, there are various standard starting points from which other results are derived. Sometimes, however, these usual assumptions are not adequate either to prove or disprove a particular statement. This is made concrete with an *independence result*, which demonstrates that the statement and its opposite are both equally compatible with the axioms.

The first significant result of this kind was that showing that the **parallel postulate** is independent of Euclid's other axioms. After Kurt Gödel proved his **incompleteness theorems** in 1931, the possibility of finding independence results within mainstream mathematics became genuine. Might it turn out for example that the Riemann hypothesis is independent of the usual axioms of mathematics? No such result has appeared within number theory yet, but through phenomena such as **Friedman's TREE sequence**, incompleteness is slowly encroaching on mainstream mathematics.

As the foundations of mathematics came under more scrutiny in the 20th century, two important independence results were proved by Paul Cohen: that of the **axiom of choice** and the **continuum hypothesis**.

Generalization

Why do mathematicians prefer to work in generalized, abstract settings such as **groups**, instead of the familiar number systems? One reason is that generalization brings power. When you can prove that something is true in the much more general context of a group, it must automatically be true in every number system, and the many other examples of groups which occur throughout mathematics. It must even hold in as yet undiscovered groups.

Some mathematical subjects push this philosophy further than others. The groups and fields studied in algebra, which are already abstractions of ordinary number systems, can then be generalized further, in one direction to the *structures* of **model theory**, or in another to the *objects* of **category theory**. Perhaps the greatest triumph of this approach came in algebraic geometry. First, geometric shapes such as circles were generalized to **varieties** given by polynomial equations, and then to the abstract heights of **schemes**. This paved the way for breakthroughs such as Deligne's proof of the **Weil conjectures**.

Abstraction

In different ways, all abstract theories of mathematics seek to identify and analyse features of the mathematical landscape. They do this by identifying an important phenomenon, such as multiplication, and then axiomatizing it. This strips away all the extraneous noise that a particular multiplicative system may carry, and allows the phenomenon under investigation to be studied in isolation.

As well as bringing the power of generalization, abstraction allows for greater precision. Proving theorems in more general settings gives a clearer insight as to *why* they are true. **Reverse mathematics** takes this idea to its limit. Once found, these different abstract approaches can then be recombined in subtle ways. A **Lie group**, for example, combines the abstract concepts of a **group** and a **manifold**.

Classifications

Among the most prized of all theorems are *classifications*. While the philosophies of generalization and abstraction encourage mathematicians to consider ever broader classes of object, a classification theorem goes the other way. The first such theorem was that of the **Platonic solids**. Plato began with an abstract definition, namely that of a *regular polyhedron* as a shape which satisfies some general requirements (that every face should be a regular polygon, and that all are identical). Then he proved that the *only* shapes which satisfy this condition are the tetrahedron, cube, octahedron, dodecahedron and icosahedron.

Classification theorems carry huge weight in mathematics, as they replace abstract theorizing with concrete examples, but at no cost of generality. Now, if we are confronted with an unidentified regular polyhedron, we do not need clever arguments based on symmetry groups, we can simply consider Plato's familiar shapes in turn.

For this reason, classification theorems can be major mathematical events, which mark the end of centuries of enquiry. Notable examples include the classification of finite simple groups, the classification of surfaces, the classification of simple Lie groups, Shelah's classification of countable first-order theories, the classifications of frieze and wallpaper groups, and the geometrization theorem.

Hilbert's problems

In 1900, David Hilbert gave an address to the International Congress of Mathematicians in Paris, in which he listed 23 problems which would set the course of mathematics for the twentieth century:
1 Establish the truth or otherwise of Cantor's **continuum hypothesis**.
2 Prove that the axioms of arithmetic are consistent. (This was addressed by **Gödel's second incompleteness theorem** and Gentzen's **proof theory**.)
3 Find two tetrahedra A and B, so that A cannot be chopped up into a family of identical, smaller tetrahedra, and these rearranged to produce B. (This was the most straightforward of Hilbert's problems, and was achieved in 1902 by Max Dehn.)

4 Systematically construct new **non-Euclidean** geometries through analysis of **geodesics**. (This problem is generally considered too vague to be answerable. But our knowledge of non-Euclidean geometries is certainly well developed.)

5 Is there a difference between **Lie groups** which are assumed differentiable and those which are not? (The answer is essentially 'no'. For large classes of Lie groups, differentiability is automatic.)

6 Fully axiomatize physics. (Our best efforts so far are the **standard model of particle physics** and **Einstein's field equation**, but the search for a Theory of Everything goes on.)

7 Understand **transcendental number theory**. The precise problem Hilbert posed was answered by the *Gelfond–Schneider theorem*, but the subject as a whole remains mysterious.

8 Prove the **Riemann hypothesis**.

9 Generalize Gauss' **quadratic reciprocity law**. (Several deep generalizations have been found, and **Langland's program** involves searching for more.)

10 Find an algorithm for solving **Diophantine equations**. This task was proved impossible in **Matiyasevich's theorem**.

11–13 Problems in **Galois theory**, which for the most part remain unresolved.

14–17 Problems in **algebraic geometry**, which have been partially answered.

18 i) Are there only finitely many *space groups* in each dimension? (See **n-dimensional honeycombs**.)

ii) Is there an anisohedral space-filling polyhedron? (See **Heesch's tile**.)

iii) Prove the **Kepler conjecture** for sphere packing.

19–21 and **23** Problems in the theory of **partial differential equations**, which have partially been resolved.

22 A problem in **differential geometry**, to do with the different ways a single surface can be described. It has largely been settled.

The Clay Institute millennium problems

In the year 2000, the Clay Institute assembled a list of major open problems, echoing David Hilbert's from a century earlier. Each of their seven millennium problems was assigned a $1,000,000 prize. They are:

1 **The Birch & Swinnerton-Dyer conjecture**

2 **The Hodge conjecture**

3 **The Poincaré conjecture**

4 **The Navier–Stokes equations**

5 **P = NP?**

6 **The Riemann hypothesis**

7 **The Yang–Mills Problems**

To date only the Poincaré conjecture has been settled. It was proved in 2003 by Grigori Perelman. He declined both the prize and the Fields Medal.

The Fields Medal

The Swedish chemist Alfred Nobel died in 1896, leaving most of his money to found the *Nobel Prizes* in literature, physics, chemistry, physiology and medicine, and peace. In 1968, the Swedish central bank endowed a new Nobel Memorial Prize in economics. There has been some speculation about why Nobel did not include mathematics. A legend says that he found himself the rival of a famous mathematician (possibly Gosta Mittag-Leffler) for a woman's love. A more prosaic theory is that he was simply uninterested in the subject.

In 1924, the International Congress of Mathematicians sought to fill this gap, with the Canadian mathematician John Charles Fields contributing the funds for the prize. The first *Fields Medal* was awarded in 1936, and since 1966 four have been awarded every four years.

The Fields Medal is traditionally awarded to mathematicians under the age of 40. When Andrew Wiles completed his proof of Fermat's last theorem he was therefore ineligible. He was, however, awarded a commemorative silver plaque in 1998. The medal itself is gold, and depicts the Greek mathematician Archimedes together with a Latin quotation from him: *Transire suum pectus mundoque potiri* ('Rise above oneself and grasp the world').

MATHEMATICS AND TECHNOLOGY

Computers

The history of computation is intimately bound up with that of mathematics. Not only were many of the innovators of computation also mathematicians, but the earliest computers were designed for mathematical purposes, namely to mechanize arithmetic. It was the great Gottfried Leibniz who first saw the potential of binary strings to encode information, back in the 17th century. Leibniz also designed an early calculating machine, the *stepped reckoner*, which could add, subtract, multiply and divide.

Charles Babbage is often cited as the father of the computer, for the *difference engine* he designed in the early 19th century. This was a programmable device which could calculate polynomial functions. Babbage was also a professor of mathematics at Cambridge. In the 20th century, **Turing machines**, conceived for purely theoretical reasons by Alan Turing, went on to form the basis on which digital computers would run algorithms. A few years later, Claude Shannon's **information theory** identified and analysed the language that these machines would use to communicate, paving the way for the internet revolution. Today the boundary between computer science and mathematics is porous, with subjects such as **complexity theory** and **cryptography** straddling the two.

Computational mathematics

A common belief about mathematicians and technology is summed up by an old joke about a university chancellor. He hated the physics and chemistry faculty, because they were constantly requesting money for expensive equipment. But he liked the mathematicians because all they ever needed were pens, papers and dustbins (garbage cans). (His favourite faculty was philosophy, because they didn't even need dustbins.) Perhaps there was once a grain of truth to the idea that mathematicians need few tools beyond their brains, but the dawn of the personal computer has had major implications for mathematics. It is striking that Gaston Julia discovered **Julia sets** in 1915, but it took another half century for these beautiful fractals to be seen by means of computer.

There are now several integrated computer packages available for mathematics, including *Mathematica, Maple, Mathcad, Maxima, Sage* and *MATLAB*. These are not only sophisticated numerical calculators and graphics packages, but they can manipulate symbols in very advanced ways. Algebraic procedures which previously took a long time, and ran the risk of human error, can now be automated. Such programs are invaluable tools for mathematical education, and are assuming an ever more important role in research. Computer simulations are essential in much of applied mathematics (such as fluid dynamics). Other mathematical technologies include distributed computing projects, and proof-checking programs.

Mathematics in the information age

Mathematics once had a reputation of being fusty and inaccessible; the internet has gone a long way to opening the subject up. A great deal of mathematics is now available online. Sites such as Wolfram Mathworld (mathworld.wolfram.com), Wikipedia (wikipedia.org) and The On-Line Encyclopedia of Integer Sequences (www.research.att.com/~njas/sequences/) are searchable depositories of large amounts of information.

Increasingly mathematicians use the web to disseminate their own work, and discuss the work of others. It has become common practice to upload papers onto personal websites or arXiv.org, as well as submitting them to peer-reviewed journals. The mathematical blogosphere is growing rapidly, and websites such as *Math Overflow* form online discussion rooms for professional researchers. Innovations such as polymath projects are beginning to turn this interaction into active collaborative research.

Distributed computing

In the internet age, you do not have to be a scientist to help with scientific research. There are numerous projects which run on the idle time of home computers and games consoles around the world, from analysing the structure of protein molecules to scanning the skies for evidence of extraterrestrial life. As computers get faster, and more people around the world connect to the internet, distributed computation has become a hugely powerful resource. Among these projects are an increasing number of mathematical investigations.

Distributed mathematics projects

Within mathematics, the best-known distributed computing project is GIMPS (the Great Internet Mersenne Prime Search), which dominates the search for **large primes**.

Primegrid investigates other aspects of prime numbers, including long arithmetic sequences of primes, and large primes of specific forms. Primegrid incorporates the *Twin Prime* Search.

ABC@home searches for abc triples, to investigate the **abc conjecture**.

Zetagrid, completed in 2005, investigated the **Riemann hypothesis**, verifying it for the first 200 billion zeroes of the **Riemann zeta function**.

NFS@home tackles the **integer factorization problem**, particularly for numbers of the form $b^n \pm 1$ for b between 2 and 12, as n becomes large.

distributed.net focuses on mathematics related to cryptanalysis, including work on the **RCA factoring challenge** and **Golomb rulers**.

Mathematical collaboration

The stereotype of the mathematician as solitary genius has always been unfair. Collaboration between mathematicians is as old as the subject itself, and historically many researchers have travelled extensively to work together. Pythagoras is reputed to have journeyed to Egypt, Judaea, Phoenicia, and even India, to learn his craft. This tendency was taken to its limit by Paul Erdős, who lived a life of perpetual travel, co-authoring papers with over 500 people (see **Erdős numbers**).

Polymath projects

In January 2009, the Fields medallist Timothy Gowers posed a problem from **Ramsey theory** on his blog, and invited anyone who wanted to join in to help prove it. This was the start of the first *polymath project*, inspired by the success of open source projects such as Wikipedia.

The idea was brand new. Mathematicians have always collaborated, but usually only in twos or threes. Asking the question 'Is massively collaborative mathematics possible?', Gowers wrote, 'if a large group of mathematicians could connect their brains efficiently, they could perhaps solve problems very efficiently as well'. So a small network of blogs and wikis was created to connect brains, discuss strategy, share ideas and collect results.

The experiment was a success; through the input of around 27 people around the world, the theorem was proved, and indeed extended, within two months. Several further polymath projects have since been run. It is hard to imagine that such projects will not form an increasingly powerful force for research in the years ahead.

Inspiration and perspiration

Many people have theorized about the psychology of mathematics, including the relationship between occasional momentary flashes of inspiration, and the many hours of sheer hard work. Wittingly or not, mathematicians themselves have helped cultivate this mystery, by following the lead of

Carl Friedrich Gauss, who always strived to present as streamlined and highly polished proof as possible. His work was logically impeccable, but left no hint of how it had been arrived at. As Gauss said, 'no self-respecting architect leaves the scaffolding in place after completing the building'.

A fresh perspective on this question comes through polymath projects, which leave behind them a complete, unabridged record of the thought processes which led to the proof, complete with mistakes and cul-de-sacs. As Timothy Gowers and Michael Nielsen wrote in *Nature*, 'It shows vividly how ideas grow, change, improve and are discarded, and how advances in understanding may come not in a single giant leap, but through the aggregation and refinement of many smaller insights'.

Proof-checking software

Technically speaking, a mathematical *proof* is a list of statements, each a consequence of what has gone before, forming a logically watertight sequence. In principle then, a computer should be able to follow, and verify, a mathematical proof. The challenges are considerable. Despite this definition, most proofs aimed at a human readership tend to resemble a semi-formal argument, with appeals to the reader's intuition and experience.

Nevertheless, it is possible to translate proofs into a format comprehensible to *proof-checking software*. Since 1973, the Mizar project has been assembling a database of classical mathematical results, whose proofs are translated into the Mizar programming language, and then checked by computer. Its database contains many thousand theorems, including the **fundamental theorem of algebra** and **Bertrand's postulate**.

This technique becomes really valuable with proofs which are too long to be easily digestible by humans. A triumph of computerized proof checking came in 2004, when Georges Gonthier at Microsoft Research and Benjamin Werner of INRIA used a system called Coq to verify the **four colour theorem**. At time of writing, a major enterprise in this area is the ongoing *Flyspeck* project, aiming towards a fully formalized and verified proof of **Kepler's conjecture**.

Automated theorem proving

Proof-checking software validates proofs which are written by humans. It is another matter entirely for a computer to come up with a proof itself.

Automated theorem proving is a major avenue in artificial intelligence research. It has produced several mathematically significant results. A breakthrough occurred in 1996, when the Equational Prover program, under the guidance of William McCune, discovered a proof of Robbins' conjecture. This algebraic conjecture was made in 1930s by Herbert Robbins and had defied the attentions of a generation of human mathematicians.

Closely related to theorem proving programs are systems for *software creation*. These are programs which design algorithms to solve particular problems. Again, they can outperform humans in certain cases, and intelligent systems are today employed by NASA as well as several commercial technology companies.

PHILOSOPHIES OF MATHEMATICS

The frivolous theorem of arithmetic

Almost all natural numbers are very, very, very large.

Coined in 1990 by Peter Steinbach, this joke has a serious point behind it. Those numbers which are accessible to humans with our finite minds form a tiny and utterly unrepresentative sample of the set of **natural numbers**. This remains true even factoring in the most powerful distributed computing projects.

Even if we consider the numbers up to some inconceivably large limit, such as **Graham's number**, this sample is still infinitely outnumbered by the numbers beyond it. In fact, our intuition for how most numbers behave is limited, since we rarely encounter them. Extrapolating from the familiar territory of the small numbers into this infinite realm is fraught with risk.

Theory and experiment

The frivolous theorem of arithmetic provides a motto in defence of the abstract approach of modern mathematics. Only by proving things axiomatically can we hope to derive statements which are true for all natural numbers. That is not to say that experimenting with small numbers has no place within mathematics; far from it. Only one counterexample is needed to refute a conjecture; and patterns among numbers can only be found by looking. In recent years computation has provided experimental evidence into many important areas in mathematics, from the Riemann hypothesis to the Navier–Stokes problem.

The frivolous theorem gives some perspective on what we are looking at, however: one snowflake on the tip of an infinitely tall iceberg.

Plato's cave

The Greek philosopher Plato held that there are two planes of existence: the transitory, imperfect, physical world that we inhabit, and the eternal, unchanging world of *forms*. In his work *The Republic*, Plato wrote an analogy to illustrate how the two realms are related. He imagined a group of prisoners who spent their lives chained up in a cave, able only to stare at the wall in front of them. Behind the prisoners is a fire, which illuminates the wall. Various objects from the outside world are brought into the cave, but the prisoners are never permitted to see them directly, only their silhouettes on the wall.

According to this analogy, these shadows represent the physical realm. They are all we can see, and we mistakenly accept them as the ultimate reality. According to Plato, however, they are only fleeting and imperfect representations of a truer reality, that of *forms*.

Platonism

Plato had more than mathematics in mind with his theory of forms, but it does have a particular resonance here. Philosophers of mathematics are particularly concerned with the existence of entities such as numbers. It is not hard to conceive of seven bananas and seven books as being two imperfect shadows of the true number seven. Plato also believed that our only access to the world of forms comes through the application of reason. This too chimes with mathematical methodology.

Whatever the ultimate validity of Platonism as a philosophy, it is fair to say that many mathematicians subscribe to it, at least as a working assumption. It offers a reassuringly unequivocal answer to the question of the existence of mathematical entities, and one which agrees with the experiences of those who spend their days in the company of abstract structures. As such, Platonism is perhaps the default position against which other philosophies of mathematics are judged.

Frege's logicism

Some statements, such as an **Aristotelian syllogism** or the statement '$A = A$', are self-evidently true, and furthermore they are true simply by virtue of their form, rather than the meaning of any of the terms used. '$A = A$' is true, whatever A is. Such statements are what philosophers call *analytic* truths. Early *logicists*, notably Gottlob Frege, had an ambitious goal. They hoped to show that all of arithmetic was the inevitable consequence of statements such as this.

The project cannot be counted a success. Any axiomatization of arithmetic, such as that in *Principia Mathematica*, must involve some axioms whose truth is not analytic (the axiom which underpins **mathematical induction** being the major sticking point).

Mathematics as logic

*The fundamental thesis ... that mathematics and logic are identical,
is one which I have never since seen any reason to modify.*—Bertrand Russell

Logicism can be interpreted more generally as the view that mathematics is reducible to logic. However, unlike Frege's original logicism, 'logic' now takes on a broader meaning and is allowed to posit axioms other than analytic truths.

This principle acted as guiding light to many mathematicians over the 20th century, with Russell and Whitehead's *Principia Mathematica* being an early triumph.

It would be difficult today to reject this neo-logicism completely, given the successes it has brought. Many objects and methods of pure mathematics can certainly be conceived as logical in nature. What is more, this realization has hugely enriched our interactions with them, through the technical branches of mathematical logic such as **computability theory**, **proof theory** and **model theory**. On the other hand, this neo-logicist principle does not address the fundamental problems of mathematical philosophy, such as the ontological status of mathematical objects; rather, it relocates them to the philosophy of logic.

Formalism

Formalism takes a reductive view of mathematics, opposed to the Platonist's. It is the view that mathematics is no more than the manipulation of certain symbols on a page, in accordance with some previously agreed rules. David Hilbert was the most famous formalist. He realized that treating mathematics as nothing more than a game of symbols was the right way to investigate the fundamental questions of consistency, completeness and decidability with which **Hilbert's program** was concerned.

On its most basic terms, formalism is hard to reject, given the successful and rigorous axiomatization of so much mathematics. Pure mathematicians do indeed spend their time manipulating symbols in accordance with certain rules. The philosophical question is whether this is *all* they are doing, or whether there is some deeper meaning to their work. It seems that some other consideration is necessary, as formalism alone cannot answer a key question: which rules should we choose? This question became acute when Kurt Gödel proved his incompleteness theorems, with their implication that no available axiomatization of the natural numbers is adequate to answer every question.

Empiricism

Rather than trying to reduce everything to the analytic truths of pure logic, or a symbol-manipulating formal game, an empirical approach to the natural numbers observes that they exist in our world as properties of physical objects. This is, after all, how every child begins mathematics: if he has two toys and is given two more toys, then he has four toys in total.

How can we investigate the behaviour of these natural numbers? The empiricist school says that we should treat this in the same way as we treat other scientific matters: we investigate, experiment, make hypotheses, test them against the evidence and draw conclusions.

Empiricism is set against formalism, and received a boost from the works of Gödel and Turing, which showed that the truth of the natural numbers lies beyond anything accessible by a formal system. The computability theorist Gregory Chaitin is an outspoken empiricist, who cites the **randomness** at the heart of mathematics as evidence that formal systems are forever doomed to fail, and that a more direct line of enquiry is needed. Axioms (and of course proof) still play a role, but they are closer in spirit to the **standard model of particle physics**. They represent our best estimate, for now. We must expect them to change and grow, as we inch towards the truth.

Constructivism

Brouwer's fixed point theorem is a curious result. It guarantees that whenever a cup of coffee is stirred there will always be one point in the same position afterwards as it was before. However, neither the statement of the theorem nor its proof provides any information whatsoever about where this point might be found. This is one example of a *non-constructive* result, and the mathematical literature is full of others. A more usual approach would be as follows: to prove that an object of a particular type exists (a point of coffee which is unmoved, or a prime number bigger than some threshold), then exhibit one directly. This is the *constructivist* approach.

Proof by contradiction offers a non-constructive alternative, which takes advantage of the **law of the excluded middle**. First assume that an object of the type you want does not exist, and then deduce a contradiction from this assumption. This is the main way that non-constructive results arise in mathematics. For this reason constructivism is closely bound up with **intuitionistic logic** (of which, somewhat ironically, L.E.J. Brouwer was a founder). The ultimate example of non-constructive mathematics is the **axiom of choice**, which is why some people have reservations about it.

Constructive mathematics

Constructive mathematics is an ongoing programme to replace non-constructive proofs and theorems with constructive equivalents, where possible. The guiding philosophy is, of course, constructivism. But there are reasons for constructive mathematics to be of interest, even to those who do not subscribe to that outlook. It is good practice to be as economical with one's starting assumptions as possible. Constructive mathematicians analyse when the law of the excluded middle, or a version of the axiom of choice, is a necessary prerequisite, rather than a convenient short-cut. This takes the subject into the deep waters of proof theory and reverse mathematics. Equally, constructive proofs are closely related to **algorithms**, and so of practical importance in computer science.

Finitism

God made the integers, all else is the work of man.—Leopold Kronecker

Kronecker was aghast at the turn that mathematics was taking in the late 19th century, most especially by the work of his own student Georg Cantor (see **Cantor's theorem**). Kronecker's quote became a motto for *finitism*, a strict take on the constructivist school, which rejects the reality of all infinite quantities. Finitists hold that infinity is only ever accessible as a *potential*. Certainly the set of natural numbers is potentially infinite. But we can never access the whole collection, only its finite subcollections. These finite sets are expressed in the physical reality of our universe, in a way that Cantor's infinite sets are not.

Ultrafinitism

Some thinkers push finitist philosophy further. If the reality of infinite sets is cast into doubt because they are not part of our physical universe or accessible to our human minds, then the same should go for enormous finite numbers such as Harvey Friedman's **TREE** 3. Perhaps the most outspoken proponent of *ultrafinitism* is Alexander Yessenin-Volpin, who is also well known as a poet, moral philosopher and human rights activist, and who spent many years as a political prisoner in the Soviet Union. For Yessenin-Volpin, even numbers such as 2^{100} are out of reach of the human mind, and therefore of doubtful validity, let alone monsters such as Friedman's.

The question is: if 2 is to be accepted, but 2^{100} is not, where is the line drawn? Harvey Friedman tells the story of when the two men, whose ideologies could hardly be more different, met:

'I raised just this objection with the (extreme) ultrafinitist Yessenin-Volpin during a lecture of his. He asked me to be more specific. I then proceeded to start with 2^1 and asked him whether this is 'real' or something to that effect. He virtually immediately said yes. Then I asked about 2^2, and he again said yes, but with a perceptible delay. Then 2^3, and yes, but with more delay. This continued for a couple of more times, till it was obvious how he was handling this objection. Sure, he was prepared to always answer yes, but he was going to take 2^{100} times as long to answer yes to 2^{100} than he would to answering 2^1. There is no way that I could get very far with this.'

PROBABIL
STATISTIC

AS WELL AS MODELLING predictable phenomena, as happens in Newtonian mechanics, mathematics also provides invaluable tools for analysing uncertainty. This is the domain of *probability*. Here simple experiments such as rolling dice and flipping coins can provide a great deal of insight into the mathematics of chance.

Every business and government in the world employs mathematics daily, in the form of *statistics*. This provides the tools for analysing any form of numerical data.

It is no surprise that probability and statistics are intimately related. Statistical data from an experiment will provide clues about its underlying *probability distribution*. Many powerful and elegant results can be proved about such

GE • FREQUENCY TABLES • MEAN FROM
NCY • INTERQUARTILE RANGE • SAMPLE
RMAN'S RANK CORRELATION • THE FIRST
S THEOREM • PROBABILITY • SUCCESSES
ULTIPLYING PROBABILITIES • MUTUALLY
THE BIRTHDAY PROBLEM • THE BIRTHDAY
YES' THEOREM • SPLITTING AN EVENT •
ROSECUTOR'S FALLACY • THE DEFENCE-
ITY INVERSION FREQUENTISM
UM NN AGR EOREM • THE
O WITH ? BU S NEEDLE
CIDENCE • PROBABILITY DISTRIBUTIONS
DISTRIBUTION • BINOMIAL TRIALS •
POISSON DISTRIBUTION • THE LAW OF
IBUTIONS • UNIFORM DISTRIBUTION
OM VARIABLE LAW OF

ITY AND S

distributions, such as the *law of large numbers* and the *central limit theorem*. It is striking that such mathematical calculations often violently disagree with human intuition, famous examples being the *Monty Hall problem* and the *prosecutor's fallacy*. (Some have even suggested that this trait may be evolutionarily ingrained.) To combat this tendency towards irrationality, people in many walks of life people apply techniques of *Bayesian inference* to enhance their estimation of risk.

Another important branch of applied mathematics is *cryptography*, ranging from the most ancient method of encrypting a message, a *monoalphabetic cipher*, to the modern subject of *information theory* which underpins the internet.

STATISTICS

Mean

Suppose we have been out in the field and collected a set of data:

$$3, 3, 4, 3, 4, 5, 3, 9, 4$$

Call this set A. It could be the number of leaves on dandelions in a garden, or the number of residents in houses on a street. Whatever it represents, we might wish to calculate its *average*. The mean, median, mode and mid-range are different formulations of the average. Typically they do not produce the same result.

The *mean* is the most well-known of the averages. To calculate it, we add up all the numbers ($3 + 3 + 4 + 3 + 4 + 5 + 3 + 9 + 4 = 38$), and divide by how many of them there are (in this case 9). So in this case, the mean is $\frac{38}{9} = 4.\dot{2}$.

More generally, if we have m numbers: $a_1, a_2, \ldots, a_m$ then the mean is $\frac{a_1 + a_2 + \ldots + a_m}{m}$.

The mean can have unexpected consequences. For instance, the mean number of arms of people worldwide is not 2, but around 1.999. (Most people have two arms, some have none or one, and very few have three or more.) The overwhelming majority of people, therefore, have an above average number of arms, when the mean is used. Using the median or mode, this will no longer be true.

Median

Suppose A is the set of data 3, 3, 4, 3, 4, 5, 3, 9, 4. To calculate the *median* of A we need first to arrange it into ascending order:

$$3, 3, 3, 3, 4, 4, 4, 5, 9$$

The median is now the middle number, in this case 4.

A complication arises when there are an even number of data points, in which case the median is taken as the mean of the central two. For example, suppose we want to calculate the median of the data set: 9, 10, 12, 14. There is no middle number here, but the middle two are 10 and 12. So we take the take the mean of these: $\frac{10 + 12}{2} = 11$.

Mode and mid-range

The *mode* is the value which occurs most often. In the case of the set A above, this is 3. Some sets of data do not have a meaningful mode: 1, 2, 3, 4, 5, for example, or 2, 2, 50, 1001, 1001.

The crudest form of average is the *mid-range*: the mean of the highest and lowest points. In the case of A above it is $\frac{3 + 9}{2} = 6$.

Frequency tables

The data set A discussed above was easy to handle, since it just contained nine data points. In many practical applications the data set will be much larger.

The underlying mathematics for calculating the mean, median and mode remains the same, but the manner of presenting it is slightly different.

Suppose we have the data for the number of people living in each house or apartment in a chosen town. We can set the results up in a *frequency table*:

Number of people in property	Frequency
0	292
1	5745
2	8291
3	4703
4	2108
5	961
6	531
Total	22,631

This means that there are 292 unoccupied properties, 5745 single-occupier properties, and so on, making a total of 22,631 properties in the town. The easiest average to identify is the mode, which can be read straight from the table: it is 2, as this is the number with the highest frequency.

Mean from frequency tables

To calculate the mean in the above frequency table, we need to add up the total number of people in all the houses, and divide by the number of houses. The frequency column tallies the number of houses, making a total of 22,631. But how can we find the total number of people?

In the unoccupied houses there are a total of 0 people (obviously). There are 5745 single-occupier properties, which contribute 5745 people to the total. In the two-person properties, there are a total of $8291 \times 2 = 16,582$ people, and in the three-person properties, a total of $4703 \times 3 = 14,109$ people.

So to calculate the total number of people we need to multiply together the first two columns of the table. It is convenient to abbreviate the first as n, the frequency as f. Then we can add a new column to the table, for $n \times f$:

n	f	$n \times f$
0	292	0
1	5745	5745
2	8291	16,582
3	4703	14,109
4	2108	8432
5	961	4805
6	531	3186
Total	22,631	52,859

The third column represents the total number of people in each class of house. Adding up this column we find the total number of people in the town: 52,859. So we can finally calculate the mean: $\frac{52,859}{22,631} = 2.34$ (to two decimal places).

Using the **sum** notation, we can write the mean in a new way as $\frac{\Sigma n \times f}{\Sigma f}$.

Cumulative frequency
The median in the above frequency table will be the middle property in the list, that is the 11,316th. So we need to work out which class this fits into. A convenient method is to add a third column which counts the cumulative frequency.

Number of people in house	Frequency	Cumulative frequency
0	292	292
1	5745	6037
2	8291	14,328
3	4703	19,031
4	2108	21,139
5	961	22,100
6	531	22,631

The first entry in the cumulative frequency column is 292, the same as in the plain frequency column. The next entry is different. While 5745 is the number of single occupier residences, 6037 is the number of residences with either 0 or 1 people in them. Similarly the next entry 14,328 is the number of residences with 0, 1, or 2 people in them.

Cumulative frequency is often useful. In particular it makes the median easy to find. In this case, we are looking for the 11,316th house. Since the class of two-person residences covers the 6038th house till the 14,328th, the 11,316th must be in this class, therefore the median is 2.

Interquartile range
The different averages of a set of data are ways to judge its centre. Another important aspect is its *spread*. Again there are different ways to formalize this.

The *range* is the full distance between the maximum and minimum data points. This is not a particularly useful measure as it is dictated completely by *outliers*, points that lie far outside the general trend. Another simple measure is the interquartile range.

The median is found by putting all the data in ascending order, and picking the point exactly $\frac{1}{2}$ way between the maximum and minimum. We may also look at the points $\frac{1}{4}$ and $\frac{3}{4}$ of the way between these points; these are the *quartiles*. The distance between them is the *interquartile range*. For example, if our data set is 1, 1, 3, 5, 7, 9, 10, 15, 18, 20, 50, then the median is 9, and the quartiles are 3 and 18, making the interquartile range $18 - 3 = 15$. The full range is $50 - 1 = 49$. For larger data sets we can read the quartiles off the cumulative frequency table:

Number of people in property	Frequency	Cumulative frequency
0	292	292
1	5745	6037
2	8291	14,328
3	4703	19,031
4	2108	21,139
5	961	22,100
6	531	22,631

The first quartile will be the 5658th house, which lies in the class of single-occupier properties. The third quartile is the 16,974th house, which lies in the class of three-person residences. So the interquartile range is $3 - 1 = 2$. The full range is $6 - 0 = 6$.

Sample variance

Like the interquartile range, the *sample variance* is a measure of the *spread* of a set of data. It is the most natural one to consider when we take the mean as the centre. Suppose our set X of data contains n data points, and has mean μ. We can measure the distance of any data point x from the mean as $x - \mu$. We then square this to make it positive, giving $(x - \mu)^2$.

If we do this for each value of x and then take the mean of the results, we get the *sample variance* of X, $\mathrm{Var}\,X$. That is:

$$\mathrm{Var}\,X = \frac{\sum(x - \mu)^2}{n}$$

For example, if our data is 3, 3, 3, 3, 4, 4, 4, 5, 9 then its mean is $4.\dot{2}$. To calculate the variance we subtract the mean from each value:

$$-1.\dot{2}, -1.\dot{2}, -1.\dot{2}, -1.\dot{2}, -0.\dot{2}, -0.\dot{2}, -0.\dot{2}, 0.\dot{7}, 4.\dot{7}$$

Now square each of these, and take the mean of the squares:

$$\frac{4 \times (-1.\dot{2})^2 + 3 \times (-0.\dot{2})^2 + (0.\dot{7})^2 + (4.\dot{7})^2}{9} = 3.284 \quad \text{(to three decimal places)}$$

The *standard deviation* of X is $\sqrt{\mathrm{Var}\,X}$, in this case $\sqrt{3.284} = 1.812$ to three decimal places. (See also **expectation and variance** of a probability distribution.)

Moments

The above procedure for calculating the variance works perfectly well, but there is a slightly quicker method. It turns out that the variance is equal to the *mean of the squares minus the square of the mean*. That is:

$$\mathrm{Var}\,X = \frac{\sum x^2}{n} - \mu^2$$

Starting with the data set 3, 3, 3, 3, 4, 4, 4, 5, 9, we calculate the mean of the squares:

$$\frac{4 \times 3^2 + 3 \times 4^2 + 5^2 + 9^2}{9} = 21.\dot{1}$$

Then we subtract the square of the mean:

$$21.\dot{1} - (4.\dot{2})^2 = 3.284 \text{ (to three decimal places)}$$

The mean $\frac{\sum x}{n}$ is also known as the *first moment*, and the mean of the squares $\frac{\sum x^2}{n}$ is the *second moment*. Statisticians also use higher moments in their analysis.

Correlation

Are smokers more likely to get cancer? Do people in rainy countries have more children? Do heavier beetles live longer? There are many occasions in which scientists want to test for a relationship between two phenomena. The statistical tool which quantifies this is *correlation*.

Suppose we have data on the weight and longevity of some beetles, and want to test if the two are related. We could start by plotting the data on a graph of weight against longevity on a graph. If the resulting points seem randomly scattered, then there is no correlation between the two factors. On the other hand, if they lie very close to a line, then the two are *strongly correlated*. Between these two situations is *weak correlation*.

Suppose that there is some degree of correlation. If an increase in weight tends to come with an increase in longevity, then they are *positively correlated*. If an increase in weight tends to come with a decrease in longevity, they are *negatively correlated*.

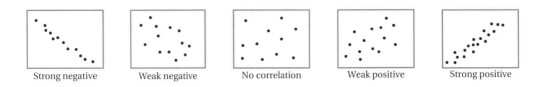

Strong negative   Weak negative   No correlation   Weak positive   Strong positive

There are various methods by which statisticians can assign a number, called a *correlation coefficient* to this situation. (A common one is Spearman's rank correlation coefficient.) In all cases, the output is a number between -1 and $+1$, where a result near -1 means strong negative correlation, a result near $+1$ means strong positive correlation, and a result of 0 means no correlation.

If our analysis does reveal positive correlation between the weight and longevity of beetles, this does not tell us whether healthier, longer-lived beetles tend to be heavier, or whether extra weight helps protect beetles from injury, or whether there is a third factor we have missed such as female beetles being both heavier and longer-lived than males. The warning is *correlation does not imply causation*.

Spearman's rank correlation

There are several approaches to testing for correlation. Charles Spearman's *rank correlation coefficient* has the advantage that it does not assume linear correlation to work.

Suppose that scientists discover a new type of luminous plant. They want to investigate whether the height of the plants is correlated with their luminosity. To illustrate the method, I will take a small sample:

Plant	Height	Luminosity
A	6.1	0.41
B	4.5	0.37
C	5.0	0.36
D	5.9	0.31
E	7.3	0.45
F	6.2	0.38

The first thing to do is rank the plants by height and by luminosity (1 being the tallest/brightest):

Plant	Height	Height rank	Luminosity	Luminosity rank
A	6.1	3	0.41	2
B	4.5	6	0.37	4
C	5.0	5	0.36	5
D	5.9	4	0.31	6
E	7.3	1	0.45	1
F	6.2	2	0.38	3

Now we can forget about the actual data and just work with the two ranks. The next thing to do is to calculate the difference (d) in ranks for each plant, and square it (d^2):

Plant	Height rank	Luminosity rank	Difference in ranks (d)	d^2
A	3	2	1	1
B	6	4	2	4
C	5	5	0	0
D	4	6	-2	4
E	1	1	0	0
F	2	3	-1	1

Next we sum up the d^2 column; in this case $\sum d^2 = 10$. Finally, we insert this into Spearman's formula for the coefficient

$$\rho = 1 - \frac{6 \sum d^2}{n(n^2 - 1)}$$

where n is the number of plants, in this case 6. (Notice that if correlation is perfect then all the ranks are the same, and the coefficient will be 1, as we would hope.) In this case, we get a rank correlation coefficient of $\rho = 1 - \frac{6 \times 10}{6(36 - 1)} = 0.71$, to two decimal places, which is moderate positive correlation.

The first digit phenomenon

Suppose we take a set of data from the real world, the account books of a company perhaps, or the heights of a mountain range. Then we tally up the number of times each of the digits 1 to 9 occur as the first digit (ignoring any leading zeros). Most people would expect that the nine digits should be equally common, each occurring as the leading digit around $\frac{1}{9}$th of the time. Remarkably, this usually fails, as two 19th-century scholars, Simon Newcomb and Frank Benford, both noticed. Benford investigated further, counting the leading digits in large quantities of data from baseball statistics to river basins. He found that the digit 1 appears as a leading digit around 30% of the time, 2 around 17% of the time, decreasing to 9 which occurs around 5% of the time.

Benford's law

Benford's law provides a formula for the first digit phenomenon. It states that the proportion of the time that n occurs as a leading digit is around $\log_{10}\left(1 + \frac{1}{n}\right)$. We have to be careful with Benford's law. Truly random numbers (that is to say *uniformly distributed* data) will not usually satisfy it (so it is of no help in selecting lottery numbers). Similarly data within too narrow a range will not obey it (not many US presidents have an age beginning with 1). Between these two extremes, however, are innumerable social and naturally occurring situations where it does apply. (Indeed, Benford's law has been of great value in fraud investigations in the USA.)

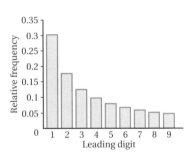

Hill's theorem

Benford's law has been justified in various ways since its discovery. Consider just the numbers 1 to 9. In this set, of course, the leading digits are perfectly uniform. Now create a new set by multiplying everything by two: 2, 4, 6, 8, 10, 12, 14, 16, 18. Now over 50% of the results have leading digit 1. This illustrates the instability of the uniform distribution on leading digits. Many arithmetical procedures will skew it towards Benford's distribution.

This is the idea, but it was only in 1998 that Theodore Hill provided the first rigorous explanation. The key observation is that Benford's law is not *base* dependent. The statement above was for data presented in base 10, but a similar thing applies in any other base b, with the average frequency of the leading digit n given by $\log_b\left(1 + \frac{1}{n}\right)$. Hill was able to show that Benford's is the only probability distribution on leading digits which satisfies this property of base-invariance.

PROBABILITY

Probability

Probability is a mathematical way to quantify the chance of an event X taking place. It works by assigning X a number called $P(X)$, which lies on a scale between 0 and 1.

In naïve terms, if $P(X) = 0$, then X is impossible: rolling 7 on an ordinary die, for example. A probability of $P(X) = 1$ corresponds to certainty. In between, the flip of a fair coin has a probability $\frac{1}{2}$ of coming out heads, and the probability of rolling a number between 1 and 5 on a fair die is $\frac{5}{6}$.

This reading of probability works well for many purposes, when there are finitely many outcomes. But it is too simplistic for some applications. It is better to interpret $P(X) = 0$ as meaning that X is *infinitely unlikely*. Suppose I (somehow) pick, completely at random, a real number between 0 and 10. What is the probability that I will select π (not to 10 or 100 decimal places, but exactly)? Since this is one of infinitely many possibilities, the probability must be 0. This situation is covered by **continuous probability distributions**.

Successes and outcomes

When you roll a fair die, there are six possible outcomes, all equally likely. Suppose we are interested in the probability of getting a 6. That means that exactly one of the six possible outcomes is classed as a success, giving a probability of $\frac{1}{6}$. The fundamental principle here is straightforward. The probability of a successful outcome is always: $\frac{\text{Number of successful outcomes}}{\text{Total number of outcomes}}$, so long as every outcome is equally likely.

If we roll two fair dice (A and B), then there are 36 different possible outcomes: 1 to 6 on A, combined with 1 to 6 on B. If we want to know the probability of rolling a total score of 7, we need to work out the number of successful outcomes. The possible successful outcomes are:

A	1	2	3	4	5	6
B	6	5	4	3	2	1

Since there are six of these, the probability of scoring 7 is $\frac{6}{36} = \frac{1}{6}$.

This simple idea can be the basis of some highly involved problems, such as the **birthday problem**. Calculating such probabilities often involves subtle combinatorics to count outcomes.

Adding probabilities

Suppose we know the probability of two events X and Y. What is the probability of the new event 'X or Y'? The rule of thumb is that 'or' means 'add the probabilities'.

For example, the probability of rolling a 4 on a fair die is $\frac{1}{6}$, and the same probability holds for rolling a 5. So the probability of rolling a 4 or 5 is $\frac{1}{6} + \frac{1}{6} = \frac{1}{3}$. Extreme caution is required here! This can easily lead to errors. Suppose I flip two coins (A and B). What is the probability that I get a head on A or a head on B? The probability that I get a head on A is $\frac{1}{2}$, and the same goes for B. Adding these two, the answer should be $\frac{1}{2} + \frac{1}{2} = 1$. This would imply that it is a certainty. Obviously, this is nonsense: I may get two tails. In general, the rule only holds when we have two **mutually exclusive events**.

Multiplying probabilities

Suppose I know the probability of two events X and Y. What is the probability of the new event 'X and Y'? Here, the rule of thumb is that 'and' means 'multiply the probabilities'. For example, if I flip two coins, the probability that they will both come up heads is $\frac{1}{2} \times \frac{1}{2} = \frac{1}{4}$.

As with adding probabilities, misapplying this rule easily produces nonsense. Suppose I flip one fair coin, once. The probability that I will get a head is $\frac{1}{2}$, and similarly for a tail. So the probability that I get a head and a tail (on the same flip) should be $\frac{1}{2} \times \frac{1}{2} = \frac{1}{4}$. This is obviously absurd, since that scenario is impossible, with a probability of 0.

The rule of thumb is valuable, nevertheless. To know when it applies and when it does not, we need to understand **independent events**.

Mutually exclusive events

Two events X and Y are *mutually exclusive* if they cannot both occur: if X happens then Y does not, and vice versa. If I roll a die, then the events of my getting a 2 and my getting a 5 are mutually exclusive.

If X and Y are mutually exclusive then to find the probability of 'X or Y', we add the probabilities:

$$P(X \text{ or } Y) = P(X) + P(Y)$$

For example, rolling a 2 and rolling a 5 on a fair die each have probability $\frac{1}{6}$. So the probability of rolling either a 2 or a 5 is $\frac{1}{6} + \frac{1}{6} = \frac{1}{3}$. However, if I roll two dice (A and B), scoring 2 on A and 5 on B are not mutually exclusive events. They can both occur. So in this case we cannot just add the probabilities.

Independent events

Two events X and Y are *independent* if they do not affect each other. Whether or not X occurs has no impact on the probability of Y, and vice versa. A classic example would be if I roll a dice and flip a coin, then rolling a 6 and flipping a head are independent events.

If X and Y are *independent*, then to find the probability of 'X and Y' we multiply the probabilities:

$$P(X \text{ and } Y) = P(X) \times P(Y)$$

In particular, the probability of getting a head and a 6 is $\frac{1}{2} \times \frac{1}{6} = \frac{1}{12}$.

Independence is not always completely self-evident. Another example is to pick a card from a pack. Then shuffle the pack, and pick again. Are the results of the two picks independent events? The answer depends on whether the first card is returned to the pack before the second is chosen. If it is, then the two are independent, if not, then the first event affects the probability of the second, so they are not independent.

It is *almost* true to say that two events cannot be both independent and mutually exclusive. There is one exception, namely if one of the events is impossible. The events of flipping a head on a coin and rolling a 7 on an ordinary die are certainly independent. They are also mutually exclusive, in the trivial sense they cannot both happen, because rolling a 7 itself can never happen.

The birthday problem
How many people do there need to be in a room so that there is at least a 50% chance that two will share the same birthday? To answer this, it is convenient to turn the question on its head, and, for different numbers of people, calculate the probability that everyone has a different birthday. When this probability first drops below 50%, we will have the answer to the problem.

The model used here comes with some built-in assumptions which should be noted. Most obviously it ignores leap years; more subtly it assumes that every day of the year is as common a birthday as every other, although this is not quite true in practice.

The birthday theorem
To solve the birthday problem, suppose first that there are just two people in the room. Then the total number of possible arrangements of birthdays is 365×365. On the other hand, if they are to have different birthdays, then the first person may have his on any day of the year (365 possibilities) and the second may have hers on any day except the first's birthday (364 possibilities). So the number of possible pairs of distinct birthdays is 365×364. The probability of one occurring is therefore $\frac{365 \times 364}{365 \times 365}$.

The same approach generalizes to when there are n people in the room. The number of total possible arrangements of birthdays is 365^n. If all the people are to have distinct birthdays, we reason as before: the first person may have her birthday on any day (365 possibilities). The next may have his on any day except the first's (364 possibilities); the third must avoid the birthdays of the first two (363 possibilities), and so on, until the nth person who must avoid the first $(n-1)$ birthdays ($366 - n$ possibilities). So the probability of this occurring is $\frac{365 \times 364 \times \ldots \times (366 - n)}{365^n}$.

This simplifies to $\frac{364!}{(365 - n)! \times 365^{n-1}}$. So, what is the first value of n for which this is below 0.5?

A little experimentation with different values of n provides the answer: 23, at which point the probability is around 0.493.

Conditional probability

In a particular city, 48% of houses have broadband internet installed, and 6% of houses have both cable television and broadband internet. The question is: what is the probability that a particular house has cable TV, *given that* it has broadband?

If X and Y are events, we write the *conditional probability* of X given Y as $P(X|Y)$. Mathematically, it is defined as follows:

$$P(X|Y) = \frac{P(X \& Y)}{P(Y)}$$

(This only makes sense when $P(Y) \neq 0$.)

In the above example, we take X to be the event that the house has cable TV, and Y to be the event that it has broadband. Notice that we do not need to know $P(X)$ to calculate the answer: $P(X|Y) = \frac{0.06}{0.48} = 0.125$, or 12.5%.

In many contexts, conditional probability is extremely useful, as it allows probabilities to be updated as new information becomes available. This is known as *Bayesian inference*.

Bayes' theorem

In 1764, an important paper by the Reverend Thomas Bayes was published posthumously. In it he gives a compelling account of conditional probabilities. The basis was *Bayes' theorem*, which states that, for any events X and Y:

$$P(X|Y) = P(Y|X) \times \frac{P(X)}{P(Y)}$$

In a sense, this formula is not deep. It follows directly from the definition of conditional probability: $P(Y|X) = \frac{P(X \& Y)}{P(X)}$, so $P(X \& Y) = P(Y|X)P(X)$. Substituting this into the definition of $P(X|Y)$ produces the result. However, this theorem has been of great use, for example in the analysis of the **problem of false positives**.

Splitting an event

Suppose we pick a person at random off the street, and want to know the probability that this person wears glasses? In our city, the only statistics available say that 65% of females and 40% of males wear glasses. We also know that 51% of the population are female and 49% male. How can we combine these to get the probability we want?

Call Y the event that our selected person is female. The event we are really interested in is X, that our selected person wears glasses. We have split this into two smaller events: $X \& Y$ and $X \&$ not Y. These two events are mutually exclusive, so:

$$P(X) = P(X \& Y) + P(X \& \text{not } Y)$$

Expressing $P(X \& Y)$ in terms of conditional probability, we get $P(X \& Y) = P(X|Y)P(Y)$ and similarly $P(X \& \text{not } Y) = P(X| \text{not } Y)P(\text{not } Y)$. Putting these into the formula above, we get:

$$P(X) = P(X|Y)P(Y) + P(X|\text{not } Y)P(\text{not } Y)$$

At this point we can use the data we have, $P(X|Y) = 0.65$, $P(X|\text{not } Y) = 0.4$, $P(Y) = 0.51$ and $P(\text{not } Y) = 0.49$. Putting this together:

$$P(X) = 0.65 \times 0.51 + 0.4 \times 0.49 = 0.5275$$

False positives

A test for a certain disease has the following accuracy: if someone has the disease, the test will produce a positive result 99% of the time, and give a *false negative* 1% of the time. If someone does not have the disease, the test gives a negative result 95% of the time, and gives a *false positive* 5% of the time. The disease itself is quite rare, occurring in just 0.03% of the population. A person, Harold, is picked at random from the population and tested. He tests positive. The critical question is: what is the probability that he has the disease?

We need to translate this into mathematics: let X be the event that Harold has the disease. Before we factor in the test data, $P(X) = 0.0003$. Let Y be the event that Harold tests positive. We can split this event to write $P(Y) = P(Y|X)P(X) + P(Y|\text{not } X)P(\text{not } X)$, which comes out as $0.99 \times 0.0003 + 0.05 \times 0.9997 = 0.0052955$.

We are really interested in $P(X|Y)$, the probability that he has the disease, given that he has tested positive, and can work this out using Bayes' theorem:

$$P(X|Y) = P(Y|X) \times \frac{P(X)}{P(Y)} = 0.99 \times \frac{0.0003}{0.0052955} = 0.056$$

So the probability that Harold has the disease, given that he has tested positive for it, is a little under 6%. The explanation for this surprising result is that the true positives form a high proportion of the very small number of disease sufferers. These are greatly outnumbered by the false positives, a fairly small proportion of a hugely larger number of non-sufferers. So, despite the seeming accuracy of the test, when randomly chosen people are tested, a large majority of the positive results will be false.

The prosecutor's fallacy

A suspect is being tried for burglary. At the scene of the crime, police found a strand of the burglar's hair. Forensic tests showed that it matched the suspect's own hair. The forensic scientist testified that the chance of a random person producing such a matching is $\frac{1}{2000}$. The *prosecutor's fallacy* is to conclude that the probability of the suspect being guilty must therefore be $\frac{1999}{2000}$, damning evidence indeed.

This is certainly incorrect. In a city of 6 million people, the number of people with matching hair samples will be $\frac{1}{2000} \times 6{,}000{,}000 = 3000$. On the basis of this evidence alone, the probability of the suspect being guilty is a mere $\frac{1}{3000}$.

The term 'prosecutor's fallacy' was coined by William Thomson and Edward Schumann, in their 1987 article 'Interpretation of Statistical Evidence in Criminal Trials'. They documented how readily people made this mistake, including at least one professional prosecuting attorney.

The defence attorney's fallacy

Thomson and Schumann also considered an opposite mistake to the prosecutor's fallacy, which they dubbed the *defence attorney's fallacy*.

In the above example, the defence attorney might argue that the hair-match evidence is worthless, since it only increases the probability of the defendant's guilt by a tiny amount, $\frac{1}{3000}$.

If the hair is the only evidence against the suspect, then the pool of potential suspects before the forensic evidence is taken into account is the entire population of the city, 6,000,000, which the new evidence reduces by a factor of 3000, to 2000.

However, one would expect that this is not the only evidence, in which case the initial pool of potential suspects will be much smaller. If it is 4000, say, then the forensic evidence may reduce this by 2000, to 2, increasing the probability of guilt from $\frac{1}{4000}$ to $\frac{1}{2}$. This is valuable evidence.

The fallacy of probability inversion

The phenomenon of false positives and the prosecutor's fallacy may seem surprising. Mathematically speaking, the two situations are similar; the ultimate fallacy in both cases is of confusing $P(X|Y)$ with $P(Y|X)$. If $P(X|Y)$ is very high, people commonly assume that $P(Y|X)$ must be too. In the example above of the medical test, $P(\text{positive result}|\text{disease}) = 0.99$ while $P(\text{disease}|\text{positive result}) = 0.056$. These examples show how wrong this can be.

This fallacy is widespread (even in doctors' surgeries and law courts). Some have argued that it is not merely a common mathematical mistake, but an ingrained cognitive bias within human nature. Either way, an appreciation of this issue is essential for making sense of statistical data in the real world.

Frequentism

What is the ontological status of probability? That is, to what extent does it really exist in the world? There are two broad schools of thought: *frequentism* and *Bayesianism*.

For a frequentist, randomness is taken as an intrinsic part of reality, which probability quantifies. To say that event A has a probability of $\frac{1}{2}$ means that if the experiment was repeated many times then A would occur exactly $\frac{1}{2}$ of the time. In other words the probability of A is a measure of the frequency with which A happens, given the initial conditions. (This would only be an approximate after finitely many repetitions, but would be exact in the limit.)

As this shows, the principle does not apply very easily to one-off events, but is best suited to repetitive occurrences.

Bayesianism

In contrast to frequentists, for a Bayesian, probability does not exist in the external world. It is purely a way for humans to quantify our degree of certainty on the basis of incomplete information. In other words, probability is a subjective concept. People will make different assessments of probability, based on the different data they have available.

So if a coin flip is initially judged to have a probability of $\frac{1}{2}$ of resulting in a head, this is because we know little about it. More data about the weighting of the coin, its initial position, and the technique of the flipper would allow us to modify our probability. If we knew these things in great detail, we would be able to predict the outcome with some certainty. (The mathematician John Conway is reputed to have mastered the art of flipping coins to order.)

Bayesian inference

There is a consequence to the Bayesian view. To a Bayesian, all probability is conditional. Suppose you estimate the probability of A happening as $P(A)$. (This is really $P(A|C)$ where C represents your current knowledge, but we suppress this.) This is your *prior probability*. When some new data (B) comes to light, you need to *update* this assessment. This means using conditional probability to calculate $P(A|B)$, called your *posterior probability*.

As the fallacy of probability inversion shows, Bayesian inference can throw up counterintuitive results. Bayesian thinkers deploy this technique to improve probability assessment in a broad range of subjects, from economics to artificial intelligence.

Aumann's agreement theorem

There are three parts to Bayesian inference: a prior probability distribution, some new data, and a posterior probability distribution produced from these.

In 1976, Robert Aumann considered two Bayesian reasoners with identical prior probabilities for some event X. The two people are then provided with different pieces of data. Of course, conditional probability is likely to produce two different posterior probabilities for X. Aumann's question was what would happen if they then share their posteriors (technically, elevate them to **common knowledge**), *without* sharing their private data.

The answer is mathematically immediate, but by no means obvious: after several rounds of sharing probabilities and recalculating, they will each, eventually, arrive at the same posterior probability for X. As Aumann put it, 'people with the same priors cannot agree to disagree'.

The Monty Hall problem

Monty Hall is the former presenter of the TV Quiz *Let's Make A Deal*, in which contestants had to choose between three doors, concealing different prizes. This scenario was the inspiration for the most infamous of all probability puzzles, the *Monty Hall problem*, concocted by Steve Selvin in 1975.

There are three doors, A, B and C. Behind one of them is a brand new sports car. Behind the other two are wooden spoons. The contestant chooses one door, let's say A. In doing so he has a $\frac{1}{3}$ probability of hitting the jackpot. Next, Monty Hall, who knows where the car is, says 'I'm not going to tell you what's behind door A, not yet. But I can reveal that behind door B is a wooden spoon. Now will you keep with door A, or swap to C?'

The natural assumption is that the odds are now 50/50 between A and C, and swapping makes no difference. In fact, this is incorrect: C now has a $\frac{2}{3}$ probability of concealing a car, but A just $\frac{1}{3}$. So the contestant should switch.

To switch or not to switch?

To say that the solution to the Monty Hall problem often comes as a surprise is an understatement. Readers of *Parade* magazine felt so passionately that when Marilyn Vos Savant discussed the problem in 1990 she was inundated with complaints, including some from several professional mathematicians, protesting her public display of innumeracy.

To see why she was right, it may help to increase the number of doors, say to 100. Suppose the contestant chooses door 54, with a 1% probability of finding the car. Monty then reveals that doors 1–53, 55–86, and 88–100 all contain wooden spoons. Should the contestant swap to 87, or stick with 54? The key point is that the probability that door 54 contains the car remains 1%, as Monty was careful not to reveal any information which affects this. The remaining 99%, insteading of being dispersed around all the other doors, becomes concentrated at door 87. So she should certainly swap.

The Monty Hall problem hinges on a subtlety. It is critical that Monty knows where the car is. If he doesn't, and opens one of the other doors at random (risking revealing the car but in fact finding a wooden spoon), then the probability has indeed shifted to $\frac{1}{2}$. But in the original problem, he opens whichever of the two remaining doors he knows to contain a wooden spoon, and the contestant's initial probability of $\frac{1}{3}$ is unaffected.

Count Buffon's needle

If you drop a needle at random onto a piece of lined paper, what is the probability that it will land crossing one of the lines? That was the question investigated by Georges Leclerc (le Comte de Buffon) in 1777, who experimented by chucking sticks over his shoulder onto his tiled floor.

The answer depends on the length (l) of the needle and the distance (d) between the lines. If $l \leq d$ then the answer turns out to be $\frac{2l}{\pi d}$. (The case $l > d$ is slightly more complex.) So if the needle is 1 cm long, and the lines are 2 cm apart, then the answer comes out very neatly as $1/\pi$, which provides what is known as a *Monte Carlo* method for calculating π: perform the experiment as many times as you like, and then divide the total number of drops by the number of times the needle lands on a line, to obtain an approximation to π.

In 1777, Comte de Buffon also attempted to apply conditional probability to the study of philosophy, by trying to quantify the likelihood that the sun will rise tomorrow, given that it has risen for the last n days.

The law of truly large numbers

The following useful principle was first named by Persi Dioaconis and Frederick Mosteller in 1989, although the phenomenon had been known for a long time:

> *With a large enough sample, any outrageous thing is likely to happen.*

Lotteries usually do not have good odds; if I buy a ticket, I have a very low chance (perhaps one in ten million) of winning. If I do win, I will be astonished that such an unlikely event has happened. But viewed from the lottery's perspective, if several million tickets are sold, there is a good chance that *someone* will win.

A still more outrageous example is the case of the US woman who won the New Jersey lottery twice. From the winner's perspective this is a truly incredible occurence. Dioaconis and Mosteller framed a broader question: 'What is the chance that some person, out of all the millions and millions who buy lottery tickets in the United States, hits a lottery twice in a lifetime?' Following Stephen Samuels and George McCabe, they report the answer: 'practically a sure thing'.

Coincidence
The law of truly large numbers is a useful tool for understanding coincidence, when highly improbable events take place. Suppose we call an event *rare* if it has a probability of less than one in a million. In 1953 Littlewood observed that, in the USA's population of 250 million people, hundreds of people should experience rare events every day. Scaling this up to the entire world, even supremely unlikely occurrences of one in a billion should be expected to occur daily.

Other sources of apparent coincidence derive from our poor intuition about probability. The surprising solutions to the birthday problem and the phenomenon of false positives show how unreliable this can be.

PROBABILITY DISTRIBUTIONS

Random variables
Elementary probability works by counting the number of successes and outcomes. This is a valuable method, but in more sophisticated scenarios it will not suffice. If we replace the fair die with a biased one, the probability of rolling a 6 will no longer be $\frac{1}{6}$.

To refine the picture, we need a little terminology. The *sample space* is the collection of all possible outcomes of an experiment. In the case of a roll of the die, this will be the numbers 1 to 6. A *random variable* is an assignment of probabilities to these outcomes. (Technically it is a **function**, from the sample space to the numbers between 0 and 1.)

There are all sorts of possible functions. The random variable which assigns number 6 probability 1, and the numbers 1 to 5 all probability 0, corresponds to a die which is certain to show a 6. The random variable which assigns each number a probability of $\frac{1}{6}$ corresponds to a fair die. These random variables have different *probability distributions*.

Probability distributions
There are a large variety of probability distributions, but they come in two fundamentally different forms. In a *discrete* distribution, the outcomes are separated from each other, as in the example of a die, which

can take the value 4 or 5, but not $4\frac{1}{2}$. In a *continuous* distribution this is not the case (for example, a person's height may take any value between 4 and 5 feet).

Given a probability distribution X, two important pieces of data are where it is centred, and how widely spread it is. These are measured by two numbers: the *mean $E(X)$* and *variance $V(X)$*, respectively. The E in $E(X)$ stands for the *expectation* or *expected value* of the distribution, an alternative (and slightly misleading) name for the mean.

In experiments, the **mean** and **sample variance** of a set of data should correspond to the theoretical mean and variance of the underlying distribution. This will happen ever more closely as larger sets of data are used, a consequence of the **law of large numbers**.

Expectation and variance

Given a *discrete random variable X*, the *mean* is defined as the sum of all the possible outcomes multiplied by their respective probabilitites. If X represents the distribution of the roll of a fair die, the possible outcomes are the numbers 1 to 6, each with probability $\frac{1}{6}$. So the mean is $\frac{1}{6} \times 1 + \frac{1}{6} \times 2 + \frac{1}{6} \times 3 + \frac{1}{6} \times 4 + \frac{1}{6} \times 5 + \frac{1}{6} \times 6 = 3\frac{1}{2}$.

The formal definition of the mean (*expectation*) is $E(X) = \Sigma x \times P(X = x)$, where x ranges over all possible outcomes, and $P(X = x)$ is the corresponding probability.

For *continuous* random variables, this becomes $E(X) = \int x f(x) dx$ where f is the probability density function (see **continuous probability distributions**).

Suppose that $E(X) = \mu$. Then the *variance $V(X)$*, measuring the spread of the probability distribution, is defined as $V(X) = E(X - \mu)^2$. An easier way to calculate it is $V(X) = E(X^2) - E(X)^2$. For a die roll:

$$E(X^2) = \frac{1}{6} \times 1^2 + \frac{1}{6} \times 2^2 + \frac{1}{6} \times 3^2 + \frac{1}{6} \times 4^2 + \frac{1}{6} \times 5^2 + \frac{1}{6} \times 6^2 = 15\frac{1}{6}$$

So:

$$V(X) = 15\frac{1}{6} - \left(3\frac{1}{2}\right)^2 = 2\frac{11}{12}$$

The spread is also measured by the *standard deviation σ*, which is defined by $\sigma = \sqrt{V(X)}$. For a die roll this is $\sqrt{2\frac{11}{12}}$, around 1.7.

Bernoulli distribution

One of the simplest probability distributions the *Bernoulli distribution*, can describe the tossing of a biased coin, or other experiments with two possible outcomes with unequal probabilities. An experiment, called a *Bernoulli trial*, has two outcomes: success, with probability p, and failure, with probability $1 - p$. It is common to identify success with the number 1, and failure with 0. Then the *Bernoulli distribution* assigns 1 probability p, and 0 probability $1 - p$. (Every other number is assigned probability 0.)

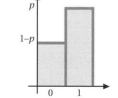

The Bernoulli distribution may not seem much on its own, but by combining Bernoulli trials together, more sophisticated distributions can be built. An important example is the **binomial distribution**.

The expectation of the Bernoulli distribution is p, and its variance is $p(1 - p)$.

Binomial trials

Suppose we roll 100 fair dice. What is the probability of getting exactly 17 sixes? Since the dice are fair, this problem can be solved by counting successes and outcomes, as follows. If we specify 17 dice, then the probability that those dice show sixes is $\left(\frac{1}{6}\right)^{17}$ (since the 17 rolls are independent). We also require that the other 83 dice do not show sixes: this probability is $\left(\frac{5}{6}\right)^{83}$. So the probability that our 17 named dice, and only these, roll sixes is $\left(\frac{1}{6}\right)^{17} \times \left(\frac{5}{6}\right)^{83}$.

For each possible choice of 17 dice, we get this probability. To answer the original question, therefore, we need to multiply this probability by the number of possible choices of 17 dice from the 100 (since the resulting events are all mutually exclusive). The number of choices of 17 from 100 is given by the **combination** $\binom{100}{17}$. So the answer to the question is $\binom{100}{17} \times \left(\frac{1}{6}\right)^{17} \times \left(\frac{5}{6}\right)^{83}$ (which comes out at around 0.1).

This is an example of a *binomial trial*, and is described by a binomial distribution.

Binomial distribution

A *binomial distribution* is specified by two numbers: n, the number of identical Bernoulli trials performed, and p, the probability that each such one ends in success. In the above example, $n = 100$ and $p = \frac{1}{6}$, and we write $X \sim B\left(100, \frac{1}{6}\right)$ to mean that X is a random variable with this binomial distribution.

More generally, if $X \sim B(n, p)$, then X measures the total number of successes from n independent Bernoulli trials with probability p. Then, for each number k between 0 and n, the probability that $X = k$ is:

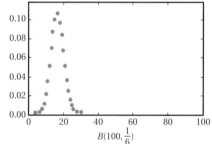

$B(100, \frac{1}{6})$

$$P(X = k) = \binom{n}{k} p^k (1 - p)^{n-k}$$

The expectation of X is $E(X) = np$, and the variance is $V(X) = np(1 - p)$.

Poisson processes

Between 9a.m. and 5p.m., an office telephone receives on average three calls per hour. In any particular hour it might receive no calls, or 1, 2, 3, 4, 5, 6, … calls. There is no obvious cutting-off point, but higher numbers of calls become progressively unlikely. If we assume that the calls come randomly, with no difference across the day, or between days, then this is an example of a *Poisson process*.

In general, a Poisson process models the number of occurrences of a random phenomenon, within a given time period or region of space.

Poisson distribution

We would like a random variable which assigns probabilities to all the possible outcomes 0, 1, 2, 3, 4, … of a Poisson process. The distribution commonly used is the *Poisson distribution*, discovered in 1838 by Siméon-Denis Poisson, via the law of rare events.

One number is needed to specify a Poisson distribution, usually called the *intensity*, λ. In the above example, $\lambda = 3$. We write $X \sim \text{Po}(3)$ to express that X is a random variable with this distribution.

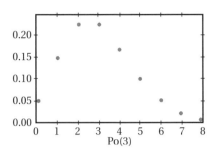

Then the probability that the phoneline receives 0 calls in a particular hour is $P(X = 0) = e^{-3}$. The probability that it receives one is $P(X = 1) = 3e^{-3}$, and for two calls, it is $P(X = 2) = \frac{3^2 e^{-3}}{2}$. The probability that it receives k calls is $P(X = k) = \frac{3^k e^{-3}}{k!}$.

More generally, if $X \sim \text{Po}(\lambda)$, then:

$$P(X = k) = \frac{\lambda^k e^{-\lambda}}{k!}$$

The expectation and the variance of a Poisson distribution with parameter λ are given by $E(X) = V(X) = \lambda$.

The law of rare events

Suppose a factory produces peanut-flavoured ice-cream. The average number of peanuts in each tub is 18. A natural way to model this situation would be using the Poisson distribution, $X \sim \text{Po}(18)$.

There is another approach, however. Suppose there are 36,000 peanuts in the ice-cream vat, and the probability that a particular peanut will end up in a particular tub is 0.0005. Then we could model the scenario as a binomial distribution, $X \sim B(36,000, 0.0005)$.

The first model is certainly more convenient, but the second may be more accurate. We need not worry too much though. The law of rare events guarantees that these two models will produce very similar answers. Technically, the *law of rare events* says that, if n grows large and p becomes small, and they do this in such a way that the average number of successes $np = \lambda$ remains constant, then the distribution $B(n, p)$ becomes increasingly close to a Poisson distribution, $\text{Po}(\lambda)$. Indeed, this is how Siméon–Denis Poisson discovered his distribution.

Continuous probability distributions

If I pick someone at random from my city, and measure their height, then distributions such as the binomial and Poisson distributions cannot apply. The problem is that the list of possible heights do not form a *discrete* list like the whole numbers: 0, 1, 2, 3, … but a *continuous* range. Continuous probability distributions apply in situations such as this. Each comes with a curve, called the *probability density function*, which represents the distribution.

In a discrete distribution, to calculate the probability that X lies in a certain range, we add up all the corresponding probabilities. In the continuous case we **integrate** the curve over that range. So the probability that X lies between 4 and 6 is represented by the area beneath the curve, in this range.

For a discrete distribution, we demand that all the probabilities add up to 1. In the continuous case, we require that the area under the whole curve should be 1.

The simplest continuous distribution is the uniform one. Occupying pride of place at the heart of modern probability theory is the normal distribution.

Uniform distribution

Imagine a spinning top whose circumference is 4 cm long and marked from 0 to 4. We give it a spin, and record the exact point on the circumference which comes to rest on the ground. This could be any number between 0 and 4 (not necessarily a whole number). Assuming the top is fair, this experiment can be modelled by a random variable X with a *uniform distribution*.

A uniform distribution is determined by its two endpoints, in this case 0 and 4. Between the two, all intervals of the same length are equally likely. So the probability of the cylinder landing between 1.1 and 1.3 is the same as between 3.7 and 3.9, for example.

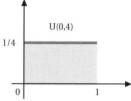

In general, if $X \sim U(a, b)$, the probability density function has a constant value of $\frac{1}{b-a}$ between a and b, and 0 elsewhere. In the above example, this gives the graph a height of $\frac{1}{4}$. The archetypal example is the *standard* uniform distribution, $U(0, 1)$. The general uniform distribution has expectation $\frac{a+b}{2}$ and variance $\frac{1}{12}(b-a)^2$.

Normal distribution

Outside mathematics, the normal distribution is commonly known as the *bell curve*. It was first introduced in Abraham De Moivre's 1756 work *The Doctrine of Chance*, on the back of Leonhard Euler's work on the exponential function. Two numbers are required to specify a normal distribution: its mean μ, which pinpoints its centre, and its variance σ^2, which determines its spread. A broad range of phenomena can be modelled by random variables X with $X \sim N(\mu, \sigma^2)$, for suitable choices of μ and σ^2. The height of adult women, test scores and the orbiting speeds of planets are all examples.

The fundamental shape of the bell curve is given by the equation $y = e^{-x^2}$. However, this needs to be rescaled to ensure the area underneath the curve is 1, to move the centre to μ, and stretch it according to the value of σ^2. Putting these together, we get the probability density function, $y = \frac{1}{\sigma\sqrt{2\pi}} e^{-\frac{(x-\mu)^2}{2\sigma^2}}$.

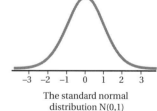

The standard normal distribution N(0,1)

In the case of the *standard* normal distribution, $N(0, 1)$, this is simplified slightly: $y = \frac{1}{\sqrt{2\pi}} e^{-\frac{x^2}{2}}$.

This standard version is the only one we really need, since every normal distribution can be *standardized*. If $X \sim N(\mu, \sigma^2)$, it is standardized by defining $Y = \frac{X-\mu}{\sigma}$. Then $Y \sim N(0, 1)$.

The normal distribution is the mother of all probability distributions, in a precise sense given by the **central limit theorem**.

Independent, identical random variables

Many situations in probability theory involve a sequence of independent, identical random variables $X_1, X_2, X_3, X_4, \ldots$ A typical example would be when a single experiment is performed repeatedly, in such a way that each round is identical to the previous rounds, but its outcome is independent from them. An example is the repeated rolling of a die. (Picking a card from a pack would only qualify if the card is replaced and the pack shuffled each time.)

A common thing to do is to take the sample mean of the first n rounds of the experiment. We do this by defining a new random variable

$$Y_n = \frac{X_1 + X_2 + \cdots + X_n}{n}$$

where, Y_n represents the mean of the outcomes of the first n experiments (such as the average score of the first n rolls of the die). **The law of large numbers** and **the central limit theorem** both describe this random variable.

Law of large numbers

Roll a fair die 10, 100 or 1000 times, and take the mean score in each case. What would you expect to find? The *law of large numbers* predicts that, as the sample gets ever larger, we should expect the mean of the sample result to get ever closer to the theoretical mean of 3.5.

Although informal versions of this had been known for many years, Jacob Bernoulli was the first person to frame it as a rigorous theorem, in 1713. The theorem refers to a sequence $X_1, X_2, X_3, X_4, \ldots$ of independent, identical random variables, each with mean μ. Then we define new random variables:

$$Y_n = \frac{X_1 + X_2 + \cdots + X_n}{n}$$

The law of large numbers asserts that, as n grows large, the random variable Y_n gets ever closer to the fixed number μ.

This law is refined by the central limit theorem. However, this law has broader applicability, since the central limit theorem requires an additional assumption about the variance of the random variables X_i.

Central limit theorem

In 1733, Abraham de Moivre used a normal distribution to model the total number of heads in a long sequence of coin flips. Something seems wrong with this: coin flips are discrete, not continuous. The binomial distribution, and not the normal, is the appropriate model.

However, de Moivre had not made an elementary error. Rather this was the first inkling of a fundamental result in probability theory: the central limit theorem.

Suppose an experiment is modelled by a random variable X with mean μ and variance σ^2. It does not matter what the distribution of X is, aside from these facts. It could be uniform, Poisson, or some as yet undiscovered distribution. The *central limit theorem* says that, if this experiment is repeated many times, the average result is approximately given by a normal distribution.

More precisely, repetitions of the experiment correspond to a sequence of independent, identical random variables $X_1, X_2, X_3, X_4, \ldots$ Then we take the sample averages:

$$Y_n = \frac{X_1 + X_2 + \cdots + X_n}{n}$$

The central limit theorem tells us that, as n gets bigger and bigger, the distribution of Y_n is approximately $N\left(\mu, \frac{\sigma^2}{n}\right)$.

A more precise statement comes from standardizing Y_n, by defining $Z_n = \frac{Y_n - \mu}{\sigma/\sqrt{n}}$. Now, as n gets larger, the random variables Z_n get ever closer to the standard normal distribution $N(0, 1)$.

Gambler's fallacy

The *law of averages* may be well known outside of mathematics, but you will find no theorem of this name in any book on probability theory. Appeals to this law, when they are valid, are likely to refer to the law of large numbers. When they are erroneous, they are usually instances of the *gambler's fallacy*. Suppose a gambler has seen black win at roulette six times in a row. He might consider that red is now 'due', and therefore more likely to occur next spin.

If successive experiments (spins of the coin or wheel) are taken to be independent, which they certainly should be, the gambler's fallacy is false by definition. It is often made as a botched invocation of the law of large numbers. The error is that this law makes probabilistic predictions about average behaviour, over the long term. It makes no predictions about the results of individual experiments.

Of course, a long run of identical outcomes may still be worth considering as evidence that the experiments are not as they seem (either the probability is not as claimed, or the events are not really independent).

STOCHASTIC PROCESSES

Stochastic processes

Start in the middle of a road, and flip a fair coin. If it comes up heads walk 1 metre north; if it comes up tails, walk 1 metre south. Then flip again. Where will you end up after 10 or 100 flips? This experiment is an example of a *random walk*.

For a 2-dimensional random walk, consider Manhattan, where the streets are laid out as a square grid. At each crossroads you take an equal probability of walking north, south, east or west for one block. (We model the city as an infinite grid, ignoring the possibility that you might reach the edge.)

Similarly we can define a random walk on a 3-dimensional lattice, or indeed on any infinite **graph**. (For finite graphs the problem is trivial.)

Random walks are the simplest examples of *stochastic processes*, processes which develop over time according to probabilistic rules, rather than along predetermined lines. More sophisticated examples include Markov chains and Brownian motion.

Pólya's random walks

The term *random walk* was coined by George Pólya, who in 1921 analysed the 1- and 2-dimensional cases described above. His question was as follows: pick a point on the graph at the beginning. Now, what is the probability that the random walker will reach it eventually? A simpler question, which amounts to the same thing, is: what is the probability that the walker will eventually return to his starting point?

Pólya showed that the answer is 1, in both cases, making it a virtual certainty. The 1-dimensional case is sometimes known as *gambler's ruin*. A gambler playing a fair, random game against a casino one chip at a time, has a probability of 1 of losing all his chips eventually. This may seem unsurprising. But Pólya showed that this fails in higher dimensions. A random walk on a 3-dimensional lattice has a lower probability of returning to its starting point, subsequently pinned down to around 0.34. Rather than covering the whole lattice, higher-dimensional random walks exhibit a striking fractal appearance as they grow.

Markov chains

At each stage of a Manhattan random walk, you flip a coin to decide in which direction to go next. In probability theory, these coin flips are modelled by a random variable of a simple kind. A *Markov chain* is a sequence of random variables, like a random walk. The difference is that these random variables may be of more sophisticated types, such as a random walk on a grid which contains random teleporters or other booby traps. Conceived by the 19th-century probabilist Andrey Markov, the defining characteristic of a Markov chain is that the probability distribution at each stage depends only on the present, and not on the past. (In a random walk, all that matters is where you are now, not how you got there.)

Markov chains are an excellent framework for modelling many phenomena, including population dynamics and stock-market fluctuations. To determine the eventual behaviour of a Markov process is a deep problem, as Pólya's 3-dimensional random walk illustrates.

Kinetic theory of heat

Robert Brown was a botanist, and a pioneer of microscopy within the biological sciences. In 1827 he turned his microscope on a primrose pollen seed suspended in water. Floating inside were tiny specks of matter, darting around in a very haphazard manner. This subsequently became known as *Brownian motion*. Brown initially thought the particles were tiny living organisms, but further investigations revealed that the same characteristic irregular movement occurred in finely powdered rock left in water.

In 1905, Albert Einstein realized that the particles were being bumped around by water molecules, too tiny to see. Significantly, the hotter the water was, the faster the visible particles moved. Einstein recognized this as powerful indirect evidence for the molecular theory of heat. As we now know, heat energy in matter is nothing more than the combined **kinetic energy** of its constituent molecules.

Brownian motion

To flesh out the details of Einstein's work on the kinetic theory of heat, a mathematical model of a particle's Brownian motion was needed. Since each change of course of the particle is random, and independent of its previous motion, it resembles a stochastic process such as a Markov chain. However, in random walks and Markov chains, time comes in *discrete* steps. Each step in the process takes place after a fixed period of time. In Brownian motion, the particle is constantly changing direction. The path looks like a random walk, zoomed out so that the individual legs of the journey shrink to zero.

A random walk is modelled by a sequence of random variables: $X_1, X_2, X_3, X_4, \ldots$, that is, a family (X_i) where i varies over the natural numbers. In contrast, Brownian motion can be modelled by a *continuous* family of random variables, that is, (X_i) where i varies over the real numbers.

What happens as this system develops? Einstein showed that after any length of time, the position of the particle is modelled by a 3-dimensional normal distribution (its position in each dimension is normally distributed, and the three are independent).

CRYPTOGRAPHY

Monoalphabetic encryption

One of the simplest ways to encipher a message is to rearrange the alphabet before writing it. For example, here is an encryption system based on the ordering of letters on a computer keyboard:

a	b	c	d	e	f	g	h	i	j	k	l	m	n	o	p	q	r	s	t	u	v	w	x	y	z
Q	W	E	R	T	Y	U	I	O	P	A	S	D	F	G	H	J	K	L	Z	X	C	V	B	N	M

We can use this key to encrypt our message. Before enciphering, the message is known as *plaintext* (and will be written in lower case letters). Suppose our plaintext reads 'meet me in the park at three a.m.'. We produce the *ciphertext* (which will be written in capital letters) by substituting the letters according to the table above: 'DTTZ DT OF ZIT HQKA QZ ZIKTT Q.D.'.

Once we have sent this to our contact, she can decipher it using the same key. Of course, there is no reason to be limited to the letters of alphabet. Any 26 symbols will do equally well. To make a sentence less guessable, we might also want to omit the punctuation marks and spaces.

a	b	c	d	e	f	g	h	i	j	k	l	m	n	o	p	q	r	s	t	u	v	w	x	y	z
$\sqrt{}$	$=$	$\cap$	$\div$	$\aleph$	$\exists$	Σ	$\neq$	$\rightarrow$	$\int$	$\in$	∞	x	$\cup$	$\pm$	e	$\frac{dy}{dx}$	Π	l	π	i	$\forall$	θ	$\subseteq$	$\varnothing$	$\times$

$$`\rightarrow \neq \sqrt{\phantom{}} \forall \aleph \pi \neq \aleph x \sqrt{\phantom{}} \pi \aleph \Pi \rightarrow \sqrt{\phantom{}} \infty l \varnothing \pm i \Pi \aleph \frac{dy}{dx} i \rightarrow \Pi \aleph`$$

Cryptanalysis

Imagine that you intercept an encrypted message intended for your enemy:

> WKRKRPUEBRXEUGRJURJFBGDRFRBKGFRGBGURPBJRXFOKGRPURZUGRXAKIJRPK
> AOFXOFVUDIJUXKIFJUGKRAKQQKZULXKIJEKGRFERBDFQUSFDPUZBQQPFTUFZPB
> RUEFJGFRBKGBGPUJPFGLXKIJEKLUZKJLBDDBDXYPIDDPUZBQQWBTUXKIFVUXR
> PUOFRUJBFQDFJUBGMKSGIOMUJDBSUBWPRABTULUDRJKXRPBDOUDDFWUFAR
> UJXKIPFTUOUOKJBDULBRLKGKREKGRFEROUFWFBGUGLKAOUDDFWU

How might you try to crack it? This is a question of *cryptanalysis*.

Let us assume that the sender used monoalphabetic encryption, the fundamental technique discovered by the ninth-century scientist Abū al-Kindī. The basis of *frequency analysis* is the observation that the letters of the alphabet are not equally common. The first step is to analyse the most common letters which occur in the ciphertext:

U	R	F	K	B	G	D	J	P
36	29	26	25	23	19	18	17	15

The basic idea is to try to replace these with the most frequent letters which arise in English, which are (in order) ETAOINSHRDLU.

Frequency analysis

In the above example, we might start by replacing the commonest two letters in the message, U and R, with the commonest two in English, *e* and *t*. This takes us to:

> WKtKtPeEBtXEeGtJetJFBGDtFtBKGFtGBGetPBJtXFOKGtPetZeGtXAKIJt
> PKAOFXOFVeDIJeXKIFJeGKtAKQQKZeLXKIJEKGtFEtBDFQeSFDPeZBQQ
> PFTeFZPBteEFJGFtBKGBGPeJPFGLXKIJEKLeZKJLBDDBDXYPIDDPeZB
> QQWBTeXKIFVeXtPeOFteJBFQDFJeBGMKSGIOMeJDBSeBWPtABTeLeLDtJ
> KXtPBDOeDDFWeFAteJXKIPFTeOeOKJBDeLBtLKGKtEKGtFEtOeFWFB
> GeGLKAOeDDFWe

Now we can open up a second line of attack. At several places, the plaintext letter t is followed by the encrypted letter P. From our knowledge of English, it seems likely that P represents h. This gives us:

WKtKtheEBtXEeGtJetJFBGDtFtBKGFtGBGethBJtXFOKGthetZeGtXAKIJt
hKAOFXOFVeDIJeXKIFJeGKtAKQQKZeLXKIJEKGtFEtBDFQeSFDheZBQ
QhFTeFZhBteEFJGFtBKGBGheJhFGLXKIJEKLeZKJLBDDBDXYhIDDheZ
BQQWBTeXKIFVeXtheOFteJBFQDFJeBGMKSGIOMeJDBSeBWhtABTeLe
DtJKXthBDOeDDFWeFAteJXKIhFTeOeOKJBDeLBtLKGKtEKGtFEtOeFW
FBGeGLKAOeDDFWe

Through a combination of analysis of the frequencies of different letters and combinations of letters, intelligent guesswork, and trial and error, further progress should be possible.

It is appropriate to include a warning here: frequency analysis is not an exact science and works better with longer sections of text. This short example is somewhat artificial but is intended to illustrate the basic techniques.

ETAOIN SHRDLU
Frequency analysis relies on knowing the relative frequencies of different letters in the English language. Of course these are averages, and will not hold exactly in any particular text.

Letter	Average number of occurrences per 100 characters	Letter	Average number of occurrences per 100 characters
e	12.7	m	2.4
t	9.1	w	2.4
a	8.2	f	2.2
o	7.5	g	2.0
i	7.0	y	2.0
n	6.7	p	1.9
s	6.3	b	1.5
h	6.1	v	1.0
r	6.0	k	0.8
d	4.3	j	0.2
l	4.0	x	0.2
u	2.8	q	0.1
c	2.8	z	0.1

The phrase ETAOIN SHRDLU lists the first 12 letters in order of frequency. It was well known in the age of linotype printing presses, where the letters on the keyboard were arranged in approximate order of frequency. Sometimes the phrase would accidentally appear in newspapers.

Frequency analysis does not just use individual letters. Some combinations of letters (such as 'th') are more common than others ('qz').

Digraph	Average number of occurrences per 2000 characters	Digraph	Average number of occurrences per 2000 characters	Digraph	Average number of occurrences per 2000 characters
th	50	at	25	st	8.5
er	40	en	25	io	18
on	39	es	25	lc	18
an	38	of	25	is	17
re	36	or	25	ou	17
he	33	nt	24	ar	16
in	31	ea	22	as	16
ed	30	ti	22	de	16
nd	30	to	22	rt	16
ha	26	it	20	ve	16

Polyalphabetic encryption, codes and spelling mistakes

There are several techniques to make monoalphabetic encryption more difficult to crack. One is to use *polyalphabetic encryption*, where each letter is enciphered in more than one way. So, the key might use a 52-letter alphabet, with each letter of plaintext enciphered by a choice of two symbols. For extra complexity, we could also introduce *dummy symbols* which have no meaning, and will be deleted by our contact, but may confuse any interceptor:

a	b	c	d	e	f	g	h	i	j	k	l	m	n	o	p	q	r	s	t	u	v	w	x	y	z	dummy
S	L	6	M	D	R	{	E	Q	W	\	A	@	B	K	7	J	3	T	C	O	?	G	4	P	H	X
Z	!	(	5	%	*	1	)	−	9	+	F	£	$	2	N	Y	~	I	;	8	^	U	#	,	V	}

Another age-old technique is to introduce higher levels of encryption. So far we have been encrypting messages by using different symbols to stand for the *letters* of our message. This is a *cipher*. A *code* is when *words* are similarly disguised. So for example,

bank	dollar	car	policeman
OSTRICH	MARBLE	DRUM	GRAVY

A third method is to introduce deliberate spelling mistakes into the message, to further throw out frequency analysis. All of these produce messages which are much harder to crack than plain monoalphabetic encryption:

'{DXC}C;EX28HXS$}B5@ZX3~!ATRX~K£X}@KXI;3,(E{}P?CX2K1~S^,P$5X38}£2O;V,}M%'

One-time pad
The disadvantage of monoalphabetic encryption is that the same letter is encrypted in the same way each time, leaving it open to frequency analysis. This problem is ameliorated through polyalphabetic encryption, spelling mistakes and other devices. But, ultimately, an expert cryptanalyst may overcome these hurdles, particularly when armed with a computer to test out different possibilities.

The *one-time pad* is an alternative method which works with a string of letters as its key. Suppose in this case that the key begins 'mathematical'.

First the letters of the plaintext and key are each converted into numbers, according to their alphabetical positions.

Plaintext:

a	b	o	r	t	m	i	s	s	i	o	n
1	2	15	18	20	13	9	19	19	9	15	14

Key:

m	a	t	h	e	m	a	t	i	c	a	l
13	1	20	8	5	13	1	20	9	3	1	12

The ciphertext is then obtained by adding together the two numbers in corresponding positions, and converting them back to letters. Where the result exceeds 26, 26 is first subtracted. That is to say, addition is performed modulo 26 (see **modular arithmetic**).

Ciphertext:

14	3	9	26	25	26	10	13	2	12	16	26
N	C	I	Z	Y	Z	J	M	B	L	P	Z

Our contact can decipher the message by reversing this process (as long as she has the key). This form of encryption is equivalent to enciphering successive letters according to different monoalphabetic ciphers. Its high level of security, and its name, derives from the fact each key is used for encrypting just one message, and is then discarded. So agents would have matching pads, with a key on each page, and turn a new page with each message.

In principle, the one-time pad is absolutely unbreakable, as Claude Shannon proved in 1949. There are many possible plaintext messages of the same length, and without the key there is no way for an enemy cryptanalyst to decide between them. (User error may always present a chink in the armour, of course. In the above example, the key is very guessable.)

Public keys

Although the one-time pad is theoretically unbreakable, it nevertheless has an Achilles heel: it is extremely expensive in keys. The sender and receiver need a new key for every message. Exchanging these keys inevitably involves risk. In the one-time pad and other traditional ciphers, encryption and decryption are symmetric procedures. In particular, the sender and receiver require access to the same key.

In public key cryptography, this symmetry is broken. Now, the key comes in two parts: a private part which is kept by the owner and never shared, and a public part, freely accessible to all.

Anyone can encrypt a message using the public key, and send it to the owner. Only the owner has the power to decipher it, however, as this requires the private key.

As an analogy, imagine that a business makes available an unlimited number of identical unlocked padlocks. There is only one key, which they keep. Anyone wishing to send them an item can place it in a box, and lock it with one of their locks. Only the business can then unlock the box.

Public key cryptography is the backbone of modern internet security. The key consists of two large prime numbers, say p and q. These are kept private, while their product $p \times q$ is made public. The security of this system relies on the inherent difficulty of reversing this process, known as the **integer factorization problem**.

Shannon's information theory

Claude Shannon's 1948 paper 'A Mathematical Theory of Communication' was a classic of the post-war era, and marked the birth of the subject of *information theory*. This paper has been of incalculable significance since the second half of the 20th century. In it, Shannon considered the process of encoding, transmitting and deciphering information, but his contribution was not a new method of encryption.

To begin with, he pioneered the use of **binary** as the natural language of information. Then, for the first time he analysed the theoretical basis of information transmission. He probed its limits, analysing the maximum rates at which a system can transmit data. The answer depends on the source of information, and in particular on a quantity called its *entropy*. This precisely quantifies the unpredictability of successive bits (binary digits) in a stream of binary code, by modelling it as a **Markov process**. Shannon's entropy was later found to be equivalent to Kolmogorov complexity.

The two cases that Shannon considered are *noiseless* systems and *noisy* systems (which errors can creep into). In the latter case, he considered the theoretical limits of the powers of error-correcting codes.

Kolmogorov complexity

In modern technology, information is encoded into binary sequences. (ASCII is a code for translating letters and various other symbols into binary, for example.) Some sequences are more complicated than others. The sequence 111111… is easily described, containing very little information. It would be a waste of disc space to store a string of a million 1s. To save space, archiving software can dramatically *compress* this sequence, repackaging the information as instructions for writing out one million 1s.

In the 1960s, Ray Solomonoff and Andrey Kolmogorov used this idea as a way to quantify the information content of a string of bits. The *Kolmogorov complexity* of a string is the minimum length to which it can be compressed. Strings which carry a lot of information are incompressible and thus have high complexity. Strings with little information, such as the million 1s, can be hugely compressed and therefore have low complexity. Kolmogorov complexity is essentially equivalent to Shannon's notion of the *entropy* of a string of bits.

Error-correcting codes

When sending information along a noisy channel, errors can creep in. *Error-correcting codes* are ciphers which allow the message to survive some level of corruption.

The simplest method is plain repetition: instead of sending COME NOW, we send CCCOOOMMMEEE   NNNOOOWWW. If one digit is corrupted, the blocks of three come to our aid: III   CCCAAANNQNNNOOOTTT.

If there is more than one error, we may be in trouble again. We could use longer repeating blocks, say 100 of each letter, but this will start to slow the process down. Shannon showed, remarkably, that this trade-off between accuracy and speed is not inevitable. With some mathematical technology, it is possible to find codes which are both quick and as accurate as required. Even if 99% of the message is corrupted, it may still be possible for it to be reliably deciphered.

One method is based on **Latin squares**, which are natural error correctors. If an entry in a Latin square is corrupted it is easily identifiable and correctable, by checking all the rows and columns. More sophisticated methods employ the algebraic structures of **finite fields**.

MATHEMA
PHYSICS

*Philosophy is written in this grand book, the universe, which
stands continually open to our gaze. But the book cannot be
understood unless one first learns to comprehend the language
and interpret the alphabet in which it is composed. It is written
in the language of mathematics.* GALILEO GALILEI

Galileo's words are as valid today as when he wrote
them, at the genesis of the first detailed physical theory
Newtonian mechanics. Even today, Newton's theory remains
adequate for many purposes, but cannot cope with the motion
of light on the astronomical scale. This was addressed by
Albert Einstein's theory of *special relativity*, bringing
many unexpected consequences, including the celebrated

WTON'S SECOND LAW • NEWTON'S THIRD
MENTUM • CONSERVATION OF MOMENTUM
DIES WITH CONSTANT VELOCITY • FALLING
TION • KINETIC ENERGY • CONSERVATION
WAVES • FREQUENCY • AMPLITUDE • AM
RUMENTS • HARMONICS • SCALAR AND
OR CALCULUS • GRAD • DIV • CURL • THE
AT EQUATION • SOLUTIONS TO THE HEAT
EULER'S FLUID FLOW
TIONS • THE NAVIER-
ATIONS • SPECIAL
NCE • GALILEAN RELATIVITY • SPACE IS
THE ECLIPSES OF IO • SPEED OF LIGHT
KOWSKI SPACETIME • THE HYPERBOLIC

TICAL

equivalence of mass and energy. Missing from this story, however, was an account of gravity. This Einstein tackled in his second theory: *general relativity*.

On the subatomic scale, too, Newtonian mechanics broke down and again light was the obstacle. An old question asked whether light consists of waves or particles. The eventual answer was very troubling: both. A completely new model of matter was built to describe this, *quantum mechanics*.

Since the early 20th century, the challenge has been to find a new model uniting general relativity and quantum mechanics. Although this dream remains unfulfilled, the approach of *quantum field theory* has brought huge advances.

NEWTONIAN MECHANICS

Newton's laws

Newtonian mechanics concerns the behaviour of objects which are subjected to *forces* pulling or pushing them around. A boy kicking a football is one example of a force, gravity causing the boy to fall down is another. Newton's second and third laws model these situations. The first law addresses something more basic: what happens to an object which is *not* subject to a force?

This question is not as easy as it sounds. Since the time of Aristotle, the belief had been that such an object would gradually slow down as its 'inertia' waned, until it became stationary. It was Galileo Galilei who first corrected this misconception. Galileo's principle became *Newton's first law*.

Newton's first law

A moving object will *not* slow down in the absence of any force. Imagine a stone sliding across a perfectly smooth frozen lake. There is no force acting on it, and it simply continues to slide, at a constant speed, in a fixed direction. (In reality, the stone will eventually slow down, of course, because ice is not perfectly smooth, and so there will be a small force acting, namely friction.)

Galileo's principle, also known as Newton's first law, says that an object will remain at rest, or travelling at constant speed in a fixed direction, unless disturbed by a force. The law can better be seen in the near vacuum of deep space. Here, a rock moving though space in a given direction will simply continue on its path indefinitely, until disturbed.

Newton's second law

When you push a box, it moves. But Newton's first law tells us that objects can move even without forces acting. So what difference does the force really make?

The answer is that, without any force acting, an object can travel in a straight line at a constant speed. A force will cause it to speed up, slow down or change direction. In other words, forces produce *acceleration*.

How much acceleration? This depends on another factor. A golf shot produces a significant acceleration in a golf ball (over a very short time). To produce the same acceleration in a house-brick would require a massively larger force. To produce a given acceleration in an object, the force required is proportional to the object's mass. The heavier the object, the greater the force needed. In equation form, we have

$$F = ma$$

where F is the force measured in *Newtons*, m is the object's mass (in kilograms) and a is the resulting acceleration (in metres per second per second, or m/s^2).

Newton's third law

In billiards or pool, the cue ball is used to move the other balls. When it hits a stationary ball, during the short time that they are in contact, the cue ball exerts a force on the target ball, causing it to accelerate according to Newton's second law. However, the cue ball does not simply continue on its former path. If the target ball moves left, then the cue ball will be deflected right. In what pool players call a 'stun shot', the cue ball hits the target head-on, and then decelerates to stationary.

According to Newton's first law, the cue ball must also have been subjected to a force. *Newton's third law of motion* says that these two forces are of equal magnitude, and in opposite directions. The general statement is when an object *A* exerts a force *F* on object *B*, then *B* exerts a force $-F$ on *A*.

Equal and opposite reactions

Newton's third law is often quoted, slightly misleadingly, as *'every action has an equal and opposite reaction'*. This is reasonable when we contemplate pool balls colliding, but seems to run counter to intuition at the level of human affairs. When we hit a nail with a hammer, we do not ordinarily think of the stationary nail as exerting a force on the moving hammer. Nevertheless the nail does causes the hammer to decelerate.

More complex situations can be confusing. If a man is pushing a car along a flat road, and the car applies an equal force to the man, why does it not push him backwards along the road? In such scenarios there are often more forces at work than are immediately apparent. The man is not only pushing the car forwards, he is also pushing back with his feet, and taking advantage of the equal and opposite force exerted on him by the ground. On sheet ice, he could not do this. Here, if he gave the car a push, he would indeed find himself sliding backwards.

Momentum

The *momentum* of an object is its mass multiplied by its velocity. Often the letter *p* is used for momentum (because *m* is already taken for mass) and *v* stands for velocity. So the defining formula is $p = mv$. If a soccer ball weighing 0.5 kg is travelling at 6 m/s, then it has a momentum of $0.5 \times 6 = 3$ kg m/s.

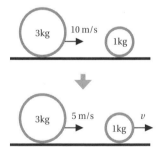

Momentum is important in Newtonian mechanics, because when two or more particles collide, the total momentum after the collision must equal the momentum before it. This property is known as *conservation of momentum*. Suppose two balls collide. The smaller has mass 1 kg and is initially stationary. The heavier has mass 3 kg and collides into the other at a velocity of 10 m/s. If the larger ball's velocity is decreased to 5 m/s, what is the velocity of the smaller ball after the collision?

The total momentum before the collision is given by $1 \times 0 + 3 \times 10 = 30$ kg m/s. If the velocity we want to find is *v*, then the total momentum after is $1 \times v + 3 \times 5 = v + 15$. Since momentum is conserved, it must be that the two are equal, and $v + 15 = 30$. So $v = 15$ m/s.

Conservation of momentum

A *closed system* is a collection of objects which interact with each other but are insulated from all external forces and objects. In such a system, the total momentum of the system remains fixed and unchanging.

This useful fact is a direct consequence of Newton's laws. The simplest case is where two particles, of fixed mass m_1 and m_2, inflict forces F_1 and F_2 on each other. By Newton's third law, it must be that $F_1 = -F_2$. If the two accelerations are a_1 and a_2, then, by Newton's second law, it follows that $m_1 a_1 = -m_2 a_2$, and so $m_1 a_1 + m_2 a_2 = 0$.

Suppose the velocities of the two particles are v_1 and v_2. To get from the acceleration to the velocity, we need to integrate (see **rates of change of position**). If we integrate the above equation, we get $m_1 v_1 + m_2 v_2 = C$, for some constant of integration C. This says that, even as F_1, F_2, a_1, a_2, v_1 and v_2 all vary, the total momentum $m_1 v_1 + m_2 v_2$ remains fixed at the value C.

Bodies with constant displacement

The simplest type of motion is where an object's displacement is constant, that is to say, it is stationary. Suppose that a lump of matter sits 5 metres away from the origin. We use t to represent the time since the stopwatch was started, and s to represent the object's **displacement**. Then, for any value of t, we have $s = 5$. There is nothing more to say, since $\frac{ds}{dt}$, $\frac{d^2s}{dt^2}$, and all higher derivatives are all equal to zero.

Bodies with constant velocity

Suppose a cyclist is travelling along a straight line with a constant velocity of $3 \, \text{m/s}$. If s represents her displacement from the origin, then her velocity is given by $\frac{ds}{dt}$. In this case $\frac{ds}{dt} = 3$. Integrating this, we get $s = 3t + C$ for some constant of integration C. What is the meaning of this number C? Setting $t = 0$ reveals the answer: it is the cyclist's initial displacement.

Usually we will orient ourselves so that the origin is the starting position. This amounts to the **boundary condition** that $s = 0$ when $t = 0$. Hence $C = 0$. Now we have the equation $s = 3t$. This will tell us the cyclist's displacement at any time. For example, after 60 seconds, her displacement is $180 \, \text{m}$.

More generally, if the constant velocity is given by $\frac{ds}{dt} = v$, then $s = vt$, or $v = \frac{s}{t}$, which happily matches the usual definition of velocity as distance divided by time.

Falling bodies

An important type of motion is that of an object moving under constant acceleration. A common example is of an object falling under gravity. The moral of **Galileo's cannonballs** is that, if we ignore the effects of air resistance, any two objects, even of widely differing masses, will fall to earth at the same rate. This constant acceleration is determined by earth's gravity, and known as g (approximately $9.8 \, \text{m/s}^2$).

So, if a body is falling without air resistance, then:

$$(1) \quad \frac{d^2s}{dt^2} = g$$

Integrating this, we get an expression for the object's velocity: $\frac{ds}{dt} = gt + C$. Assuming that the object is dropped from rest, when $t = 0$, $\frac{ds}{dt} = 0$, and so $C = 0$. So:

$$(2) \quad \frac{ds}{dt} = gt$$

Integrating this again, we get $s = \frac{g}{2} t^2 + D$. If we have organized matters so that the object was dropped from the origin, then $D = 0$, and we get:

$$(3) \quad s = \frac{g}{2} t^2$$

So, after 10 seconds, the object has a velocity of 98 m/s (by equation 2) and has dropped 490 metres (by equation 3). This is one example of a body with constant acceleration.

Bodies with constant acceleration

If an object has constant acceleration, say of a, then:

$$(1) \quad \frac{d^2s}{dt^2} = a$$

Integrating this we get $\frac{ds}{dt} = at + C$. Now if we say that the object has an initial velocity of u, then when $t = 0$, $\frac{ds}{dt} = u$, meaning that $C = u$. Often people write v for the velocity instead of $\frac{ds}{dt}$, so we get:

$$(2) \quad v = at + u$$

It is important to recognize here that t and v are variables, while a and u are fixed constants.

Integrating equation 2 again we get $s = \frac{1}{2} at^2 + ut + D$. If we assume that the object starts at the origin, then $D = 0$. So:

$$(3) \quad s = \frac{1}{2} at^2 + ut$$

Equations 2 and 3 are often used for calculating the velocity and displacement of an object under constant acceleration, after a certain amount of time. It can also be useful to connect v and s directly, without needing to go via t. A little manipulation of equations 2 and 3 produces:

$$(4) \quad v^2 = u^2 + 2as$$

Kinetic energy

In Newtonian mechanics, a moving body has a certain amount of energy by virtue of its motion. If a mass m is moving at a velocity v, its *kinetic energy* is $K = \frac{1}{2}mv^2$. So a car weighing 1000 kg which is travelling at 15 m/s has a kinetic energy of $\frac{1}{2} \times 1000 \times 15^2 = 112{,}500$ J (J standing for 'Joule', the international unit of energy). In the study of mechanics we prefer to consider idealized masses consisting of a single point, rather than realistic masses such as cars, to avoid extraneous factors such as internal moving parts.

Conservation of kinetic energy

A useful property of kinetic energy is that, like momentum, it is conserved during idealized collisions. If there are two bodies of mass 1 kg and 2 kg, travelling towards each other at speeds of 10 m/s and 20 m/s, respectively, then their total kinetic energy is $\left(\frac{1}{2} \times 1 \times 10^2\right) + \left(\frac{1}{2} \times 2 \times 20^2\right) = 450$ J. After they collide, this kinetic energy should remain the same. So, as an easy example, if the heavier body is at rest after the collision, and the lighter one is travelling with speed v then we must have $\frac{1}{2} \times 1 \times v^2 = 450$, and we can deduce that $v = 30$.

(The conservation rule will not hold exactly in the real world, where some energy will always be lost as heat and sound.)

Potential energy

It requires energy to lift a heavy object, because you are fighting against the gravitational force. By doing this you are imbuing the object with *gravitational potential energy*. This energy remains stored in the object, and can be converted to kinetic energy by dropping it. The formula for an object's gravitational potential energy is $V = mgh$ where m is its mass, g is the earthly gravitational constant, and h is the height to which it has been lifted. Potential energy is not limited to gravity, but can exist in relation to other forces, such as electromagnetism.

An important principle in mechanics is that energy is conserved within any system. Ignoring the leaking of energy through heat, sound and light, an object may convert kinetic energy to potential (by rolling up a hill for example) or vice versa (by rolling down it), but the total quantity of energy $V + K$ should remain constant.

WAVES

Waves

Many physical phenomena come in *waves*. Common examples are sound, light and the ripples on a pond. In each case, some property perpetuates through a medium over time, and does so in a repetitive way.

The simplest waves are 1-dimensional, such as those carried along a violin string. Sound and light travelling through space spread out in three directions, while the displacement waves on a pond's surface are 2-dimensional. (These can be modelled by **scalar fields**.) But the 1-dimensional model remains relevant in each case.

Waves are modelled mathematically as *periodic* functions, meaning functions which repeat themselves. The purest is the **sine wave**, but any more complex waveform (such as that produced by a musical instrument) can be built by **Fourier analysis**. This subject also clarifies how waves interfere with each other, simply through the addition of their corresponding functions.

The two fundamental attributes of any wave are its frequency and its amplitude.

Frequency

A wave such as sound or light is subject to the ordinary formula: speed $= \frac{\text{distance}}{\text{time}}$. The distance from the start of one cycle to the start of the next is called the *wavelength*. If this has a value of L, the wave has a speed v, and it takes a time of t to complete one cycle, then this produces the relationship $v = \frac{L}{t}$.

The *frequency* (f) of a wave is the number of complete cycles per second (measured as Hertz, Hz). This means $f = \frac{1}{t}$, where t is as in the above formula. So $v = L \times f$.

For sound waves, frequency is interpreted by our brains as *pitch*. A taxi's squeaky brakes have a high frequency of around 5000 Hz (meaning that 5000 cycles are completed every second), while blue whales sing at around 20 Hz, around the lower limit of the human ear. The upper limit is approximately 20,000 Hz (though bats can hear up to 100,000 Hz).

For visible light, frequency determines colour, the visible spectrum being from 4.3×10^{14} Hz to 7.5×10^{14} Hz.

In many cases (such as sound and light) the wavelength and the frequency can vary, while the speed is fixed by the ambient conditions. If we rescale, so that this fixed speed $v = 1$, this sets up an inverse relationship between wavelength and frequency: $f = \frac{1}{L}$.

Amplitude

Frequency describes how bunched up the peaks of a wave are. The closer together they are, the higher the frequency. The height of those peaks is given by the *amplitude*. (More precisely, the amplitude is half the height from a peak to a trough, or the distance from the centre to a peak.)

The amplitude of a sound wave determines its *volume*. On a stringed instrument, the harder the string is plucked, the greater the amplitude of the resulting wave, and the louder the sound. (In fact, this story is slightly more complicated, as humans have a tendency to hear high-frequency noises as psychologically louder than low ones. So we do not usually use metres as the measure of loudness, but *decibels*, which also incorporate some measure of frequency.)

The amplitude of a wave of visible light defines its *brightness*, and the amount of energy it carries.

AM and FM

Before the dawn of digital radio in the 1990s there were two different ways in which sound could be encoded into electromagnetic waves for broadcast. These were *amplitude modulation* (*AM*), first used in 1906, and *frequency modulation* (*FM*) in 1933. The difference is that the AM encodes the information into

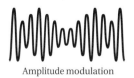

Amplitude modulation

the amplitude of the electromagnetic wave (keeping the frequency constant), while FM keeps the amplitude constant, and encodes the information into the frequency. Generally FM is better resistant to noise, with a bigger range to play with (AM has to contend with the limits to the range of amplitudes which it is practical to broadcast and receive).

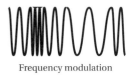

Frequency modulation

Digital radio
Like FM, digital radio uses frequency modulation to carry information. The difference between this and standard FM is that the sound is first encoded as a stream of bits (see **information theory**), and these bits are broadcast via frequency modulation. Each digital station essentially needs just two frequencies, to represent 0 and 1, meaning that more stations can be accommodated in a narrower range of frequencies. Because **error-correcting codes** are built into the data stream, digital radio is better resistant to noise than either AM or ordinary FM.

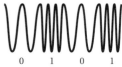

Stringed instruments
Players of stringed instruments, such as banjos, cellos and pianos, create sounds by plucking, bowing or hitting the strings. This causes the string to vibrate, and the resulting pitch (or *frequency*) of the note created is determined by the string's length, mass and tension. A double bass produces a deeper pitch than a ukulele, because its strings are longer and heavier. When you press on the string, you effectively shorten it, raising the pitch. As you loosen a tuning peg, the tension lessens, causing the pitch to drop.

The principal note that the string produces is called its *fundamental frequency* or *first harmonic*. This is produced by a wave which is exactly double the length of the string. But there are other pitches produced too, namely the higher harmonics.

Harmonics
The *first harmonic* on a vibrating string is the fundamental frequency (or *root*), a wave where only the ends of the string are stationary.

The *second harmonic* leaves an additional stationary point in the middle of the string. (Instrumentalists produce this by gently touching the centre of the string.) This has wavelength half that of the first harmonic. Halving the wavelength corresponds to doubling the frequency, so the first harmonic has double the frequency of the fundamental note. To human ears, this sounds an octave higher than the root.

The *third harmonic* has stationary points one third and two thirds of the way along the string. Its wavelength is a third of that of the fundamental, and its frequency is three times higher. In musical terms, this sounds an octave plus a fifth above the root.

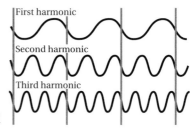

The *fourth harmonic* divides the string into four parts, producing a frequency four times that of the first, and a sound two octaves higher.

FIELDS AND FLOWS

Scalar and vector functions

The commonest form of function takes in a number (x) as input and gives out another $u(x)$ as output. In three dimensions, however, the input might be a triple of numbers (x, y, z), representing the coordinates of a point in 3-dimensional space. The output $u(x, y, z)$ might describe the temperature at the specified point, for example. Such a function is called a *scalar field*.

Temperature is a *scalar* quantity, as it has magnitude but no direction. In many settings we want a **vector** as the output of our function. Forces and velocity are examples of vector quantities, as they have both a magnitude and a direction.

Interpreted as a velocity, the vector $\begin{pmatrix} 1 \\ 0 \\ -1 \end{pmatrix}$ means that a particle is travelling at $1\,\text{m/s}$ in the positive direction of the x-axis, and $1\,\text{m/s}$ in the negative direction along the z-axis. (This produces a total magnitude, of $\sqrt{2}\,\text{m/s}$ by Pythagoras' theorem.)

Often bold letters such as **u** are used for such vector functions. For example,

$\mathbf{u}(3, 2, -1) = \begin{pmatrix} 1 \\ 0 \\ -1 \end{pmatrix}$ means that the function **u** assigns the vector $\begin{pmatrix} 1 \\ 0 \\ -1 \end{pmatrix}$ to the point $(3, 2, -1)$.

Such vector-valued functions are important in many applications of mathematics, for example in modelling fluid flow. At every point (x, y, z) there is a vector $\mathbf{u}(x, y, z)$ which describes the velocity of the fluid at that point. So if $\mathbf{u}(1, 2, 3) = \begin{pmatrix} 0 \\ 0 \\ 0 \end{pmatrix}$, then the fluid is static at the point $(1, 2, 3)$. If $\mathbf{u}(0, 4, -1) = \begin{pmatrix} 2 \\ 0 \\ 0 \end{pmatrix}$, then the fluid is travelling rightwards at $2\,\text{m/s}$ at the point $(0, 4, -1)$.

More generally we write u_x, u_y, u_z to stand for the components of **u** in the x, y and z directions respectively. So $\mathbf{u}(x, y, z) = \begin{pmatrix} u_x \\ u_y \\ u_z \end{pmatrix}$.

Vector fields

Fluid flow is an example of a *vector field*. This is a function which assigns a vector to every point in space (not to be confused with the algebraic structure called a **field**). Usually, we require the vector field to be *smooth*, so that small movements in space produce small changes in the corresponding vectors.

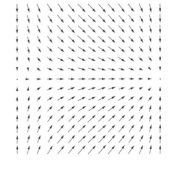

Just as with scalar functions, we are interested in the rate of change of vector fields. This is the topic of vector calculus. In the case of fluid flow, the vector field is governed by the Euler and Navier–Stokes equations. Vector fields may also be interpreted as forces, such as in Maxwell's equations for electromagnetic fields.

The **hairy ball theorem** says that every smooth vector field on a sphere must be equal to zero somewhere.

Vector calculus

To investigate scalar and vector fields, we want to look at the way they vary from point to point, and moment to moment. This involves calculus. But with three spatial coordinates, and one of time, by which we might want to differentiate, the notation can become very messy and the equations very long.

In the 19th century, mathematicians doing vector calculus started to use a new symbol ∇, called 'del' or 'nabla' (not to be confused with a capital Greek delta Δ) as a useful short-hand for the following operator:

$$\nabla = \begin{pmatrix} \dfrac{\partial}{\partial x} \\[6pt] \dfrac{\partial}{\partial y} \\[6pt] \dfrac{\partial}{\partial z} \end{pmatrix}$$

Almost by treating ∇ as an ordinary vector, several important quantities can be expressed: notably div, grad, curl and the Laplacian.

Grad

Suppose that f is a scalar field, which assigns a number to every point of 3-dimensional space. (An example is temperature.) Then:

$$\nabla f = \begin{pmatrix} \dfrac{\partial f}{\partial x} \\[6pt] \dfrac{\partial f}{\partial y} \\[6pt] \dfrac{\partial f}{\partial z} \end{pmatrix}$$

This is a vector, called the *gradient* of f (commonly abbreviated to 'grad f'). Like the ordinary derivative, ∇f expresses the rate at which f increases. The vector ∇f points in the direction of greatest increase, and its magnitude quantifies the rate of increase.

Div

Although ∇ is an operator, we can pretend it is an ordinary vector, and take a sort of 'dot product'.

If $\mathbf{u} = \begin{pmatrix} u_x \\ u_y \\ u_z \end{pmatrix}$ is a vector field, then:

$$\nabla \cdot \mathbf{u} = \frac{\partial u_x}{\partial x} + \frac{\partial u_y}{\partial y} + \frac{\partial u_z}{\partial z}$$

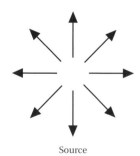

Source

This is pronounced 'div **u**', short for the *divergence* of **u**. This is now a scalar quantity, which quantifies the total inflow and outflow at each point.

If $\nabla \cdot \mathbf{u} > 0$ at some particular point, this means that the net effect of flow at that point is outwards. That is to say, the point acts as a *source* of flow. If $\nabla \cdot \mathbf{u} < 0$, then the net flow is inwards at that point, meaning that it acts as a *sink*.

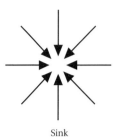

Sink

Curl

Just as the divergence is obtained by taking the dot product with ∇, so the *curl* of a vector field **u** is defined by taking the **cross product**.

If $\mathbf{u} = \begin{pmatrix} u_x \\ u_y \\ u_z \end{pmatrix}$, then:

$$\nabla \times \mathbf{u} = \begin{vmatrix} \dfrac{\partial u_z}{\partial y} - \dfrac{\partial u_y}{\partial z} \\ \dfrac{\partial u_x}{\partial z} - \dfrac{\partial u_z}{\partial x} \\ \dfrac{\partial u_y}{\partial x} - \dfrac{\partial u_x}{\partial y} \end{vmatrix}$$

Rotational vector field

The curl is another vector field, which quantifies the extent and direction of the *rotation* of **u**. If $\nabla \times \mathbf{u} = 0$, then **u** is said to be *irrotational*.

The curl is an important tool in applied mathematics, notably in **Maxwell's equations** for electromagnetic fields.

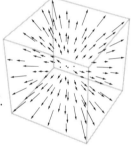

Irrotational vector field

The Laplacian

Applying the del operator twice we can also use it as short-hand for second derivatives. An important case is the *Laplacian* ∇^2, named after its inventor Pierre–Simon Laplace. If f is a scalar field, then:

$$\nabla^2 f = \nabla \cdot (\nabla f) = \frac{\partial^2 f}{\partial x^2} + \frac{\partial^2 f}{\partial y^2} + \frac{\partial^2 f}{\partial z^2}$$

This is a scalar quantity. The Laplacian at a point expresses the average value of f in a small neighbourhood of that point (technically the limit of the average as the neighbourhood shrinks to zero). It features in Laplace's equation, the heat equation and in the quantum mechanical **Hamiltonian**, among many other places in mathematics.

A vector form of the Laplacian is central to the **Navier–Stokes equations**. Starting with a vector field **u** as above, we define:

$$\nabla^2 \mathbf{u} = \begin{pmatrix} \nabla^2 u_x \\ \nabla^2 u_y \\ \nabla^2 u_z \end{pmatrix}$$

Writing this out fully really demonstrates the space-saving benefits of the ∇ notation!

Laplace's equation

The simplest equation which can be built from the Laplacian is:

$$\nabla^2 f = 0$$

This is *Laplace's equation*.

According to this equation, *f* is a scalar field such that around every point the average value is zero. In one dimension, this is only satisfied by fields corresponding to straight lines. In higher dimensions there are subtler possibilities, the *harmonic functions* studied in Hodge's Theorem.

Hodge's theorem

In the 1930s, William Hodge performed some exceptionally deep analysis on the Laplace equation and its possible solutions, or *harmonic functions* as they became known. In particular, Hodge showed that solving this equation on a manifold is equivalent to calculating certain *homology groups* for it (see **algebraic topology**).

This profound and unexpected connection had major repercussions from topology to group theory, and was described by Hermann Weyl as 'one of the great landmarks in the history of science in the present century'. The breakthrough presented Hodge with the right language to formulate the famous **Hodge conjecture**. It also provides the template for solutions of more complicated equations such as the heat equation, in which the scalar field is no longer fixed, but now additionally depends on time, becoming a *flow*.

The heat equation

In 1811, Joseph Fourier published a theory on the flow of heat in a solid. According to this, heat is a scalar quantity. The heat *f* at a point depends on the time *t*, as well as the point's coordinates in 3-dimensional space (x, y, z). So we model *f* as a function of four variables: *t* and *x*, *y*, *z*. By considering the way that heat diffuses over time, Fourier derived the fundamental partial differential equation for modelling the flow of heat:

$$\frac{\partial f}{\partial t} = \nabla^2 f$$

The physical interpretation of this equation is that the rate at which *f* changes at a single point is determined by the average temperature of the points around it.

Solutions to the heat equation

We cannot hope for a unique solution to the heat equation, $\frac{\partial f}{\partial t} = \nabla^2 f$. The general theory of partial differential equations is firmly set against this; but, more than this, common sense dictates that the way that heat flows must depend on the initial distribution of heat around the solid.

The simplest solution is $f(t, x) = e^{-t} \sin x$. This is a 1-dimensional solution, in which *y* and *z* are ignored (equivalent to modelling the flow of heat in a 1-dimensional wire). This solution corresponds to an initial configuration of a sine wave, which evenly dissipates over time. This illustrates the important fact that functions which solve the heat equation tend towards a state of equilibrium over the long term. Another 1-dimensional solution is $f(t, x) = e^{-4t} \sin 2x$ and $f(t, x) = e^{-n^2 t} \sin nx$ generally.

To deal with more complicated initial conditions, we can build solutions from series of these solutions, just as in **Fourier analysis**. Indeed the study of heat was Fourier's initial motivation.

Flows

The importance of Fourier's heat equation $\frac{\partial f}{\partial t} = \nabla^2 f$ carries far beyond models of heat flow. This equation occupies a foundational place in several subjects, including modelling stock options in financial mathematics. Along with Laplace's equation, it forms the basis upon which other partial differential equations are built, especially those modelling any form of *flow*. A notable example is the Navier–Stokes equation.

The Laplace and heat equations are not limited to ordinary 3-dimensional space, but can apply on any **manifold**. This is useful for modelling diverse physical phenomena, and also within pure mathematics. *Ricci flow* is a striking example of an abstract flow on a manifold, the analysis of which led to the proof of the **geometrization theorem** and **Poincaré conjecture**.

Fluid dynamics

How do fluids flow? This continuous form of mechanics is far more difficult to model than the discrete particles of Newtonian mechanics. In the 18th century, Leonhard Euler considered this question. Euler's formula for fluid flow is essentially a statement of Newton's second law, transferred to an idealized fluid.

His work was built on during the 19th century, separately by Claude-Louis Navier and George Stokes. Both worked with essentially the same mathematical model of fluid flow.

As well as assuming that the fluid is *incompressible*, Euler assumed that it has no *viscosity*, that is to say, that it can move freely over itself without any internal frictional forces. Navier and Stokes introduced a new constant v to quantify the viscosity of the fluid. Mathematically, this made the resulting equations even more difficult to solve. This is the celebrated **Navier–Stokes problem**.

The fluid model

We assume that our fluid is distributed throughout some region. Mathematically, the central quantity in which we are interested is a vector **u**, describing the velocity of the fluid at a particular point (x, y, z) and at some moment t, say

$\mathbf{u} = \begin{pmatrix} u_x \\ u_y \\ u_z \end{pmatrix}$. The value of each of u_x, u_y, u_z will each depend on the time t as well as the spatial

coordinates (x, y, z). In mathematical terms, this constitutes a **vector field** $\mathbf{u}(t, x, y, z)$.

The principal factor affecting the flow is the pressure of the liquid, p. This may also vary from point to point and moment to moment. So we write it as a function $p(t, x, y, z)$. Assuming that there are no other forces acting on the liquid, what Euler's work and that of Navier and Stokes provides is a formula relating **u** and p.

In their analysis, a fundamental assumption is that the fluid is *incompressible*. When a force is applied, the liquid may move in some direction but cannot contract or expand to fill a different volume. This amounts to asserting that $\nabla \cdot \mathbf{u} = 0$. In other words no point may act as either a source or sink of flow at any time.

Euler's fluid flow formula

Suppose for a moment that we are interested in a fluid flowing in a 1-dimensional space, so we only need to consider u_x, with $u_y = u_z = 0$. Euler's formula for fluid flow says that the flow must satisfy:

$$\frac{\partial u_x}{\partial t} + u_x \frac{\partial u_x}{\partial x} = -\frac{\partial p}{\partial x}$$

For a 3-dimensional flow, acceleration in the x-direction is instead governed by the equation:

$$\frac{\partial u_x}{\partial t} + u_x \frac{\partial u_x}{\partial x} + u_y \frac{\partial u_x}{\partial y} + u_z \frac{\partial u_x}{\partial z} = -\frac{\partial p}{\partial x}$$

Euler also derived two other similar equations, for acceleration in the y- and z- directions. In these, u_y and u_z replace u_x as the object to be differentiated on the left-hand side, with $\frac{\partial p}{\partial y}$ and $\frac{\partial p}{\partial z}$ respectively replacing $\frac{\partial p}{\partial x}$ on the right.

These three equations can be expressed more succinctly as one equation, using vector calculus:

$$\frac{\partial \mathbf{u}}{\partial t} + (\mathbf{u} \cdot \nabla)\mathbf{u} = -\nabla p$$

When the liquid is subject to an additional external force $\mathbf{f}$ (such as gravity), the equation becomes $\frac{\partial \mathbf{u}}{\partial t} + (\mathbf{u} \cdot \nabla)\mathbf{u} = \mathbf{f} - \nabla p$.

Navier–Stokes equations

In 1822, the engineer and mathematician Claude-Louis Navier first amended Euler's fluid flow formula to allow the liquid to be *viscous*. This means that as it flows over itself there are frictional forces which hinder movement. Honey is an example of a highly viscous fluid. Navier arrived at the fundamental equations for describing this situation, although his mathematical argument was not quite correct. Working independently a few years later, George Stokes was able to provide a correct derivation of these fundamental equations, from Newton's second law.

The critical ingredient which Navier and Stokes added was a constant v to quantify the *viscosity* of the fluid. If no external force is acting on the fluid, the Navier–Stokes equations then say that its velocity $\mathbf{u}$ should satisfy the following:

$$\frac{\partial \mathbf{u}}{\partial t} + (\mathbf{u} \cdot \nabla)\mathbf{u} = v\nabla^2\mathbf{u} - \nabla p$$

Here $\nabla^2\mathbf{u}$ is the **Laplacian** of $\mathbf{u}$. In the presence of an external force $\mathbf{f}$, the equation becomes $\frac{\partial \mathbf{u}}{\partial t} + (\mathbf{u} \cdot \nabla)\mathbf{u} = \mathbf{f} + v\nabla^2\mathbf{u} - \nabla p$.

The Navier–Stokes problem

The Navier–Stokes equations are a triumph of the power of mathematics to model nature. They have received extensive, detailed experimental verification, in a wide variety of contexts. They occupy a central position in fluid dynamics, and their study has led to technological advances from aircraft wings to artificial heart valves.

It is all the more surprising, then, that it has never been established that these equations have mathematical solutions. To be more precise, it is possible to find mathematical formulas for **u** which satisfy an Euler or Navier–Stokes equation over a short period of time. But often these solutions break down at a later stage, ceasing to be smooth functions (an impossible condition for a physical fluid).

No-one has yet been able to find a single formula which remains valid for all values of t, which solves either the Euler or Navier–Stokes equations (although their 2-dimensional analogues can be solved). This is certainly frustrating, since evidence from computer simulations (as well as the natural world) suggests that there should be a great many such solutions.

In 2000, the **Clay Institute** announced a prize of $1,000,000 for a solution of a Navier–Stokes equation, as one of their millennium problems.

Electromagnetic fields

In the early 19th century, the phenomenon of *electromagnetism* became a major focus of scientific research. It began when Hans Christian Ørsted noticed that the electrical current in a battery disrupted a nearby compass. Investigating further, it was found that rotating a wire in a magnetic field would cause current to flow through it. In 1831, Michael Faraday successfully exploited this mechanism to build the first electric generator. Reversing this idea, he was also able to build an electric motor for converting electric energy into mechanical energy.

These discoveries rely upon the intertwining of magnetic and electric fields. If an electric field **E** causes current to pass through a wire, this sets up a magnetic field **B** wrapping around the wire, at right angles to it. The precise geometry is subtle; pinning it down exactly involved the work of physicists such as André-Marie Ampère, and mathematicians such as Carl Friedrich Gauss. In 1864 James Clerk Maxwell was finally able to write four partial differential equations which perfectly capture the geometry of these two fields. This combined *electromagnetic force* is now counted as one of the **fundamental forces of nature**.

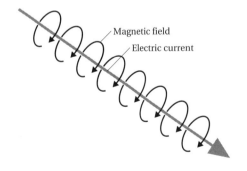

Magnetic field

Electric current

Maxwell's equations

James Clerk Maxwell's equations relate two vector fields: a magnetic field **B** and an electric field **E**. Their exact geometry depends on the ambient matter and the distribution of electrons around it. This can be quantified by two pieces of data: a scalar field ρ, called the *charge density*, and a vector field **J**, called the *current density*. The speed of light c is also involved. After rescaling through a careful choice of units, Maxwell's equations may be written using vector calculus as:

$$\nabla \cdot \mathbf{E} = \rho$$
$$\nabla \times \mathbf{E} = -\frac{\partial \mathbf{B}}{\partial t}$$
$$\nabla \cdot \mathbf{B} = 0$$
$$\nabla \times \mathbf{B} = \mathbf{J} + \frac{1}{c^2}\frac{\partial \mathbf{E}}{\partial t}$$

SPECIAL RELATIVITY

Inertial frames of reference

In deep space, thousands of light-years from the nearest star, float two rocks. Rock 1 is stationary, and rock 2 drifts past it at 5 m/s. Or was it the other way around? Perhaps rock 2 was stationary, and rock 1 floated by at 5 m/s. Or possibly they sailed past each other, each travelling at 2.5 m/s. Or maybe, even, rock 1 was whizzing along at 996 m/s, and was overtaken by rock 2 flying at 1001 m/s.

How can we tell which of these descriptions is right? The answer is that we cannot. Each of the above is a perfectly legitimate characterization of the situation. Which one we need depends on our choice of *inertial frame of reference*.

Because neither rock has any force acting on it, both of them are said to have *inertial motion*, and, according to Galilean relativity, all such are fundamentally equivalent. If we have a particular interest in rock 1, it might make sense to consider that as stationary, and rock 2 as moving. Making this decision fixes our inertial frame of reference, and the velocities of other bodies can then be measured relative to it.

Galilean relativity

We are familiar with a certain level of relativity, according to which different *positions* are equivalent. Imagine two identical sealed rooms in different parts of a city. There is no conceivable experiment by which you could determine which you are in. The laws of physics are identical in the two cases. The only difference is in their relative locations: from the first room, the second is 6 miles north. Equivalently, from the second, the first is 6 miles south.

As Galileo noted in 1632, inertial frames of reference extend this equivalence to different *velocities*. If the rooms are on two spacecraft, each moving freely through deep space at different constant speeds, then again there is no conceivable experiment that could determine which you are in. The two scenarios are identical, even if one is travelling at 10,000 m/s, relative to the other.

This idea does not come naturally to humans, since we spend our lives firmly rooted in one inertial frame (as far as our senses can detect, anyway). But it follows from Newton's first law. However, this equivalence does *not* extend to acceleration. If one room is in an accelerating train, and the other is stationary, then an experiment (such as dropping a ball) can distinguish the two.

Space is relative

The implication of Galilean relativity is that space is *relative*. If there is no possible way to distinguish between objects which are stationary and those which are travelling with constant velocity, then we have to acknowledge that the notion of absolute stationarity is neither useful nor meaningful.

A consequence is that we have no way to cling onto individual points in space. If a cup is on a table in an inertial frame, if no-one touches it, is it in the same position 10 seconds later? Relative to the chosen inertial frame it is; relative to another it is not.

But which is true, *really*? It sounds like a simple question: is the cup occupying the same region of space as before, or is it not? However frustrating it may be, there is no valid answer which can be given. For some time, physicists, including Newton, theorized the existence of an all-pervading *ether*, which would have fixed one inertial frame as the true measure of stationarity. The demise of ether theory with **the Michelson–Morley experiment** took with it the last hope for a universal default frame of reference.

Spacetime

The attempt to unify the three dimensions of space and one of time into a 4-dimensional geometric *spacetime* dates back to at least Joseph-Louis Lagrange in the late 18th century. A first attempt to formalize this might involve quadruples of coordinates (t, x, y, z) where t represents time, and (x, y, z) are the ordinary 3-dimensional spatial coordinates. We could write this more briefly as $(t, \boldsymbol{x})$, where $\boldsymbol{x}$ is short-hand for (x, y, z).

This is naïve, because it fails to take into account that space is relative. If we pick a point in space, say $\boldsymbol{A}$, then $(2, \boldsymbol{A})$ and $(15, \boldsymbol{A})$ represent the same point $\boldsymbol{A}$ thirteen seconds apart. But this is exactly what Galilean relativity warns us against! Galilean spacetime is a solution to this problem.

Galilean spacetime

Spacetime should not come equipped with a built-in notion of 'the same point at different times', but should be flexible enough to allow us to specify possible inertial frames by which to judge this.

Mathematicians have a device, called a *fibre-bundle*, for arranging this. It looks very much like the naïve spacetime, with one-dimension of time, and three of space. The difference is that there is no pre-assigned correspondence between the different layers of

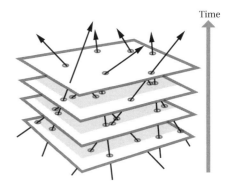

Time

3-dimensional space. Various paths through these slices are equally valid as constant velocities, called *world-lines*, represented as straight arrows. A family of parallel arrows corresponds to picking an inertial frame and measuring everything relative to it.

The eclipses of Io

When you light a candle, does the light take time to travel to the corners of the room, or does it get there instantaneously? With no obvious way to answer this question, scientists have debated and disagreed over the centuries. Galileo (1564–1642) believed that light travels at a finite speed, and tried to demonstrate this experimentally. All he could conclude was that, if he was right, light's speed must be extremely high.

Among his more successful work was the discovery of *Io*, the nearest moon of Jupiter in 1610. Later in the 17th century Ole Rømer found evidence that light travels at a finite speed, by observing Io, which orbits Jupiter once every 42.5 hours. For some of that time it is eclipsed from our view by Jupiter. Rømer noticed that the time between successive eclipses varied, depending on the relative motion of earth and Jupiter. As they neared each other, this time decreased; as they drew apart, it increased. This could only be because light takes longer to reach earth from Jupiter the futher apart the two planets are.

Speed of light

The eclipses of Io were the first solid evidence that light moves at a finite speed. Subsequent experiments have pinned this down to 299,792,458 m/s, in a vacuum. (This number is exact, since a *metre* is now defined as the distance travelled by light in $\frac{1}{299,792,458}$ of a second.)

The letter c is usually used to stand for the speed of light. The term 'light' is a little misleading; it is the speed of all electromagnetic radiation in a vacuum, of which visible light is just a small sliver.

The Michelson–Morley experiment

Once scientists knew that light travels at a finite speed, it was expected that this would slot neatly into Galilean relativity. If all velocities are relative, and all inertial frames equivalent, then this should apply equally to the speed of light (c). If you were travelling with speed c, alongside a beam of light, it would appear stationary, like two cars travelling side by side at the same speed. If you exceed c, the beam of light would seem to move backwards (just as happens when one car overtakes another). Two beams of light fired at each other should have a relative velocity of $2c$.

An experiment in 1887 by Albert Michelson and Edward Morley investigated the existence of luminiferous ether (see **space is relative**). As well as providing convincing evidence against the ether theory, the Michelson–Morley experiment was the first hint of a very unexpected fact indeed: the speed of light does *not* obey the principle of Galilean relativity. Subsequent results backed this up. From different inertial frames, the speed of light appears the same, fixed and unchanging, absolute and not relative. Even if you could travel at $\frac{1}{2}c$, light would not seem any slower!

To explain this paradox required a complete reworking of the basic conception of time and space.

Minkowski spacetime

Galilean spacetime, in which all inertial frames of reference are equivalent, cannot handle the constancy of the speed of light. Around 1907, Hermann Minkowski built on earlier work of Henrik Lorentz, Henri Poincaré and Albert Einstein, to formulate a new spacetime to cope with this. Through some inventive mathematics, notably the *Lorentz transformation*, every point of spacetime could be endowed with a double cone, with one half pointing to the past, and the other to the future.

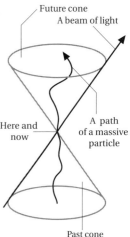

These *light cones* describe the history of every possible flash of light passing through the point. Paths inside the cones represent lower speeds, the possible world-lines of a massive particle passing through the point. Outside the cone are points inaccessible to both light and matter, as travel that is faster than light would be needed to reach them.

The hyperbolic geometry of light

Minkowski spacetime is endowed with a new notion of *special* relativity. The basic principle is that, below the speed of light, Galilean relativity holds. But, once c is reached, it breaks down. More precisely, within the light cone, every future-pointing arrow is as good as every other. Despite appearances, there is no unique line passing through the centre of the cone (so there is no notion of being stationary). Also, however close to the boundary an arrow appears to be, from its own perspective the edge is no closer. (This corresponds to the speed of light remaining constant, no matter how fast you travel.)

This is made rigorous by endowing the insides of the cone with **hyperbolic geometry**. This has the properties we want: on the Poincaré disc, there is no well-defined centre, and the boundary of the disc is infinitely far away from each point inside the disc.

Time is relative

A consequence of special relativity is that the concept of *simultaneity* becomes blurred. *Past* and *future* are preserved by the light cones: event A is in the future of event B, if it is inside its future light cone. Similarly B is in the past of A.

But if A and B are outside each other's light cones, they cannot influence each other by any means. Are they simultaneous?

A natural way to answer this would be to wait for a third event C which is in the common futures of A and B, and then calculate the times that have elapsed since both. If these times are the same, we judge A and B to have been simultaneous. However, because of the constancy of the speed of light, this answer will not be unique. It will depend on the choice of C. For this reason, we must give up on the idea of absolute time.

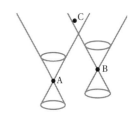

Mass is relative

Suppose an object is travelling at 99% of the speed of light and is subject to a large force causing it to accelerate further. Newtonian mechanics predicts that after some time it should exceed the speed of light (according to $v = u + at$, see **bodies with constant acceleration**).

In relativity theory, this is forbidden. So the object's acceleration must be lower than predicted by Newton's second law, $F = ma$. As the object gets nearer the speed of light, its acceleration must decrease.

If F is constant, and a is decreasing, the mass m must be increasing. So as an object approaches the speed of light, its mass m seems to increase.

There are now two values of mass that we might want to assign to an object. One is the observed mass m, as it seems to a stationary observer. As the object's speed gets close to the speed of light, m tends towards infinity. The second is the *rest mass*, the mass of the object in its own inertial frame, where it is judged stationary. The rest mass is usually denoted m_0.

Special relativity

1905 was Albert Einstein's *annus mirabilis*, or miraculous year. As well as completing his doctoral thesis on molecular physics, he published four seminal scientific papers: one on quantum mechanical **photons**, one on the **kinetic theory of heat**, and two on relativity theory. 'On the Electrodynamics of Moving Bodies' introduced his theory of special relativity, building on earlier work of Lorentz and Poincaré. This picture was completed in 1907, when Hermann Minkowski introduced his mathematical model of spacetime. The theory was quickly welcomed by the scientific community.

Einstein showed how to rework important parts of physics, notably Maxwell's equations, to allow for the constancy of the speed of light. In doing so, he replaced the familiar notions of time and space with time-like, space-like and light-like paths.

Time-like, space-like and light-like paths

If A and B are two events (which we think of as points in Minkowski spacetime), then there are three possibilities:

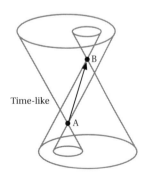

Time-like

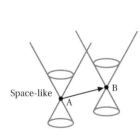

Space-like

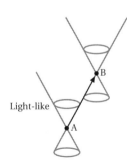

Light-like

1 There is a *time-like path* between them. This means that they lie within each other's light cones, and there is an inertial frame which carries one to the other. So they can be conceived as two events happening at the same point of space, at different times.

2 There is a *space-like path* between them. They lie outside each other's light cones, and cannot influence each other. Since time is relative, they can be conceived as two events happening at the same moment, in different places.

3 There is a *light-like path* between them. This means that they lie on the boundary of each other's light cones. Light can pass from one to the other, but matter cannot.

Causality is determined by events which lie on time-like and light-like paths.

Energy is relative

The Newtonian concept of **kinetic energy** ceases to be absolute, even in Galilean relativity. Since all velocities are equivalent, there is no meaningful number which can be assigned to v in the formula $\frac{1}{2}mv^2$.

In special relativity, this gets worse. Since mass is relative, the value of m is also dependent on the choice of inertial frame. As a result, we have to conclude that kinetic energy is also relative, and depends on the choice of inertial frame. This realization prompted Einstein to revisit the whole concept of energy, in his final paper of 1905.

$E = mc^2$

Albert Einstein's final paper of 1905 was entitled 'Does the Inertia of a Body Depend on its Energy Content?' It was only three pages long, and consisted of one fairly simple mathematical deduction. But its stunning conclusion would again shatter our understanding of the workings of the physical universe.

It concerned the question of energy in special relativity. Just as the mass of an object is relative, so is its energy. The faster an object travels, the greater both its mass and energy become. Einstein realized that these two quantities increase in fixed proportion to each other. This ratio is expressed by the number c^2, where c is the speed of light.

Conversely, as the object slows down, its relativistic mass decreases. But it never reaches zero. There is a fixed lower limit, given by its mass at rest (m). Similarly, Einstein found, the energy of an object decreases with its speed as expected. But again, there is a baseline which is never crossed: its energy at rest (E).

What is the meaning of this energy at rest? The unavoidable conclusion was that any object has energy, simply by virtue of having mass, and the two are related by perhaps the most famous equation of all:

$$E = mc^2$$

Amid all the relativistic quicksand, these three numbers E, m and c are absolute, not depending on any choice of inertial frame. Einstein's equation therefore provided a new firm foothold from which to view the universe.

The equivalence of mass and energy

In Einstein's equation $E = mc^2$, the number c is a fixed constant. Physicists often choose to rescale and work with a different choice of units, where $c = 1$. Then Einstein's formula can be rephrased as:

$$E = m$$

The message here could not be clearer: energy and mass are equivalent. More precisely, mass is energy, frozen into material form. Einstein realized that the ultimate test of this bold claim would come from nuclear reactions.

These come in two main forms: in *radioactive decay*, an atom of radioactive material flies apart to leave a lighter atom (or, in the case of *nuclear fission*, more than one such atom). In nuclear *fusion*, two hydrogen atoms collide to form one heavier helium atom. Crucially, in both processes, the end result weighs less than the initial ingredients. A small amount of mass has been lost, and released as energy.

The study of such reactions subsequently provided detailed experimental support for Einstein's formula. With our usual units of seconds and metres, the number c^2 expresses the exchange rate between mass and energy. Because c^2 is so large (89,875,517,873,681,764), a tiny drop in mass can buy an enormous amount of energy. In 1945, this was demonstrated to dreadful effect at Hiroshima and Nagasaki. It also underlies the energy on which we all ultimately depend: that of the sun.

The symmetry groups of spacetime

Different conceptions of spacetime carry different principles of equivalence. In the most naïve version (plain 4-dimensional Euclidean space) all that is required is that every point is the same as every other, the laws of physics do not vary from place to place or moment to moment. The statement is similar to the statement that the four corners of a square are identical. It translates as saying that whenever we pick two points, there should be a symmetry of the structure of spacetime taking one to the other.

Just as for a square, these symmetries combine to form a **symmetry group**. In this case, we get the full symmetry group of 4-dimensional Euclidean space. Known as E(4), this is itself a 10-dimensional object. More sophisticated models of spacetime get correspondingly more sophisticated groups, starting with the *Galilean group*: the symmetry group of Galilean spacetime.

The Poincaré group

What is the symmetry group of Minkowski spacetime? Geometrically, this is an intricate question. Light cones must be preserved, but all directions inside light cones are equivalent. On the other hand, because time is relative, the symmetries no longer have to preserve simultaneity. It was Henri Poincaré who found the answer, in a supreme piece of analysis. The resulting *Poincaré group* is a 10-dimensional **Lie group**, which continues to play a central role in our understanding of the universe. In particular, in relativistic quantum mechanics an individual particle is determined by a **representation** of the Poincaré group.

GRAVITY

Galileo's cannonballs

Galileo Galilei is said to have climbed the leaning tower of Pisa, armed with two cannonballs of different masses. By dropping them, he demonstrated once and for all that objects of different masses fall at the same rate. Whether Galileo actually performed this particular experiment is a matter of doubt. Nevertheless, he did make the discovery, and it was enough to contradict the view of gravity that had predominated since Aristotle, that heavier bodies fall faster.

Galileo was a supporter of the ideas of Nicholas Copernicus, who a century earlier had posited that the sun, and not the earth, is at the centre of the solar system. Building on Galileo's work, Robert Hooke and Isaac Newton developed a theory of *universal gravitation*, according to which gravity is not limited to earth, but an attractive force between all masses. This insight brought the necessary tools to flesh out the Copernican theory. The backbone of their theory is the inverse square law.

Newton's inverse square law

In ordinary earthly life, we think of gravity as being constant, and producing a fixed acceleration. Taking a broader perspective, this is not true. When Neil Armstrong stepped out of the Apollo 11 lunar module, he did not hurtle towards earth at $9.8 \, \text{m/s}^2$. We know that more massive objects create greater gravitational fields, and the earth is 80 times more massive than the moon, so why did he not?

The answer is that, as you move away from a planet or star, its gravitational effects diminish. At what rate does this happen? Johann Kepler thought that the gravitational effect between two objects was inversely proportional to the distance between them.

This was not quite right. In Isaac Newton's *Philosophiæ Naturalis Principia Mathematica* (known as the *Principia*) the correct answer was provided. Newton said that at a distance r away from a mass, the gravity was proportional to $\frac{1}{r^2}$.

Since then the details have been filled in: the gravitational force between two objects of mass m_1 and m_2 is $\frac{Gm_1m_2}{r^2}$, where G is the *universal gravitational constant*, of around 6.67×10^{-11}.

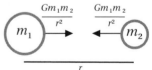

If we take m_1 as the mass of the earth ($5.97 \times 10^{24} \, \text{kg}$), and r as the radius of the earth ($6.37 \times 10^6 \, \text{m}$), then $\frac{Gm_1}{r^2}$ comes out at around 9.8, as we expect.

The two-body problem

Two bodies in space will attract each other according to Newton's inverse square law. We idealize these bodies as points in space endowed with mass. If they are stationary to start with, the two bodies will simply fall towards each other and collide.

If they are moving to start with, more complex outcomes are possible. Examples are two stars cycling around interlocking ellipses, an asteroid which spirals around a star before plunging into it, or a single-apparition comet, which does a parabolic U-turn around a star and then flies off. These *two-body problems* amount to systems of **differential equations** with **boundary conditions** given by the initial positions and velocities; in all cases the problem can be solved easily, unlike the three-body problem.

The three-body problem

In 1887, King Oscar II of Sweden announced a prize to celebrate his 60th birthday. A keen supporter of mathematics, he offered 2500 crowns to the mathematician who could solve the *three-body problem*. This is the same as the two-body problem, with an extra body added. But this extra mass makes the question incomparably more complicated.

It had come out of the works of Newton, who had considered the motion of the earth, moon and sun. He wrote that when more than two masses interact, 'to define these motions by exact laws admitting of easy calculation exceeds, if I am not mistaken, the force of any human mind'.

Henri Poincaré devoted much attention to the problem, and analysed exactly what would be needed to solve the problem. It involved ten separate integrals; but Poincaré believed that these would not be solved exactly without profound mathematical advances. With no complete solution on the horizon, Poincaré was awarded King Oscar's prize.

A huge development in the subject was Sundman's series, but the three-body problem (and *n*-body for $n \geq 4$) remains an active topic of research today.

Sundman's series

In 1912, to the mathematical community's amazement, an astronomer at the University of Helsinki, Karl Sundman, provided a complete solution to the three-body problem. He found an infinite **power series** in $t^{\frac{1}{3}}$ which converges and perfectly describes the arrangement of the three bodies at time t. An astonishing achievement though it is, Sundman's is not the last word on the three-body problem. The problem is that although his series converges it does so extremely slowly. To describe the system over any significant length of time, around 10^{10^8} terms would have to be added together, making this an impractical procedure in most cases.

Gravitational equivalence

There remains something rather mysterious about Galileo's cannonballs. According to Newton's second law, the effect of a force on an object should depend on its mass: $F = ma$. Galileo's discovery seemed to contradict this. How could this be?

The answer is that, uniquely in the case of gravity, the magnitude of the force *also* depends on the object's mass: $F = mg$, where g is the earthly gravitational constant. These two considerations exactly cancel each other out, to give $a = g$.

This *equivalence principle* between gravity and acceleration would not receive a satisfactory explanation until Albert Einstein began to develop his theory of general relativity, in 1907.

Einsteinian frames of reference

To simulate life in zero gravity, trainee astronauts often travel in special aeroplanes. When their plane goes into freefall, the sealed room inside acquires the conditions of zero gravity. In other words, it resembles an inertial frame.

Of course the room frame is not inertial, since it is not moving at constant velocity, but accelerating at 9.8 m/s². However, the acceleration and gravity cancel each other out exactly.

This insight is fundamental to Albert Einstein's theory of general relativity. An *Einsteinian frame of reference* is one that is freefalling under gravity. In a revolutionary move, freefall acceleration replaced constant velocity as the fundamental form of motion.

From the perspective of the trainee astronaut in his room, the effects of gravity depend on the acceleration of the plane. When it is flying level, gravity is as we feel it on earth. If the aeroplane points vertically downwards and switches on its afterburner to descend even faster than gravity, the astronaut would feel that 'up' and 'down' had flipped. Only during freefall is life inside the room free of all gravitational effects.

General relativity

Special relativity is adequate to describe physics in the absence of gravity. In this situation all inertial frames travelling below the speed of light are equivalent. In general relativity, Einsteinian frames of reference take centre stage. The principle of equivalence states that all Einsteinian frames are equivalent. There is no conceivable experiment that can distinguish one from any other.

What is more, it is impossible to separate frames which are subject to gravity, from those which are subject to acceleration. If you are travelling in deep space with no gravity and your spacecraft accelerates at 9.8 m/s², life inside would be indistinguishable from that in earthly gravitation. In general relativity, gravity and acceleration are one and the same.

Gravitational tides

Only in the gravity-free environs of deep space do inertial and Einsteinian frames coincide; in an empty universe, special and general relativity are the same. The idea of general relativity is that if an entire room is in freefall then, *relative to the walls and the floor*, the contents of the room are totally free from the effects of gravity. The complicating factor is that gravity is not a constant force. It varies, depending on your distance from the earth, or other nearby mass. In our hypothetical falling room, the floor will be subject to a stronger gravitational force than the ceiling, because it is slightly nearer the earth. From the perspective of a point in the centre, the floor will be pulled *down* and the ceiling will be pulled *up*.

Furthermore, because gravity is directed towards the centre of the earth rather than downwards in parallel lines, the walls will be pulled slightly *inwards*. If we imagine that our

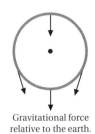

Gravitational force relative to the earth.

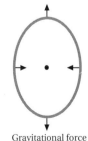

Gravitational force relative to the centre of the room, with its deforming effect.

room begins as a sphere, and is made of some compliant material, as it falls it will be deformed to a prolate spheroid (see **ellipsoid**).

In the case of the planet earth, the effect of this *gravitational tide* is imperceptible. Near stronger centres of gravity, it will be more dramatic. In the extreme example of a **black hole**, if someone is unlucky enough to fall in, the difference between the gravitational force on their head and that on their feet would pull them into spaghetti.

Einsteinian spacetime

In general relativity, Einsteinian frames take the place of inertial frames. So, in *Einsteinian spacetime*, the path traced out by a particle (its *world-line*) should be straight if it is in gravitational freefall. What can we make of gravitational tides? The world-lines of nearby particles seem to veer apart in a way that straight lines do not.

The best solution is to see spacetime itself as curved, with the world-lines of free particles being given by its **geodesics**. Not every geodesic is a viable world-line, only those which pass through the future light cones at each point. (There may be other geodesics which represent faster-than-light travel.)

The falling room in spacetime. The black lines represent geodesics.

Einsteinian spacetime, the setting for general relativity, is therefore a deformed version of Minkowski spacetime. This provides an elegant description of gravity: it is the curvature in spacetime. In the absence of any other force, particles travel along a geodesic. Where there is no gravity, spacetime is flat and these will be genuine straight lines, exactly as in Minkowski spacetime.

Einstein's field equation

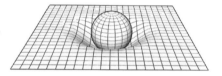

If Einsteinian spacetime is curved, what causes this curvature? The answer is whatever causes gravity: the presence of mass, or equivalently energy.

In 1915, Einstein published a system of equations which describe exactly how the presence of mass deforms spacetime. The equations are best expressed in the language of *tensor calculus*, a technically demanding extension of **vector calculus**. With this machinery in place, and working in the appropriate units, Einstein's equations can be condensed into one field equation:

$$G_{ab} = T_{ab}$$

Here G_{ab} is the *Einstein tensor*, which measures the curvature of a region of space. On the other side T_{ab} is the *stress–energy tensor*, which quantifies the amount of energy or mass in the region. The solutions to this equation are the possible geometries of spacetime in the theory of general relativity.

Black holes

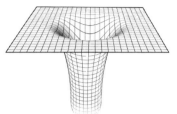

Einstein's field equation embodies his theory of general relativity. Unfortunately, its superficial simplicity belies a considerable complexity. The equation is very difficult to solve, a fact which Einstein himself found disheartening. One possible solution does present itself, namely the **Minkowski spacetime** of special relativity. This corresponds to a universe devoid of all matter and gravity, and therefore completely flat. Over the 20th century, other solutions have been found, and these match well with astronomical observations, providing a strong evidence base for general relativity.

An unusual solution was found in 1960 by Martin Kruskal, building on earlier work by others. In this solution, there was a region of spacetime so steeply curved that even light could not escape. *Black holes* such as this were considered geometric anomalies until 1971, when the astronomer Charles Thomas Bolton studied the star system Cygnus X-1 and realized that the large star HDE226868 was locked in orbit with another very massive, but invisible object. Further calculations showed that this could be nothing but a black hole.

Subsequently dozens more suspected black holes have been found, and it is believed that many stars collapse to form black holes after their death.

QUANTUM MECHANICS

Young's double-slit experiment

Around 1801, the physicist Thomas Young placed a light source behind a barrier pierced with two slits. On the other side was a screen. The pattern that Young saw would provide an important insight into the nature of light. The two dominant theories about light at that time were the particulate theory and the wave theory. If light consists of particles, then the number of particles reaching any point on the screen should be simply the number which arrive through the first slit plus the number which arrive via the second. This should produce a smooth picture.

However, if light is a wave, then Young should see something else. When two waves meet, they may reinforce each other at some places, and cancel each other out at others. The result is an *interference pattern*: brighter in some places, but darker in others. When Young performed the experiment, he saw an interference pattern on the screen. This provided powerful evidence for the wave theory of light. Over a hundred years later, the particle theory of light was to reappear under the guise of photons, when Young's experiment would take on a new significance.

Photons

Isaac Newton, among others, had believed that light consists of particles. His 'corpuscular' theory was killed off by Young's double-slit experiment, but was revived in the early 20th century.

In 1900, Max Planck analysed *black-body radiation*, the electromagnetic radiation produced inside a black box as it is heated up. Previous attempts to understand this situation had predicted that the energy inside the box should become infinite, as light of every frequency was emitted. Planck found a way to side-step this problem, by supposing that energy was emitted in small clumps, or *quanta*. This bold assumption led Planck to a formula which matched well with experimental measurements.

In 1905, Albert Einstein took Planck's argument further and used it to explain the mysterious *photoelectric effect*. In 1887, Heinrich Hertz had shone light at a metallic surface and noticed that it caused electrons to be emitted from the metal. In 1902, Philipp Lenard made the puzzling observation that the energy of these electrons was not affected by changes in the intensity of the light, as would be expected under a wave model of light. Einstein was able to explain this effect, using Planck's quanta of light, later dubbed *photons*. It was for this work at the birth of quantum theory that Einstein was awarded the Nobel Prize, in 1921.

The quantum double-slit experiment

With the re-emergence of photons, the results of Young's double-slit experiment posed a fresh conundrum.

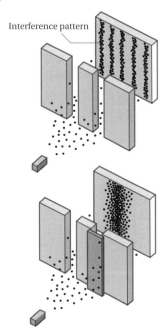

Interference pattern

With more sophisticated technology than Young had available, it was possible to refine the experiment by emitting particles from the source, one at a time. One particle could be released, and then detected at one point on the screen. Then the next would be released. Over time, a truly astonishing picture emerged.

With just one slit open, the final locations of the particles were distributed smoothly. But, with both slits open, interference patterns again appear. Opening the second slit seems to *prevent* particles from taking a previously permitted route through the first!

This experiment has been replicated with several different types of particles, including electrons, neutrons and even molecules as large as *Buckminster–Fullerene* composed of 60 carbon atoms. It provides strong evidence that matter comes neither in classical waves, nor Newtonian particles, but has a different quantum character. This has become known as *wave–particle duality* and is modelled by probability amplitudes.

Wave–particle duality

A *particle* is precisely located at one location in space. A *wave*, on the other hand, fills the whole of the accessible region of space, having different intensities at different points. (Mathematically this is modelled as a **scalar field**.) According to the theory of quantum mechanics, developed in the early 20th century, each particle comes with an associated wave, which propagates over 3-dimensional space. The value of the wave at a particular place embodies the *probability* of finding the particle at that point.

In the absence of a better term, we use the word *particle* for these dual wave–particle entities, but it should always be borne in mind that their behaviour is far from that of familiar Newtonian particles. The **de Broglie relations** translate between the language of waves and particles.

Quantum mechanics

If quantum mechanics hasn't profoundly shocked you,
you haven't understood it yet.—Niels Bohr

At first sight, wave–particle duality may not appear too strange. Suppose we place a ball at the top of a hill, and want to know how far it will have rolled after 5 seconds (treating this as a question in Newtonian mechanics). Even using state-of-the-art equipment, we can never know the starting position of the ball with perfect precision, so there is a range of possible initial states. From this range we could concoct a probability distribution for the position of the ball after 5 seconds, to fulfill a predictive role similar to that of the quantum wave.

There are two major differences between these scenarios. The first is the nature of the wave. As the double-slit experiment shows, quantum wave functions *interfere*. Ordinary probability distributions such as that for the ball cannot do this. Their values are always positive numbers, so they can never cancel each other out. This suggests that the quantum wave has an objective physical existence, while the probability distribution for the Newtonian ball is a purely human abstraction.

The second is that uncertainty about the state of a quantum particle is not simply a failure of our measuring equipment, but is an innate property of the system. **Heisenberg's uncertainty principle** imposes limits beyond which we can never hope to pin a quantum particle down.

In quantum mechanics, particles are modelled by probability amplitudes, governed by Schrödinger's equation.

The Planck constant

When Max Planck first investigated black-body radiation, he made the radical supposition that light energy comes in discrete clumps, later dubbed *photons*. More precisely, if the light waves have frequency f then each photon has energy hf, where h is a particular fixed number, now known as the *Planck constant*. Numerically, h is approximately 6.626×10^{-34}.

The Planck constant is now considered one of the fundamental constants of nature. It is because h is so tiny that we do not see quantum phenomena in ordinary life.

Planck's constant is crucial for moving between the languages of waves and particles, via the de Broglie relations. Often we work with Paul Dirac's variant on the Planck constant, written as $\hbar$ (pronounced 'h bar'). This is defined as $\hbar = \frac{h}{2\pi}$.

De Broglie relations

Max Planck and Albert Einstein had arrived at the conclusion that light exhibits particle-like as well as wave-like behaviour. Going the opposite way, the French aristocrat Louis de Broglie suggested that ordinary matter should have wave-like as well as particle-like properties.

His prediction was supported by subsequent investigation, including two-slit experiments performed with molecules in place of beams of light. De Broglie formulated two fundamental rules for translating between the languages of particles and waves. If a wave has frequency f, then each corresponding particle should have energy E, where:

$$E = hf$$

(Here h is Planck's constant.) The second formula relates the momentum p of a particle to the wavelength λ of the corresponding wave:

$$p = \frac{h}{\lambda}$$

Wave functions and probability amplitudes

Neither classical waves nor Newtonian particles are adequate to describe quantum phenomena such as that revealed by the double-slit experiment. The distribution of the particles on the screen seems to be best modelled as a probability distribution. However, classical probability distributions do not interfere as waves do. A new mathematical device was needed.

A *wave function* is a function ψ which takes as inputs the coordinate vector $\mathbf{x} = \begin{pmatrix} x \\ y \\ z \end{pmatrix}$ of a point in 3-dimensional space, at time t. Then the output $\psi(\mathbf{x}, t)$ is a complex number, called a *probability amplitude*.

If we fix $t = 5$, say, then the function $\mathbf{x} \to \psi(\mathbf{x}, 5)$ provides a complete description of the state of the particle 5 seconds after the clock was started. Included in this is the probability of finding the particle at the point $\mathbf{x}$.

So, a probability amplitude is rather like a *probability density function* (**see continuous probability distributions**). The difference is that probability density functions are always real-valued functions, whereas ψ is complex.

If we take the modulus of the probability amplitude, and square it, we do get an ordinary probability density function $\mathbf{x} \to |\psi(\mathbf{x}, 5)|^2$. This gives the probability of finding the particle at the point $\mathbf{x}$.

The total probability of finding the particle somewhere must be equal to 1. So a defining characteristic of a wave function ψ is that it is normalized, meaning that $\int |\psi(\mathbf{x}, t)|^2 \, d\mathbf{x} = 1$ for every value of t.

Observables

The state of a quantum particle is given by a wave function ψ. One reason for quantum theory's unfamiliar feel is that the outputs of this function are complex numbers rather than real numbers. It is not immediately obvious what statements such as

$\psi(\mathbf{x}, 5) = \frac{i}{4} + \frac{1}{3}$, say about the particle at the point $\mathbf{x}$ at 5 seconds. (Here, $\mathbf{x} = \begin{pmatrix} x \\ y \\ z \end{pmatrix}$ is a coordinate vector in 3-dimensional space.)

However, this function does encode the familiar properties of the particle, such as position, momentum and energy, even if they take on radically new flavours. In particular, these properties do not have unique values, but are probabilistic in nature. Such properties are called *observables*.

The simplest observable is *position*. This has a straightforward formula, given just by the coordinates of the point: **x**. Accompanying this is a probability distribution which gives the particle's probability of having this position, at time t. This is given by the real number $|\psi(\mathbf{x}, t)|^2$.

Quantized observables

Starting with a particle's wave function ψ, we can extract information about its **momentum p**, another important observable. Momentum is a vector quantity, with components in each direction: x, y and z. Focusing on the x-direction, the defining formula of momentum is:

$$p_x = -i\hbar \frac{\partial}{\partial x}$$

This looks extremely strange! We expect momentum to be a number, but this formula defines it as something else, a differential operator. However, in certain instances we can identify this with a numerical value. For example, suppose $\psi = e^{i6x}$ (this is not quite a permissible wave function since it is not normalized, but it illustrates the point). Then:

$$p_x \psi = -i\hbar \frac{\partial}{\partial x}(\psi) = -i\hbar \frac{\partial}{\partial x}(e^{i6x}) = -i\hbar \times i6e^{i6x} = 6\hbar\psi$$

Comparing the first and final terms of this equation, it makes sense to assign a numerical value to the momentum p_x, namely $6\hbar$.

This argument only worked because of the choice of ψ. In many cases we will not get a numerical value for momentum. Momentum is the first example of a *quantized* observable, meaning that it is an operator which crystallizes to a numerical value only under certain special circumstances.

Quantum momentum

The formula for momentum in the x-direction is given by:

$$p_x = -i\hbar \frac{\partial}{\partial x}$$

Similar formulas hold for the y and z components of momentum.

Using **vector calculus**, we can write these three formulas as one, to give the definition of the *momentum operator*:

$$p = -i\hbar\nabla$$

Just as for the position **x**, we may also extract a corresponding probability distribution which gives the probability of the particle having momentum of a particular value.

There are two ways of viewing a wave function. One is for the function ψ to take position as the primary consideration, and deduce information about momentum from it. It is equally possible to take the opposite perspective. These two viewpoints are related by the **Fourier transform**, and provide a beautiful symmetry to the mathematics of quantum mechanics.

Non-commuting operators

Position and momentum are the two principal observables of a quantum particle. Mathematically, they are given by two operators: $\mathbf{x}$ for position and $-i\hbar\nabla$ for momentum. Focusing on the x-direction, these are x and $-i\hbar\frac{\partial}{\partial x}$.

However, the order in which these are applied makes a difference. A simple application of the **product rule** says that $\frac{\partial}{\partial x}(x\psi) = 1\psi + x\frac{\partial}{\partial x}(\psi)$. Rearranging this, we get $\frac{\partial}{\partial x}(x\psi) - x\frac{\partial}{\partial x}(\psi) = \psi$. Since this is true for any ψ, we might abbreviate it as $\frac{\partial}{\partial x}x - x\frac{\partial}{\partial x} = 1$.

Of course, momentum in the x-direction is not given by $\frac{\partial}{\partial x}$, but by $-i\hbar\frac{\partial}{\partial x}$. Writing p_x for this momentum operator and x for the position operator, we find that:

$$p_x x - x p_x = -i\hbar$$

In particular then $p_x x \neq x p_x$, which says that the two operators do not *commute*. Similar formulas hold in the y-and z-directions.

Heisenberg's uncertainty relations

The non-commuting of the position and momentum operators was first noticed by Werner Heisenberg in 1925. At first it may have seemed no more than a minor technical inconvenience. Within a year, Heisenberg had realized its extraordinary repercussions, with major consequences for experimental physics and the philosophy of science.

Focusing on the x-direction, the position is simply given by the coordinate x. The particle does not have a unique position, however; it is smeared out across all the possible values of x. How wide this spread is can be quantified by a number, Δx (technically the standard deviation of the positional probability distribution).

Similarly, the momentum of the particle in the x-direction is given by p_x. This is also not uniquely defined, and the number Δp_x determines how widely spread the momentum is. *Heisenberg's uncertainty relation* says that:

$$\Delta x\, \Delta p_x \geq \tfrac{1}{2}\hbar$$

Here $\hbar$ is the reduced Planck constant. Similar inequalities hold along the y-and z-directions.

Heisenberg's uncertainty principle

Heisenberg's uncertainty relations have a profound implication. If the position x is extremely localized in space, it follows that Δx is very small. If so, then for the inequality $\Delta x\, \Delta p_x \geq \tfrac{1}{2}\hbar$ to hold, it must be that Δp_x is a large number, and therefore the momentum is widely spread out.

The converse is also true: if the momentum is narrowed down to a small number, then it must be that the position is widely spread. In other words, it is impossible to pin down both the position and the momentum, simultaneously.

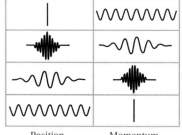

Position   Momentum

The Hamiltonian

Apart from position and momentum, another important observable is the total *energy* of the system. Just as for momentum, this is quantized and given by an operator. This is known as the *Hamiltonian, H*.

In Newtonian mechanics, the kinetic energy of a particle is $E = \frac{1}{2}mv^2$, while its momentum is $p = mv$. Rearranging these, we find $E = \frac{1}{2m}p^2$.

If we translate this formula into quantum mechanics, we need to apply it to the quantum momentum operator p. In this case we get:

$$E = -\frac{\hbar^2}{2m}\nabla^2$$

If the particle is *free*, that is, not subject to any external force, then kinetic energy is the particle's only form of energy, and the Hamiltonian H is defined by $H = E$.

If there are external forces acting, the particle will have additional **potential energy**. At the point **x** and time t, say, this energy is given by $V(\mathbf{x}, t)$. Then the total energy of the system is given by the Hamiltonian:

$$H = E + V$$

Schrödinger's equation

The state of a quantum system is described by a wave function ψ. As with other branches of physics, a central question is how this changes over time. This change of ψ is expressed by the partial derivative $\frac{\partial \psi}{\partial t}$.

The fundamental equation of quantum mechanics was discovered by Erwin Schrödinger, in 1926. It is:

$$i\hbar\frac{\partial \psi}{\partial t} = H(\psi)$$

Here $i = \sqrt{-1}$, $\hbar$ is the reduced Planck constant, and H is the Hamiltonian. This says that the way the wave function changes is determined solely by its energy.

At its most basic level, quantum mechanics investigates the possible solutions to this partial differential equation. From a mathematical perspective, this is not too problematic. Mathematicians are well used to analysing such equations. (Certainly it is easier to tackle than the **Navier–Stokes' equations**).

The major difficulty is instead a conceptual one, in interpreting the mathematics in terms of physical processes, far removed from the familiar Newtonian behaviour of the human scale. This is a problem with which physicists and philosophers continue to grapple.

The measurement paradox

Wave–particle duality is central to quantum mechanics, and it certainly has some strange consequences. The basic principle is that the wave function ψ permeates all of space and evolves according to Schrödinger's equation. The function ψ encodes the probability of finding the particle within any given region (as well as probabilities for its momentum, energy, and so on). This theory can predict the outcomes of experimental observations with impressive accuracy.

The problem is this: if such an observation is being made, and the particle is located at a specific point, then the probability of it being found at another point at the same moment disappears. So the original wave function ψ is no longer a valid description of the state of the particle. It is difficult to avoid the conclusion that whenever someone (or perhaps something) takes a *measurement*, the quantum system mysteriously jumps from being smoothly spread out, to crystallizing at a specific position.

This is known as the *collapse* or *decoherence* of the wave function. The *measurement paradox* is that measurement apparently triggers this decoherence. Scientists continue to debate the meaning of this phenomenon. However, the evidence for it is compelling and includes the double-slit experiment with extra measurements.

The double-slit experiment with extra measurements

The measurement paradox is not a conclusion that scientists would concede without serious evidence for it. Some such evidence is provided by revisiting the quantum double-slit experiment. What might happen if extra sensors are added to the two slits, to measure which route a photon takes? The answer is that the interference patterns disappear. It seems that the additional measurement causes the particle to decohere at one slit. There is then nothing to prevent it going on to reach a previously forbidden section of the screen. The probability of it passing through the second slit has reduced to zero, and so no interference will occur.

Schrödinger's cat

Erwin Schrödinger was one of the mathematical pioneers of quantum theory, but he was unsettled by the measurement paradox, and particularly by the interpretation placed upon it by the Copenhagen school of Bohr and Heisenberg. To illustrate its philosophical problems, Schrödinger performed a thought-experiment. Imagine that a live cat is placed in a sealed box. Alongside it is a particle of radioactive material and a Geiger counter attached to some 'diabolical mechanism' for killing the cat. Over the course of an hour, there is a 50% chance that an atom of the material will decay and set off the Geiger counter, which in turn triggers the murderous device. From outside the box, we have no idea whether or not this has happened, until we open it an hour later.

The measurement paradox would suggest that the wave function of the particle should remain spread out, embracing both the possibility of radioactive decay and no radioactive decay. This situation endures until the box is opened, at which stage it crystallizes into a firm state of either having decayed or not.

The implications for the cat are that it is neither firmly alive nor dead, but smeared out in a quantum living–dead state, until the box is opened. Schrödinger considered this conclusion ridiculous, but its correct resolution continues to be debated.

Quantum systems

As well as highlighting the philosophical conundrums posed by quantum mechanics, Schrödinger's cat illustrates another phenomenon of quantum mechanics. In the universe, particles do not exist in isolation, but interact with each other as part of larger systems. Two particles A and B might collide head on, or slightly deflect each other, or travel straight past each other, for example.

This is true for classical Newtonian particles and quantum particles. The difficulty is that quantum wave functions may be blurred across all three possibilities. When this happens the wave function of A becomes *entangled* with that of B. Two entangled wave functions can no longer be considered separately, but only as two aspects of a combined wave function for the two particle system. When measured, this system decoheres as a whole.

The Schrödinger equation is not limited to the wave functions of single particles, but governs those of larger systems too, including potentially the wave function of the entire universe. (It requires a lot of technical work to reinterpret this equation in a suitable higher-dimensional *Hilbert space*, to coordinatize all the particles, and possible correspondences between them.)

The EPR paradox

God does not play dice.—Albert Einstein

There is nothing to prevent two particles, having become entangled, from then parting and travelling far away from each other. In 1935, Albert Einstein, Boris Podolsky and Nathan Rosen (EPR) wrote a paper drawing out some of the seemingly paradoxical consequences of entanglement over large distances.

Suppose particles A and B became entangled and then fly apart. Now Albert will perform some measurement on particle A, and then Boris on particle B. If Albert measures the position of A, this collapses the wave function of the pair, pinning down the position of B. If Boris then measures B's position, his result is nearly a foregone conclusion.

Suppose instead that Albert measures A's momentum instead of position. Then, according to Heisenberg's uncertainty principle, the positional distributions of both A and B must become broadly spread out. So Boris' positional measurement gains a much wider range of possible readings.

The upshot of this is Albert's choice of reading fundamentally alters the positional probability distribution associated to Boris' particle, possibly a large distance away.

Entanglement

The EPR paradox seems to suggest that some sort of 'quantum information' passes between the two particles, and does so instantaneously. Even more striking EPR-scenarios have subsequently been dreamt up, involving measurements whose outcomes are just 'yes' or 'no', rather than spread over a probability range. In each case, the measurement on one particle has a definite, quantifiable impact on the second, irrespective of the distance between them.

Albert Einstein considered quantum mechanics an incomplete theory because of this problem. But in 1964, John Bell showed that it is unavoidable. There is no mathematical way to consider the two particles independently; they are inseparable. It follows, unsettlingly, that quantum mechanics is fundamentally *non-local*.

These predictions certainly run counter to common sense. Even worse, they threaten the inviolability of the speed of light. However, they have been confirmed experimentally, not over interplanetary distances, admittedly, but up to several kilometres.

QUANTUM FIELD THEORY

The quantum field

Quantum mechanics is a powerful theory for modelling the behaviour of subatomic particles and is supported by a large body of experimental evidence. On its own, however, it is not a complete description of our universe. The missing ingredients are accounts of the fundamental **forces of nature**. One of the greatest challenges in science is to construct a combined model of all four. The approach that emerged in the early 20th century was *quantum field theories*.

In quantum mechanics, objects acquire wave–particle duality, which is governed by the probabilistic wave function. Quantum field theories aim to model an underlying medium through which this travels: the *quantum field*. Particles are no longer taken as fundamental objects, but are excitations of this field. This approach provides mathematical room for manoeuvre, allowing a greater range of obervables to be modelled independently. These can represent quantizations of the forces of nature.

Quantum field theory

A quantum field theory is a mathematical model of the quantum field. Such theories are extremely challenging from a mathematical perspective. Nevertheless, considerable progress was made over the 20th century, in a sequence of increasingly sophisticated field theories, most significantly **quantum electrodynamics**, **electroweak theory** and **quantum chromodynamics**.

Some of these theories received spectacular experimental support, notably in the discovery of *antimatter* and *quarks*. The underlying mathematics remains poorly understood, however, striking examples being the **Yang–Mills problems**.

Meanwhile, the ultimate goal of a *Theory of Everything*, incorporating quantum explanations of all four forces of nature, remains elusive.

The forces of nature

Physicists believe that there are four fundamental forces at work in the universe:

1 *The electromagnetic force*. Comprehensively studied in the 19th century, the classical theory of electromagnetism is summed up in Maxwell's equations for electromagnetic fields.

2 *The strong nuclear force*. Two protons carry the same positive electric charge, so, according to the theory of electromagnetism, they should repel each other. Yet they coexist peacefully

inside the nucleus of an atom. How can this be? The answer is that there is another force which attracts them, with enough muscle to overpower the electromagnetic repulsion. This is the *strong nuclear force*.

3 *The weak nuclear force.* There is another force at work within atomic nuclei, which explains why they sometimes fly apart in radioactive decay. This is the weak nuclear force, which, unlike the strong nuclear force, is repellent.

4 *Gravity.* Successive versions of quantum field theory have succeeded in bringing together the three forces above, culminating in the **standard model of particle physics**. But gravity continues to defy our best efforts. On a galactic scale this is well understood via **Einstein's field equation** for general relativity. Between this and the quantum world, there remains a great gulf.

Relativistic quantum theory

In the early 20th century, two major physical theories were developing: quantum mechanics and relativity theory. Albert Einstein was heavily involved with both, but the man who took the first significant step to uniting the two was Paul Dirac. In 1930, Dirac successfully built a quantum model of the electron, which was compatible with special relativity. Central to his work was the *Dirac equation*, a relativistic counterpart of Schrödinger's equation.

Dirac's ideas formed the basis on which further relativistic field theories would be built, most immediately quantum electrodynamics. Scientists are continuing to search for a theory which embraces general relativity. Dirac's work also made a remarkable prediction: the existence of antimatter.

Antimatter

In special relativity, Einstein's equation $E = mc^2$ predicts that energy and mass are interchangeable. In situations of high energy, massive particles such as electrons may appear spontaneously. Mass therefore is not *conserved* overall.

However, according to quantum theory, there are other properties which must be conserved. Electric charge is one example. This seems to preclude electrons springing into existence, since this would involve a charge appearing where previously there was none.

Dirac saw a way through this conflict, when he noticed that the Dirac equation for the electron allowed a second solution. In this case, the mass was the same, but electric charge was reversed. He thus predicted the existence of the *positron*. In high-energy situations, electrons and positrons could now appear in pairs, as their charges cancel each other out.

The positron was discovered by Carl Anderson in 1932, in line with Dirac's prediction. Both men won Nobel prizes for their work. Subsequently it was found that other particles too have corresponding *antiparticles* of the same mass, but with the electric charge and other properties reversed.

Quantum electrodynamics

Building on Dirac's relativistic quantum theory, in the 1940s Richard Feynman, Julian Schwinger and Shin-Itiro Tomonaga assembled a quantum field theory to describe the first of the forces of nature: electromagnetism. The resulting theory is *quantum electrodynamics* (or *QED*), which incorporates a quantized version of Maxwell's equations. QED was a stunning success, able to predict the outcome of laboratory experiments to within one part in a trillion, an unheard of level of accuracy.

Electroweak theory

In 1967, Abdus Salam, Sheldon Glashow and Steven Weinberg produced a new quantum field theory, building on quantum electrodynamics, to incorporate the weak nuclear force. Their *electroweak theory* successfully modelled both forces, on an unexpected basis. It predicted that the two are not fundamentally separate, but are different expressions of a single force.

It certainly does not seem this way on planet earth, but electroweak theory predicts that at vey high energies (such as existed at the beginning of the universe) the two forces are fully unified.

Electroweak theory was the first quantum field theory to exploit the work of Yang and Mills. This provided an invaluable technical tool, but came at a cost. The particles carrying the weak nuclear force are called the W and Z bosons. These are massive particles in contrast to the massless photon which carries the electromagnetic force. But if the two forces are ultimately one, what is this disparity? To answer this question, it was necessary to revisit the whole question of where mass comes from. Electroweak theory provided a brand new answer: the Higgs field.

The Higgs boson

What is mass? The Newtonian answer is that it is an innate property of all matter, and the mass of an object corresponds to the amount of matter contained in it. This became less convincing in the early 20th century, with the discovery of many subatomic particles with different masses, and especially **photons**, the first massless particles.

According to the electroweak theory, mass derives from the *Higgs field*. This permeates everything, and the mass of different particles is an expression of their interaction with it.

If the Higgs field exists, it should be evidenced by a corresponding particle, called the *Higgs boson*. The discovery of the W and Z bosons at the CERN particle accelerator in 1983 was a triumph for the electroweak theory. The Higgs boson, however, remains as yet unobserved.

Quantum chromodynamics

At the same time that some physicists were working on the electroweak theory, others were applying quantum field theory to the next force: the strong nuclear force.

The result was *quantum chromodynamics (QCD)*, developed in the early 1970s by Harald Fritzsch, Heinrich Leutwyler and Murray Gell-Mann. They used *QED* as a template, but there were two additional challenges to overcome.

Firstly, electromagnetism is, in a sense, 1-dimensional: particles are positively charged negatively charged or neutral. In contrast, the strong nuclear force is 3-dimensional. There are three types of 'charges' (red, green and blue), and each has an opposite (antired, antigreen, and antiblue). It should be stressed that these are just whimsical names and have nothing to do with ordinary colour. (Incidentally, the theory's name 'chromodynamics' means the *dynamics of colour*.)

A second challenge was to account for the surge of particles called *hadrons* that experimental physicists had been finding since the late 1940s. In QCD, these two phenomena were both explained through a new fundamental particle: the *quark*.

Like QED, the predictions of QCD were subsequently validated in the laboratory with astonishing accuracy.

Hadrons
Particle physics had come a long way, since the ancient elemental theory of Earth, Air, Fire, and Water. The theory of the atom developed over the 19th century with **Brownian motion** providing important early evidence. Its name comes from the Greek word *atomos*, meaning indivisible. But, around 1912, Ernest Rutherford and Niels Bohr suggested that the atom was composed of negatively charged electrons orbiting a positively charged nucleus. In 1919 Rutherford found that this nucleus was itself composed of positively charged *protons*, and in 1932 James Chadwick also found neutral particles called *neutrons* within the nucleus.

Protons, neutrons and electrons were for some time taken to be the fundamental units of matter. But in the 1940s, physicists armed with more powerful tools began to unveil a startling array of new particles. Primary among them was the large family of *hadrons*. By the mid 1960s over one hundred new particles had been identified.

Suspicions grew that hadrons (including the proton and neutron) must be composed of smaller building blocks. Quantum chromodynamics eventually provided the answer: *quarks*.

Quarks
In the early 1960s, Murray Gell-Mann and others proposed that protons and neutrons are not indivisible but are each composed of three more basic particles. These he called *quarks*, a word taken from James Joyce's novel *Finnegan's Wake*: 'Three quarks for Muster Mark'. Just as electrons are electromagnetically charged, so quarks are charged by the strong nuclear force. Each may carry a charge of red, green or blue. *Antiquarks*, meanwhile, carry an anticolour.

Quarks carry other intrinsic properties, namely electric charge, *spin*, and mass, and come in a total of six *flavours* known as *up, down, strange, charm, top* and *bottom* (there are additionally six corresponding *antiquarks*). These six varieties have been all been identified in particle accelerator experiments. The last to be seen was *top* in 1995.

The quark theory formed an important component of quantum chromodynamics. Recipes for combining quarks of different flavours produce different hadrons. For example, the proton comprises two up and one down quark, and the neutron two down quarks and one up.

Standard model of particle physics

Putting together the electroweak theory (of electromagnetism and the weak nuclear force) and quantum chromodynamics (for describing the strong nuclear force) gives the *standard model of particle physics*. Since the 1970s, this framework has represented our best understanding of the particles which make up matter. Major questions remain, however:

1. Even on its own terms, the standard model is not a complete theory. There are several numerical constants which are currently unexplained; these have to be observed from nature and written into the theory.
2. Much of the standard model has received dramatic experimental verification. A notable exception is the **Higgs boson**. Finding this elusive particle is a major goal of experimental physics.
3. The electroweak theory demonstrated that the weak nuclear force and electromagnetism are not separate but two aspects of a single fundamental force. The strong force, however, is treated separately. Many physicists believe it should be possible to unite all three, in a *Grand Unified Theory*.
4. A glaring absence is an account of the final force: gravity. A major goal of contemporary physics is to build a model of all four forces, in a so-called Theory of Everything.
5. The mathematics underlying this framework is very poorly understood indeed. A case in point is the Yang–Mills problem.

Gauge groups
It was Hermann Weyl who first realized the crucial role that **symmetry groups** would play in understanding fields such as electromagentism. Unlike the **symmetry groups of spacetime**, these *gauge groups* do not describe global symmetries but arise from the underlying algebra.

In quantum mechanics, a particle is described by a wave function ψ, with probability distribution $|\psi|^2$. Now, suppose a is any complex number where $|a| = 1$ (so a lies on the unit circle). What happens if ψ is replaced by $a \times \psi$? The resulting probability distribution remains the same, since $|a \times \psi|^2 = |\psi|^2$. Replacing ψ with $a \times \psi$ is known as a change of *phase*, and there is no conceivable measurement which could distinguish the two. Multiplying by a is therefore *gauge symmetry* of the system.

In quantum electrodynamics, the same thing holds, and the electromagnetic field is unaffected by multiplication by a. The *gauge group* corresponds to taking together all these gauge symmetries. In this case, we get the circle group, known as U(1).

Yang–Mills theory
A major technical step in quantum field theory was made by Chen-Ning Yang and Robert Mills in 1954, when they had the bold idea of replacing U(1) with a larger **Lie group**, in the first instance SU(2), a group of 2×2 complex matrices. This group allowed extra hidden symmetries within the system, but a major technical difficulty is that it may be *non-Abelian*. This means that combining two symmetries g and h could produce different results, depending on the order: $gh \neq hg$.

Yang–Mills theory is the study of non-Abelian gauge groups. It provided the scaffolding from which both the electroweak theory (with group U(1) × SU(2)), and quantum chromodynamics (with group group SU(3)) were subsequently constructed. Putting these together, the standard model of particle physics operates with gauge group U(1) × SU(2) × SU(3).

Yang and Mills wrote down two equations that any non-Abelian gauge theory should satisfy. These equations are the source of some of the thorniest questions in mathematical physics: the Yang–Mills problems.

The Yang–Mills problems

Despite the prominent role that Yang–Mills theory has played in physics over the last half century, it is remarkable that the Yang–Mills equations themselves have never been fully solved. The *existence problem* addresses this:

1 For any simple Lie group G, show that a quantum field theory can be built which has Gauge group G and satisfies the Yang–Mills equations.

A question of a more physical nature concerns an expected consequence of any Yang–Mills theory: that energy or mass cannot come in arbitrarily small quantities, but there is a cut-off point. The *mass gap problem* says:

2 Show that there is some number $\Delta > 0$ so that every excitation must have energy at least Δ.

In 2000, the Clay Institute offered a prize of $1,000,000 for a solution to these two questions. The mathematical physicist Edward Witten was involved in the selection of the Millennium problems, and wrote that the existence problem 'would essentially mean making sense of the standard model of particle physics'.

There has been a great deal of research into the Yang–Mills problems, some of which has had repercussions elsewhere in mathematics. A notable example is in differential topology, in the discovery of **aliens from the fourth dimension**.

Quark confinement

A third problem coming from Yang–Mills theory (though without the million dollar price tag) is to show that quarks are *confined*. This would explain why we can never extract individual quarks from protons or neutrons; they only come in threes, or in quark–antiquark pairs.

When two electrically charged particles draw apart, the magnitude of the force between them drops. The same happens with gravity and the weak nuclear force. This is not true of the strong nuclear force however. It remains constant over distance. The result is that when you try to separate a quark from its antiquark, for example, the energy required is so high as to bring another quark into existence to replace it. In terms of mathematics, the problem is to show that all possible particle states in QCD are *SU(3)-invariant*.

GAMES &
RECREATI

THERE CAN BE NO FIRM DIVIDING LINE between recreational mathematics and any other kind. It is thanks to the pleasure that Pierre de Fermat found in number theory that he took time out from his work as a government official and lawyer to provide us with his great insights. Today, it is true that mathematical research is largely, but not exclusively, a professional's game. At the same time, interest in mathematics reaches further beyond the walls of universities and research institutes than ever before. One topic of enduring popular fascination is from the work of the 13th-century mathematician Fibonacci. The celebrated *Fibonacci sequence* and the related

AND ON

golden section continue to inspire artists and architects as well as scientists.

The growth of *game theory* over the 20th century means that mathematics has much to contribute to ancient games such as Chess and Go. Highlights include the growth of games-playing machines such as Deep Blue, and the solution of checkers in 2007. But the importance of game theory extends far beyond board games, to any context where *strategy* is needed, from economics to artificial intelligence. Today, game theorists are even consulted on military and political strategy.

GAME THEORY

Games and strategies

The game of *noughts and crosses* (or *tic-tac-toe*) is one of the simplest there is. The action takes place on a 3 × 3 grid, with one player inserting noughts (Os) and the other crosses (Xs). They take turns, and if either player gets three in a row, she has won. After some practice, an intelligent player should never lose at this game, whether playing first or second. There is a *strategy* that either player can use to force a draw.

The central question in *game theory* is to identify when a strategy exists either to win or draw. Game theory began with reasoning about traditional board games, but has hugely outgrown these shoes. Game theorists were consulted on strategy during the Cold War, and the subject has great significance to the study of stock markets.

Implementing optimal strategies closely resembles problem solving; therefore game theory is also of considerable interest to researchers into artificial intelligence.

The blue-eyed suicides

There is an island of 1000 people, 100 of whom have blue eyes, and 900 of whom have brown. However, there are no mirrors on the island, and the local religion forbids all discussion of eye-colour. Even worse, anyone who inadvertently discovers their own eye-colour must commit suicide that same day.

One day, an explorer lands on the island, and is invited to speak before the whole population. Ignorant of the local customs, he commits a faux pas. 'How pleasant it is', he says, 'to see another pair of blue-eyes, after all these months at sea'. What happens next?

(We must assume that the islanders follow their religion unerringly. An islander could not, for example, simply decide to disobey the suicide law. Despite their crazy religion, the islanders are also assumed to be hyperlogical: if there is some way by which someone can deduce their eye-colour, they will do so instantly.)

The blue-eyed theorem

The solution to the blue-eyed suicides puzzle depends on the number of blue-eyed islanders. In the original version, there are 100 of them. But it is easier to start with the case where there is just one, call him A. Then A knows from the explorer's words that there is at least one blue-eyed islander. Since he can see none, A concludes that it must be him. He commits suicide on the first day.

Suppose now that there are two, A and B. A can see B, and so A knows that there is at least one blue-eyed islander. However, come the second day, A observes that B has not committed suicide, and deduces that B must also be able to see someone with blue eyes. Since he can see none other, he concludes that it must be him. He commits suicide on the second day, and B does likewise.

The general statement is the following: if there are *n* blue-eyed islanders, they will all commit suicide on the *n*th day. This can be proved quite simply by **induction**. So the solution to the original question is that all the blue-eyed islanders will commit suicide on the 100th day. (The next day the brown-eyed islanders will realize what has happened, and do the same.)

Orders of knowledge

The blue eyed-theorem seems bizarrely counterintuitive, since the explorer does not seem to have told anyone anything that they did not already know. He did, however, as is explained by the concept of different orders of knowledge. If everyone in a group of people knows X, then X is said to be *first-order knowledge*. If everyone knows that everyone knows X, then X is *second-order knowledge*. Another way to say this is that if everyone knows that X is first-order knowledge, that makes it second-order knowledge.

Not all first-order knowledge is second order. For example, the people in a room may each individually notice that the clock on the wall has stopped, but they will have no idea whether the others have also noticed, until someone mentions it. At this point it becomes second-order knowledge (in fact common knowledge).

Second-order knowledge is the maximum needed for everyday use; indeed it is difficult to conceive of higher orders of knowledge. Nevertheless we can generalize: X is $(n + 1)$th-order knowledge, if everyone knows that it is nth-order knowledge. Repeatedly applying this rule defines knowledge of every order.

He knows that she knows that ...

What is remarkable about the problem of the blue-eyed suicides is that it involves orders of knowledge up to 100 (admittedly under somewhat artificial assumptions). Let X be the statement 'at least one islander has blue eyes'. The case that is easiest to understand is when there is only one blue-eyed islander. Here, the explorer directly increases the stock of first-order knowledge, by telling everyone X.

When there are two blue-eyed islanders (A and B), everyone knows X, which is first-order knowledge. But X is *not* second-order knowledge, because A does not know that B knows this, until the explorer speaks.

The first case which is difficult to imagine is where there are three blue-eyed islanders, call them A, B and C. So each of them knows that there are at least two blue-eyed islanders; this much is first-order knowledge. Also, each knows that each of the others can see at least one blue-eyed islander. So X is second-order knowledge. But X is not third-order knowledge (until the explorer speaks), because A cannot know that B knows that C knows this.

In the original puzzle, the explorer's seemingly harmless words actually increase the stock of 100th-order knowledge. Amazingly, this is enough to doom the island.

Common knowledge

In ordinary language a piece of information *X* is common knowledge to a group of people if each of them knows it. Within game theory, this is first-order knowledge. *Common knowledge* has a much stronger meaning. It is required not only that everyone knows X, but that everyone knows that everyone knows X, and furthermore that everyone knows that everyone knows that everyone knows X, and so on. That is to say, common knowledge is knowledge of every order.

The typical way for a piece of information to become common knowledge is for it to be announced publicly.

Although the roots of the idea go back to the philosopher David Hume in 1740, common knowledge has been studied in detail only more recently. It is a central idea in game theory, notably in the work of Robert J. Aumann, via whom the subject entered economics. The path of a piece of data from first-order knowledge, up through the orders of knowledge to common knowledge is believed to be highly significant for financial markets.

The prisoner's dilemma
Alex and Bobby are arrested for serious fraud, and held in separate cells. The prosecutor makes the same offer to each:

1 If you confess, and your accomplice does not, then you will go free. With your testimony they will be convicted of fraud and sent to prison for ten years.

2 If your accomplice confesses and you do not, the opposite will happen: they will walk free, and you will go to prison for ten years.

3 If you both confess, then you will both be convicted of fraud, but your sentence will be cut to seven years.

4 If neither of you confess, then you will both be convicted on lesser charges, and each sentenced to prison for six months.

(We assume that neither of the prisoners has any ethical concerns, or cares for the welfare of the other; each is just interested in minimizing their own prison time.)

The optimal solution seems to be option 4, as this minimizes the total amount of prison time. However, Alex and Bobby could each reason as follows: 'whatever the other does, I am better off confessing'. If they both follow this strategy, they will arrive at option 3, arguably the worst for the pair of them.

Equilibrium
In the prisoner's dilemma, option 3 is an example of an *equilibrium*. Even if Alex is told of Bobby's strategy, he has nothing to gain from altering his own, and vice versa. Option 4 is not an equilibrium, since if Alex knows Bobby's intended strategy, he can gain by switching and deciding to betray her. Similarly, options 1 and 2 are not equilibria.

In a game with two or more players, an equilibrium is defined as a situation where no-one has any incentive to make a unilateral change of strategy, even with full knowledge of the others' intentions. The prisoner's dilemma illustrates that an equilibrium by no means necessarily represents an optimal outcome.

Nash's equilibium theorem
In his doctoral thesis, John Nash considered *non-cooperative games*, where players cannot enter into binding agreements. In 1950, he proved his celebrated theorem, that equilibria always exist in any such game, Typically,

the equilibrium involves mixed strategies in which each player assigns probabilities to his various possible moves. His proof rested on a generalization of **Brouwer's fixed point theorem**.

Nash's insight is of fundamental importance in economics, where equilibria can predict the likely development of a market. Though suffering from schizophrenia, Nash made other significant contributions to mathematics and was awarded the Nobel Memorial Prize for Economics in 1994.

Deep Blue

The first chess-playing automaton was devised in the 18th century, by Wolfgang von Kempelen. Known as 'The Turk', it gained a reputation for defeating all-comers, including Napoleon Bonaparte and Benjamin Franklin. In 1820, the workings of the Turk were finally revealed. It involved a human chess-master sitting inside a cabinet, moving the pieces with levers. As many had suspected, the Turk was an ingenious hoax.

During the 20th century, however, the chess-playing computer became a reality. After the theoretical groundwork was laid by Claude Shannon, Alan Turing and others, the first fully functional chess program was developed in 1958. In 1980, Edward Fredkin of Carnegie Mellon University offered a prize of $100,000 to the programmers of the first chess computer to defeat a reigning world champion. This prize was claimed in 1997 by programmers at IBM, whose machine *Deep Blue* defeated Garry Kasparov 3½–2½.

Games-playing machines

The descendants of Deep Blue can now, for the most part, defeat humans at Chess. Other games have similarly succumbed to the dominance of computers. Maven is a computerized Scrabble® player developed by Brian Sheppard which can defeat even world champions. In the game of Othello® (or Reversi), Michael Buro's Logistello program can defeat the best. In the ancient Chinese game of Go, however, humans continue to rule the roost, for the time being.

Game-theoretic solutions

Impressive though Deep Blue and other games-playing machines undoubtedly are, they are a long way from providing complete *solutions* to these games. A mathematical solution to Chess would involve determining whether there is a perfect, infallible strategy which a player can use to force a win or a draw, no matter how brilliant their opponent.

Complete solutions to Chess and Go remain a distant dream. The games are simply too intricate, with too many possible scenarios. In his important 1950 paper 'Programming a Computer for Playing Chess', Claude Shannon estimated the number of distinct games of Chess to be at least 10^{120}. Dwarfing the number of atoms in the universe, this is far too large for any computer to handle directly.

Go is even more complex, partly because it is played on a larger 19×19 board. There are estimated to be around 10^{768} distinct games of Go.

Some simpler games have been solved, however, notably Connect Four (in which the first player can always force a win), Gomoku (or Connect Five), Awari (an ancient African game), the ancient Roman game Nine Men's Morris, Nim and checkers.

Checkers is solved

'Checkers is Solved' was the title of a 2007 paper by a team of computer scientists led by Jonathan Schaeffer. By some distance the most complex game to have been fully solved, this was a triumph of mathematical analysis, and a milestone in the development of artificial intelligence.

What Schaeffer discovered was a strategy for playing which would never lose. If two computers played against each other, each using this perfect strategy, they would draw (much as two competent players of noughts and crosses, or tic-tac-toe, will always draw). This result had been conjectured by the top checkers players many years earlier.

Proving the result was a mammoth operation, which involved a complete analysis of all possible endgames with ten pieces on the board, and a sophisticated search algorithm to determine how starting positions in the game relate to the possible endgames. To do this entailed in Schaeffer's words 'one of the longest running computations completed to date', involving up to 200 computer processors operating continuously from 1989 till 2007.

Nim

Claude Bachet was a 17th-century European nobleman, and the first person to make a serious study of recreational mathematics. One of his finds, Bachet's game, is even simpler than noughts and crosses (tic-tac-toe). A pile of counters is placed on the table, and two players take turns to remove counters. Each can choose to remove 1, 2 or 3. The winner is the person who takes the final counter.

The outcome of the game depends on the number of starting counters (between 15 and 25 is normal), and which player goes first.

Bachet's is the simplest of the family of Nim games. In other variants such as Marienbad, there may be more than one pile of counters, and players can choose from which pile to take counters. In *misère* versions, the player who takes the last counter loses. Other complicating factors may involve increasing the number of counters a player may remove, or forbidding players from removing the same number of counters as the previous player.

These additional factors can turn Nim into a complex game of strategy. But Nim has more than recreational value, as the Sprague–Grundy theorem demonstrates.

Mex

To analyse the game of Nim, we can assign numbers called *nimbers* to the different states of the game. The key device is that of the *minimum excluded number* (or *mex*) of a set of natural numbers.

For the set {0, 1, 3} the mex is 2, as this is the smallest number missing from the set. For {0, 1, 2} the mex is 3, and from {1, 2, 3} the mex is 0.

Nimbers

We can assign a number, called a *nimber*, to every possible position in a game of Nim. A winning position will have nimber 0.

In Bachet's game, reducing the pile to zero is a winning move, so a pile of size zero has nimber 0.

We calculate the nimbers of position X as follows: list the nimbers of the positions you could move to from X. The nimber of X is then defined as the mex of this list.

A pile of size 1 can only move to a pile of size 0 which has number 0. So the number for the 1-counter position is the mex of {0} which is 1. Similarly piles of two and three counters have nimbers 2 and 3 respectively. But from four counters, the possible moves are to one, two or three, so the number will be the mex of {1, 2, 3} which is 0. This accords with the experience of playing the game, that leaving four counters is a winning move. Then we continue:

Nimbers for Bachet's game															
Counters	0	1	2	3	4	5	6	7	8	9	10	11	12	13	14	15
Number	0	1	2	3	0	1	2	3	0	1	2	3	0	1	2	3

More complicated variants of Nim have more complex patterns of nimbers. But in each case the number 0 represents a winning position, so the strategy is always to aim for these.

In any game of Nim, one or other player will always have a winning strategy.

Sprague–Grundy theorem

In Chess and checkers, each player has their own set of pieces, with which they battle their opponent's. Therefore, from the same position on the board, the moves that the two players could make are entirely different. Nim, on the other hand, is an *impartial game*. The two players play with the same pieces and, from the same position, what would count as a good move for one would be equally good for the other. The only difference between the players, then, is who goes first.

The *Sprague–Grundy theorem* was discovered independently by Roland Sprague and Patrick Grundy in the 1930s. It says that Nim is not merely an *example* of an impartial game; every impartial game is a variant of Nim, albeit in disguised form. It follows that nimbers can be used to analyse any impartial game.

The Sprague–Grundy theorem is the foundational theorem of the subject of *combinatorial game theory*, developed by Berlekamp, Conway and Guy in their book *Winning Ways for your Mathematical Plays*, and Conway's *On Numbers and Games*. In these and subsequent works, the analysis is extended from impartial to *partisan* games, where the moves available to each player may be different.

FIBONACCI

Fibonacci's rabbits

In 1202, Leonardo da Pisa (known as Fibonacci) posed himself a challenge: 'A man leaves a pair of rabbits in a walled garden. How many pairs of rabbits can be produced from that pair in a year?'

To investigate, he modelled this situation, making several simplifying assumptions:

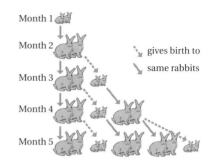

Month 1
Month 2 — gives birth to
Month 3 — same rabbits
Month 4
Month 5

1 Rabbits do not die
2 Rabbits come in pairs
3 Rabbits have two forms: baby and adult
4 Baby rabbits cannot breed
5 Baby rabbits become adult after one month
6 Each pair of adult rabbits produces one pair of babies every month

According to these, the garden contains one pair of baby rabbits in the first month. In the second, there is one pair of adult rabbits. In the third month these have reproduced, so there is one adult pair and one new baby pair, and so on.

This sequence continues 1, 1, 2, 3, 5, 8, 13, 21, 34, 55, 89, and in the twelfth month there are 144 pairs. So the answer to Fibonacci's original question is 288, subject to his assumptions. The sequence he had discovered became known as the Fibonacci sequence and is one of the most famous in science.

The Fibonacci sequence

$$1, 1, 2, 3, 5, 8, 13, 21, 34, 55, 89, 144, \ldots$$

Discovered in 1202 by Fibonacci in the course of his investigations into rabbit breeding, this sequence has intrigued people ever since and has a remarkable tendency to appear in nature. Each term of the sequence is the sum of the preceding two: $1 + 1 = 2$, $1 + 2 = 3$, $2 + 3 = 5$, $3 + 5 = 8$, and so on. So, once the first two terms are fixed (as 1 and 1) the rest of the sequence is set. We can define this as $F_1 = F_2 = 1$, and $F_{n+2} = F_n + F_{n+1}$.

In terms of rabbits, the total number of rabbits in July (F_{n+2}) is the number of adults plus the number of babies. The adults are those rabbits which have been alive since June (F_{n+1}). July's baby rabbits are equal in number to their parents, namely June's adults, who are all of May's rabbits (F_n).

A formula for F_n is given by **Binet's formula**.

Fibonacci spiral
On a piece of squared paper, draw a 1×1 square. Next to it, draw another. Adjoining these, draw a 2×2 square, and then a 3×3. Spiralling round, you can keep drawing squares whose sides are given by the Fibonacci sequence.

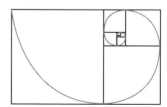

Once this is done, drawing arcs between the meeting points of the squares produces a *Fibonacci spiral*, a good approximation to a **logarithmic spiral**.

Ratios of Fibonacci numbers
We can form a new sequence from the ratios of successive Fibonacci numbers:

$$\frac{1}{1}, \frac{2}{1}, \frac{3}{2}, \frac{5}{3}, \frac{8}{5}, \frac{13}{8}, \dots$$

The interesting thing about this is that it converges. While the Fibonacci sequence grows bigger and bigger without limit, this sequence of ratios gets ever closer to some fixed number.

If a and b are successive Fibonacci numbers (a long way along the sequence), then $\frac{a}{b}$ and $\frac{a+b}{a}$ should be very close together. This is reminiscent of the definition of the **golden section** ϕ, where a line is divided into lengths a and b so that $\frac{a}{b} = \frac{a+b}{a} = \phi$. This hunch is correct, the ratio sequence does indeed tend to ϕ, intimately linking the two topics.

Binet's formula
What is the 100th term in the Fibonacci sequence? Is there a way to find it, without trudging through the first 99?

The defining relation of the Fibonacci sequence is $F_{n+2} = F_n + F_{n+1}$. There is a standard method for solving *difference equations* such as this. Doing this, with the **boundary conditions** $a_1 = a_2 = 1$, gives the solution known as *Binet's formula*. However, it was known to Leonhard Euler among others, before Jacques Binet derived it in 1843:

$$F_n = \frac{1}{\sqrt{5}}(\phi^n - (1-\phi)^n)$$

Here, $\phi = \frac{1+\sqrt{5}}{2}$ is the golden section. So the 100th Fibonacci number is $F_{100} = \frac{1}{\sqrt{5}}(\phi^{100} - (1-\phi)^{100})$, which comes out at 354,224,848,179,261,915,075.

Lindenmayer systems
Fibonacci's rabbit investigations made several assumptions which impair the effectiveness of that particular model. (The one-month gestation period is roughly correct, but rabbits can produce up to 12 babies per litter. Assumption 1 is obviously false, as is 5 since female rabbits do not become sexually mature until at least 4 months old.)

Nevertheless the Fibonacci sequence undeniably does appear in the natural world. It is a better model for the family trees of honeybees than for rabbits. If you ask how many parents, grandparents or great-grandparents an individual bee has, the answers tend to be Fibonacci numbers. (For humans, the answers are powers of 2, but the solitary queen changes this for bees.)

It is also striking that the numbers of petals on many flowers are often Fibonacci numbers, as are the number of spirals of fruitlets on pine-cones and pineapples.

There is certainly something in the simple iteration of the sequence which reflects natural growth. It is what biomathematicians term a *Lindenmayer system*, after the botanist Aristid Lindenmayer who first used them to model plant growth. Lindenmayer systems are logically stripped down **dynamical systems**. They describe many familiar fractals such as the **Koch curve** and **Cantor dust**, as well as producing excellent models of plant growth.

THE GOLDEN SECTION

The golden section

In Proposition 6.30 of his *Elements*, Euclid showed how to take a segment of line, and divide it into two lengths, so that the ratio of the whole line to the longer part is the same as that of the longer to the shorter. This special ratio is known as the *golden section* or *golden ratio*. It is denoted by the Greek letter *phi* (ϕ) after the sculptor Phidias who, in around 450 BC, first exploited its aesthetic qualities.

ϕ is intimately related to the **Fibonacci sequence**. As Robert Simpson proved in 1753, if you take a Fibonacci number, and divide it by the previous one, the result gets closer and closer to ϕ. The golden section makes many unexpected appearances within mathematics. In the theory of Penrose tilings, for example, Penrose's *rhombs* have areas in proportion given by ϕ. In a pentagram, the ratios of the major line segments are in golden proportion to each other.

The value of the golden section ϕ

Starting with a line 1 unit long, we divide it into two lengths according to Euclid's rules for the golden section, resulting in a longer part *a*, and a shorter piece *b*. First we can express *a* and *b* in terms of ϕ. Since the ratio of 1 to *a* is ϕ, this says that $\frac{1}{a} = \phi$. So $a = \frac{1}{\phi}$. Similarly if the shorter piece has length *b*, the ratio of *a* to *b* is $\frac{a}{b} = \phi$, and $b = \frac{a}{\phi}$. This means $b = \frac{1}{\phi^2}$.

Since $a + b = 1$, this means $\frac{1}{\phi} + \frac{1}{\phi^2} = 1$. Multiplying both sides by ϕ^2 and rearranging, this becomes $\phi^2 - \phi - 1 = 0$. We can now solve this with the formula for **quadratic equations**. This produces two numbers, but since ϕ must be positive, the solution is $\phi = \frac{1+\sqrt{5}}{2}$. So ϕ is an irrational, algebraic number, approximately 1.61803.

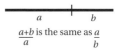

$\frac{a+b}{a}$ is the same as $\frac{a}{b}$

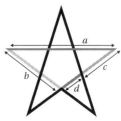

a/*b*, *b*/*c*, and *c*/*d* are golden

The golden section in the arts

Phidias' works around 450 BC, including his sculptures for the Parthenon in Athens, probably mark the start of the relationship between the golden section ϕ and the arts (although it is difficult to be certain). In 1509, the mathematician Luca Pacioli published a three-volume treatise on ϕ, *Divina Proportione* ('the Divine Proportion'), with illustrations by his friend and fellow ϕ-enthusiast, Leonardo da Vinci.

A golden rectangle is often said to be among the most aesthetically pleasing of shapes. Psychologists have tested this claim, with conflicting results. This figure does appear in the decorative arts (whether through conscious design or not). This happens most explicitly in the work of the 20th-century painter Salvador Dali, and the architect Le Corbusier, who developed a system of scales called 'The Modulor' based on ϕ.

Although the place of the golden section in art and architecture is assured, many specific examples remain controversial. Egypt's Great Pyramid of Giza, Paris' Notre Dame Cathedral and the work of the Italian renaissance architect Andrea Palladio are all cases where the involvement of the golden section is a matter of dispute.

Golden rectangle

A golden rectangle is one whose sides are in proportion given by the golden section. It has an elegant defining property: if you start with a golden rectangle, construct a square from the shorter side and remove it, what is left is a smaller golden rectangle. Golden rectangles are often said to be the most aesthetic of figures and form the basis of much art and architecture.

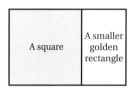

A golden rectangle

Kepler triangle

Johannes Kepler wrote: 'Geometry has two great treasures: one is the theorem of Pythagoras, the other the division of a line into mean and extreme ratio. The first we may compare to a mass of gold, the second we may call a precious jewel'. In the phrase 'mean and extreme ratio', Kepler was echoing Euclid's terminology for the golden section. In a piece of mathematical art, Kepler brought the two treasures together by constructing a right-angled triangle whose lengths are 1, $\sqrt{\phi}$ and ϕ.

Paper size

Since the 18th century, manufacturers of writing paper have appreciated the benefits of scalability. This has become even more important since the advent of the computer and home-printing. If you want to print a document, you might first want to print a draft at half size. This means that the smaller-size paper must be in the same proportions, or **similar**, to the original. It would be particularly convenient if the smaller paper was exactly half the size of the original, so that two small pages could be printed on one large one.

But this is not true for most shapes. Starting with a square and cutting it in half produces two rectangles, not two squares. What was wanted was a special rectangle which, when cut in half, produces a similar rectangle. (This makes an interesting comparison to the defining property of the golden rectangle.) Taking the shorter side as 1, and the longer as a, the requirement is that the ratio of a to 1, should be the same as the ratio of 1 to $\frac{a}{2}$. That is $\frac{a}{1} = \frac{1}{\left(\frac{a}{2}\right)}$. Rearranging, this says that $a^2 = 2$, and so $a = \sqrt{2}$.

So the sides of the sheets of paper should be in the proportion $1 : \sqrt{2}$. In the A-series, A0 is defined to be a rectangle in these proportions whose area is $1\,\text{m}^2$. A1 is formed by cutting this in half, and A2 by cutting that in half, and so on.

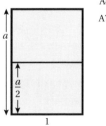

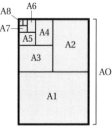

PUZZLES AND PERPLEXITIES

Magic squares

Around 2250 BC, Emperor Yu of China discovered a turtle in the Yellow River. On its shell were some curious markings. On closer inspection, they formed a 3 × 3 square, with the numbers 1 to 9 inside.

4	9	2
3	5	7
8	1	6

At least, so goes the story of the first *magic square*, known as the *Lo Shu* or *Yellow River Writing*. The magic is that each row and each column add up to the same number, 15, as do the two main diagonals.

Leaving aside rotations and reflections, the Lo Shu is the only 3 × 3 magic square. There are 880 different 4 × 4 magic squares, including Dürer's. These were listed in 1693, by Bernard Frénicle de Bessy. In 1973, Richard Schroeppel calculated that there are 275,305,224 different 5 × 5 magic squares.

It is not known how to calculate the exact number of $n \times n$ magic squares, but in 1998 Pinn and Wieczerkowski used statistical methods to estimate the number of 6 × 6 magic squares at around 1.77×10^{19}.

Dürer's magic square

In Albrecht Dürer's picture *Melancholia*, he paid tribute to his love of mathematics. As well as a mysterious polyhedron (known as his *melancholy octahedron*), the engraving also contains Europe's first magic square.

16	3	2	13
5	10	11	8
9	6	7	12
4	15	14	1

Dürer's is truly a connoisseur's square. Not only do the rows, columns and diagonals sum to 34, but so do the four quadrants, the four central numbers, the four corners and several other significant groupings. Even more, the central numbers on the bottom row date the picture: 1514. Outside these are the numbers 4 and 1, alphanumeric code for D and A, Dürer's initials.

Generalized magic squares

Magic squares are the oldest of all recreational mathematics, and it is not surprising that many variations on the theme have developed. One such is a *multiplication magic square*, where, instead of adding, the numbers are multiplied together. In this case we no longer insist that the numbers are consecutive, but they do all have to be different.

18	1	12
4	6	9
3	36	2

In 1955, Walter Horner found an 8 × 8 square which functions both as an ordinary magic square and as a multiplication magic square. Another 8 × 8 *addition–multiplication magic square*, as well as a 9 × 9 example, was found in 2005 by Christian Boyer. It is not known if there are any smaller examples.

Magic cubes

In 1640, Pierre Fermat lifted the principle of *magic* into higher dimensions. A *magic cube* is a cube where all rows, columns, pillars and the four body diagonals add up to the same number. The cube is *perfect* if the diagonals on each layer also add up to the same number, meaning that the cube is built from magic squares. It was an open question for many years what the smallest perfect magic cube was. In 2003 Christian Boyer and Walter Trump found it: a $5 \times 5 \times 5$ cube, built from the numbers 1 to 125.

Needless to say, mathematicians have not stopped at dimension 3. In the 1990s John Hendricks produced perfect magic hypercubes in dimensions 4 and 5, as well as studying magic hypercubes in higher dimensions.

Boyer's square of squares

In 1770, Leonhard Euler wrote to Joseph-Louis Lagrange, sending him a 4×4 *square of squares*, that is a magic square whose entries are not consecutive, but are all square numbers. More recently, Christian Boyer has found 5×5, 6×6 and 7×7 squares of squares. His 7×7 square is a wondrous thing, comprising the squares of consecutive whole numbers, from 0 to 48. It is unknown whether there can be a 3×3 square of squares.

In 2003, Boyer also discovered a gigantic *tetramagic cube*: an $8192 \times 8192 \times 8192$ magic cube. Astonishingly, this remains magic (and perfect) when each entry is squared, cubed or raised to the fourth power.

25^2	45^2	15^2	14^2	44^2	5^2	20^2
16^2	10^2	22^2	6^2	46^2	26^2	42^2
48^2	9^2	18^2	41^2	27^2	13^2	12^2
34^2	37^2	31^2	33^2	0^2	29^2	4^2
19^2	7^2	35^2	30^2	1^2	35^2	40^2
21^2	32^2	2^2	39^2	23^2	43^2	8^2
17^2	28^2	47^2	3^2	11^2	24^2	33^2

Knight's tours

In Chess, most pieces move either in horizontal, vertical or diagonal lines. A *knight's move* is the simplest move not covered by these possibilities. A knight can move two squares forwards or backwards, and then one right or left. Alternatively, he can move two squares left or right, and then one forward or back, making eight possible moves.

A *knight's tour* is a path that a knight can take around the chessboard, visiting each square exactly once. A *closed* tour is one where the knight loops back to his starting point. The smallest board where this is possible is the 6×6 board. Here, there are 9862 closed tours. Of particular interest are tours which have some level of symmetry. On the 6×6 board, there are five closed tours which have rotational symmetry of order 4, discovered by Paul de Hijo in 1882.

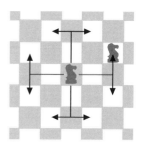

Magic knight's tours

In a knight's tour, if we label the knight's starting square 1, the next 2, and so on, then we can hope that not only will the knight visit each square exactly once, but his trail of numbers will form a magic square.

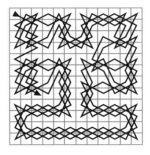

The first *magic knight's tour* was discovered in 1848 by William Beverley, on a standard 8 × 8 chessboard. In 2003, Stertenbrink, Meyrignac and Mackay used a computer to show that there are exactly 140 such tours. None of these, however, are perfect magic squares as their diagonals do not sum to the same totals as the rows and columns; there is no perfect magic knight's tour on a standard chessboard.

On a 12 × 12 board though there are perfect magic tours, including the illustrated example, discovered by Awani Kumar. Recently Kumar has extended the problem into higher dimensions, and discovered magic knight's tours of cubes, and even of **hypercubes** in up to five dimensions.

Latin squares

Despite their name, Latin squares originate in the medieval Islamic world, where they were considered mystic and engraved on amulets. The appeal is in their simplicity and symmetry. To create one, fill a 3 × 3 grid with the numbers 1, 2, 3 so that each number appears exactly once in each column and each row. The challenge extends to 4 × 4, 5 × 5 and all $n \times n$ squares.

1	2	3
2	3	1
3	1	2

These squares were picked up by Leonhard Euler who considered them 'a new type of magic square', and they are of genuine mathematical significance. The **Cayley table** of a finite group is a Latin square, for example (although the converse is not always true).

A Latin square is *reduced* if the first row and first column both consist of 1, 2, 3, …, n in the correct order. Every Latin square can be reduced by swapping around its columns and rows. There is just one reduced Latin 2 × 2 square, and similarly one 3 × 3 example. There are four 4 × 4 distinct reduced Latin squares, and, as Leonhard Euler showed, 56 5 × 5 examples. In 1900, Gaston Tarry showed that there are 9408 6 × 6 squares, which led him to the solution of the 36 officers problem.

There are several variations on the Latin theme, most notably Sudoku and Euler's Graeco-Latin squares.

Sudoku

Sudoku was first dreamt up in 1979 by Howard Garns in New York, and published in Dell *Pencil Puzzles and Word Games* magazine under the name 'Number Place'. Sudokus then became fashionable in Japan when published by Nikoli puzzle magazines, where they picked up their current name, an abbreviation of '*Suuji wa dokushin ni kagiru*', meaning 'numbers should be unmarried'. Since then, they have achieved worldwide popularity.

Underlying the puzzle is a 9 × 9 Latin square, into which the digits 1 to 9 must be written, each appearing exactly once in each row and column. The extra rule comes from the grid being subdivided into nine 3 × 3 blocks. Each of these too must contain the numbers 1 to 9.

The Sudoku begins with a few numbers already in place, these are the *clues*. The challenge is to complete the whole grid. Importantly, it is designed to have a unique solution. Typically this can be arrived at a by a process of elimination, step by step. More difficult puzzles might present the solver with a choice of ways forward, both of which need to be investigated in greater depth before either can be ruled out.

Sudoku clues

Setters of Sudoku need to ensure that their puzzle has a solution, and that it only has one solution. This is a classic **existence and uniqueness** problem. For existence, of course an empty grid (that is 0 clues) has a solution, as does a completed puzzle (81 clues). To ensure that your puzzle has a solution, you just need to avoid contradictory configurations.

It is a thornier problem to guarantee uniqueness. A basic question is: how many clues are required (that is, how many numbers present at the start)? Surprisingly, the answer is not known. The lowest number of clues which is known to generate a unique Sudoku is 17, and it is widely suspected that this is the lowest possible answer.

The 36 officers problem

In 1782, Leonhard Euler imagined 36 army officers, from six different regiments, and of six different ranks. The question he posed was whether it was possible to arrange these soldiers in a 6 × 6 grid so that each rank and regiment appears exactly once in each row and column.

What this amounts to is finding a 6 × 6 Graeco-Latin square. Euler wrote that 'after spending much effort to resolve this problem, we must acknowledge that such an arrangement is absolutely impossible, though we cannot give a rigorous proof'. In 1901 Gaston Tarry listed the 9408 possible 6 × 6 Latin squares, and showed that there was no way to combine any two of them without some pair being repeated, thereby proving Euler's conjecture.

Graeco-Latin Squares

Leonhard Euler was interested in ways to put Latin squares together. For example, can we form a 3 × 3 Latin square with the symbols 1, 2, 3 and another with A, B, C, and then put them together so that no two squares contain the same pair? If so, the result is a Graeco-Latin square.

The answer in this case is yes. But if you try the same thing for 2 × 2 squares, you will not succeed. Euler conjectured that no Graeco-Latin square can exist for squares of side 2, 6, 10, 14, 18, and so on. He was correct for 2 and for 6, as evidenced by Tarry's solution to the 36 officers problem. But in 1959 Parker, Bose and Shrikhande (known as 'Euler's spoilers') did create a Graeco-Latin square of side 10, and showed how to construct one of sides 14, 18, ... and so on, refuting Euler's conjecture.

The name comes from the fact that Euler used the Latin and Greek alphabets for the two labellings. They have broad applications for producing optimal matchings between different sets of objects, such as sports contests and experiment design.

A1	B3	C2
B2	C1	A3
C3	A2	B1

Sports contests and experiment design

Suppose two teams of five tennis players have a contest in which each player plays every player from the other team.

Labelling the players from the first team A, B, C, D, E and those from the second α, β, γ, δ, ε an optimal schedule for matches is provided by a 5 × 5 Graeco-Latin square:

	Monday	Tuesday	Wednesday	Thursday	Friday
Court 1	A v. α	B v. δ	C v. β	D v. ε	E v. γ
Court 2	B v. β	C v. ε	D v. γ	E v. α	A v. δ
Court 3	C v. γ	D v. α	E v. δ	A v. β	B v. ε
Court 4	D v. δ	E v. β	A v. ε	B v. γ	C v. α
Court 5	E v. ε	A v. γ	B v. α	C v. δ	D v. β

This principle is also important in devising scientific experiments to minimize any innate source of error. If we want to test the effectiveness of seven different machines, and we have seven different human operators, the best solution would again be to use a Graeco-Latin square. (If we had six machines, then we would encounter the 36 officers problem.)

Gardner's logician

Martin Gardner, who died in 2010, was the world's foremost expert on recreational mathematics. In over 65 books, and 25 years of columns for the *Scientific American* magazine, he brought to public attention a wealth of brain-bending puzzles, delightful curiosities, and deep mathematics including **Penrose tilings**, **fractals** and **public key cryptography**. In his earliest columns for the *Scientific American*, Gardner introduced a fictitious logician whose exploits have become classic.

The logician is travelling on an island inhabited by two tribes, one of which always lies, while the other always tells the truth. He is walking to a village but comes to a fork in the road. Not knowing which path to take, he consults a local man resting under a tree nearby. Unfortunately, he cannot tell whether the local belongs to the lying tribe or the truthful tribe. Nevertheless, he asks a single question, and from the answer he knows which way to go. What question could he have asked?

The unexpected hanging

A logician was condemned to be hanged. The judge informed him that the sentence was to be carried out at noon, one day next week between Monday and Friday. But it would be unexpected: he would not know which day it would be, before it happened. In his cell, contemplating his doom, the logician reasoned as follows:

'Friday is the final day available for my hanging. If I am still alive on Thursday afternoon, then I can be certain that Friday must be the day. Since it is to be unexpected, this is impossible. So I can rule out Friday.

'Therefore Thursday is the last possible day on which the sentence can be carried out. If I am still here on Wednesday afternoon, I must expect to die on Thursday. Again, this conflicts with the execution's unexpectedness, and is therefore impossible. I can rule out Thursday.'

Repeating the same argument, the logician was able to rule out Wednesday, Tuesday, and Monday, and went to sleep slightly cheered. On Tuesday morning the hangman arrived at his cell, completely unexpectedly. As he stood on the gallows, the logician reflected that the dreadful sentence was being carried out exactly as the judge had promised.

Logic and reality
The unexpected hanging is one of the most troubling of all logical paradoxes, as it separates pure deduction from real life in a very disconcerting fashion. As the philosopher Michael Scriven put it 'The logician goes pathetically through the motions that have always worked the spell before, but somehow the monster, Reality, has missed the point and advances still'. The paradox is of uncertain origin, but has been contemplated by logicians and philosophers including Kurt Gödel and Willard Quine. It entered popular consciousness in Martin Gardner's 1969 book *The Unexpected Hanging and Other Mathematical Diversions*.

The impossible sentence
To analyse the core of the paradox of the unexpected hanging, it is useful to shorten the judge's sentence. What happens if we reduce it to a single day? In this case, the sentence amounts to 'You will be hanged on Monday, but you do not know that'. This is already problematic. If the logician can be certain that he will be hanged on Monday, then this knowledge renders the judge's statement false. But if the judge's statement is false, this removes his basis for believing he will be hanged on Monday at all.

Like the **liar paradox**, the logician is unable to accept the judge's statement at face value, since it imposes conflicting restrictions about what is true, and what he can know to be true. But while the liar paradox is a perpetual absurdity, the passing of time changes the situation for the doomed logician, and may reveal the judge to have spoken with complete truth all along.

The bald man paradox
This is not a genuine mathematical result, but a warning that mathematics and ordinary language do not always mix easily. It uses **mathematical induction** to 'prove' something untrue: that every man is bald.

The *base case* is a man with 0 hairs on his head: he is self-evidently bald. The *inductive step* is to suppose that a man with n hairs on his head is bald. Then the next man, who has $n + 1$ hairs on his head, is either bald or not. But one single hair cannot be the difference between baldness and hirsuteness. So the man with $n + 1$ hairs must be bald too. Hence, by induction, a man with any number of hairs on his head is bald.

The resolution of the paradox comes from the fact that 'bald' is not a rigorously defined term: men are more or less bald, rather than rigidly either bald or not. A man with no hair is indisputably bald, but as individual hairs are added to his head, he becomes incrementally less bald, until around 100,000, when he is no longer bald at all.

The 1089 puzzle

Write down any 3-digit number, the only rule is that it must not read the same forwards as backwards (not 474, for example). I will pick 621. First, reverse the digits: 126. Next, subtract the smaller of these two from the larger: $621 - 126 = 495$. Now reverse the digits of this new number: 594. Add the two new numbers together: $495 + 594 = 1089$.

Whatever number you start with, the final answer is always 1089. With a bit of panache this can be exploited to give the illusion of psychic powers. (The only place to be careful is to make sure you always treat the numbers as 3-digit numbers. So if you get 099 after the second stage, this should be reversed to give 990.)

The 1089 theorem

The 1089 problem relies on a quirk of the decimal system. Why does it work? If we start with a number abc, this really means $100a + 10b + c$. If we then subtract cba (which is really $100c + 10b + a$), we get $99a - 99c$. The important thing about this number is that it is a multiple of 99. Such numbers are of the form $a9d$, where $a + d = 9$. (This is the divisibility test for 99, similar to that for 9.) Now when you add $d9e + e9d$, the units (or ones) digit will be 9 (because $d + e = 9$). Then the two 9s in the tens column produce 8, with 1 to carry. So in the hundreds column we have $d + e + 1 = 10$.

Hailstone numbers

Pick a whole number, any whole number. We will apply the following rule: if the number is even, halve it; if it is odd, triple it and add 1. So, starting with 3, we first move to 10. Applying the rule again takes us to 5, and from there to 16. Then we go 8, 4, 2, 1. Once we hit 1 we stop. We call 3 a *hailstone number* because it eventually falls to the ground, that is to say, it ends up at 1.

Some numbers have more complicated sequences. Starting at 7 we get: 7, 22, 11, 34, 17, 52, 26, 13, 40, 20, 10, 5, 16, 8, 4, 2, 1. So 7 is also a hailstone number.

The question Lothar Collatz posed in 1937 is: is every number a hailstone?

The Collatz conjecture

When Collatz first considered hailstone numbers, he conjectured that every number is indeed a hailstone. Though simple to state, this is a very puzzling conundrum indeed. Paul Erdős commented that 'mathematics is not ready for such problems'.

It would be false if one of two things happened: some sequence ends up not at 1, but in an infinitely repeating cycle, or some sequence simply grows and grows for ever.

In 1985, Jeffrey Lagarias proved that if there is a cycle, then it must have length at least 275,000. The question as a whole remains wide open, but in 2009, Tomás Oliveira e Silva verified it by computer for all numbers up to 5.76×10^{18}.

INDEX

Listings correspond to text entry headings; **bold** denotes section headings